Power System Fault Diagnosis

Power System Fault Diagnosis

A Wide Area Measurement Based Intelligent Approach

Md Shafiullah
Interdisciplinary Research Center for Renewable Energy and Power Systems, Research Institute, King Fahd University of Petroleum & Minerals, Dhahran, Saudi Arabia

M. A. Abido
Electrical Engineering Department; Interdisciplinary Research Center for Renewable Energy and Power Systems; K·A·CARE Energy Research and Innovation Center; King Fahd University of Petroleum & Minerals, Dhahran, Saudi Arabia

A. H. Al-Mohammed
Engineering & Design Department, National Grid SA, Saudi Electricity Company, Dammam, Saudi Arabia

Elsevier
Radarweg 29, PO Box 211, 1000 AE Amsterdam, Netherlands
The Boulevard, Langford Lane, Kidlington, Oxford OX5 1GB, United Kingdom
50 Hampshire Street, 5th Floor, Cambridge, MA 02139, United States

Copyright © 2022 Elsevier Inc. All rights reserved.

No part of this publication may be reproduced or transmitted in any form or by any means, electronic or mechanical, including photocopying, recording, or any information storage and retrieval system, without permission in writing from the publisher. Details on how to seek permission, further information about the Publisher's permissions policies and our arrangements with organizations such as the Copyright Clearance Center and the Copyright Licensing Agency, can be found at our website: www.elsevier.com/permissions.

This book and the individual contributions contained in it are protected under copyright by the Publisher (other than as may be noted herein).

Notices
Knowledge and best practice in this field are constantly changing. As new research and experience broaden our understanding, changes in research methods, professional practices, or medical treatment may become necessary.

Practitioners and researchers must always rely on their own experience and knowledge in evaluating and using any information, methods, compounds, or experiments described herein. In using such information or methods they should be mindful of their own safety and the safety of others, including parties for whom they have a professional responsibility.

To the fullest extent of the law, neither the Publisher nor the authors, contributors, or editors, assume any liability for any injury and/or damage to persons or property as a matter of products liability, negligence or otherwise, or from any use or operation of any methods, products, instructions, or ideas contained in the material herein.

British Library Cataloguing-in-Publication Data
A catalogue record for this book is available from the British Library

Library of Congress Cataloging-in-Publication Data
A catalog record for this book is available from the Library of Congress

ISBN: 978-0-323-88429-7

> For Information on all Elsevier publications visit our website at
> https://www.elsevier.com/books-and-journals

Publisher: Joseph P. Hayton
Acquisitions Editor: Lisa Reading
Editorial Project Manager: Andrae Akeh
Production Project Manager: Debasish Ghosh
Cover Designer: Mark Rogers

Typeset by Aptara, New Delhi, India

Contents

About the authors	ix
Acknowledgments	xi

1 Introduction — 1
  1.1 Introduction — 1
  1.2 Electric power system fault diagnosis importance — 3
  1.3 Electric power system fault diagnosis techniques — 5
  1.4 Wide area measurement system and phasor measurement units — 10
  1.5 Book organization — 16
  1.6 Summary — 18
  References — 18

2 Metaheuristic optimization techniques — 27
  2.1 Introduction — 27
  2.2 Classical optimization techniques — 27
  2.3 Metaheuristic techniques — 30
  2.4 Summary — 55
  References — 59

3 Artificial intelligence techniques — 69
  3.1 Introduction — 69
  3.2 Artificial intelligence techniques — 73
  3.3 Hybrid, ensemble, and other artificial intelligence techniques — 86
  3.4 Summary — 90
  References — 91

4 Advanced signal processing techniques for feature extraction — 101
  4.1 Introduction — 101
  4.2 Signal processing techniques — 102
  4.3 Wavelet transform — 102
  4.4 Feature extraction illustration — 107
  4.5 Summary — 117
  References — 117

5	**Improved optimal phasor measurement unit placement formulation for power system observability**	**121**
	5.1 Introduction	121
	5.2 Optimal phasor measurement unit placement formulation for power system observability	123
	5.3 Network observability and measurement redundancy illustration	126
	5.4 Transmission network observability	127
	5.5 Distribution network observability	133
	5.6 Summary	140
	References	140
6	**Transmission line parameter and system Thevenin equivalent identification**	**143**
	6.1 Introduction	143
	6.2 Transmission line parameter identification	145
	6.3 Thevenin equivalent identification	148
	6.4 Simulation results	150
	6.5 Summary	155
	References	156
7	**Fault diagnosis in two-terminal power transmission lines**	**159**
	7.1 Introduction	159
	7.2 Fault location algorithms	160
	7.3 Simulation results	171
	7.4 Summary	191
	References	191
8	**Fault diagnosis in three-terminal power transmission lines**	**195**
	8.1 Introduction	195
	8.2 Parameter estimation of a three-terminal line	196
	8.3 Adaptive fault location algorithm for three-terminal transmission line	200
	8.4 Simulation results	205
	8.5 Summary	220
	References	220
9	**Fault diagnosis in series compensated power transmission lines**	**223**
	9.1 Introduction	223
	9.2 Series capacitor locations	224
	9.3 Series capacitor schemes	225
	9.4 Phasor measurement unit-based parameter calculation of series-compensated line	227
	9.5 Fault location algorithm description	230
	9.6 Simulation results	235
	9.7 Summary	245
	References	246

10	**Intelligent fault diagnosis technique for distribution grid**	**249**
	10.1 Introduction	249
	10.2 Four-node test distribution feeder modeling	250
	10.3 Intelligent fault diagnosis approach	253
	10.4 Fault diagnosis results	253
	10.5 Optimized machine learning tools for fault location	264
	10.6 Fault diagnosis under unbalanced loading condition	276
	10.7 Summary	287
	References	287
11	**Smart grid fault diagnosis under load and renewable energy uncertainty**	**293**
	11.1 Introduction	293
	11.2 IEEE 13-node test distribution feeder modeling	294
	11.3 Load and renewable energy uncertainty modeling	299
	11.4 Fault modeling and feature extraction	304
	11.5 Fault diagnosis results and discussions	307
	11.6 Developed intelligent fault diagnosis scheme validation	326
	11.7 Summary	342
	References	342
12	**Utility practices on fault location**	**347**
	12.1 Introduction	347
	12.2 Fault location methods	348
	12.3 Local/device fault location solutions	349
	12.4 Commercially available fault location solutions	354
	12.5 Fault location detection on tapped transmission lines	357
	12.6 Overview of fault location in distribution systems	358
	12.7 Distribution management system-based fault location	360
	12.8 Advanced fault location approaches in distribution systems	362
	12.9 Examples of utility implementations	368
	12.10 Artificial intelligence deployment for fault location application	370
	12.11 Underground cable fault location	371
	12.12 Summary	392
	References	393
Appendices		**397**
	A.1 Software and hardware tools	397
	A.2 Statisctical quantities	403
	A.3 IEEE 13-node test distribution feeder data	407
	References	410
Index		**411**

About the authors

Dr. Md. Shafiullah is currently working as a faculty member in the Interdisciplinary Research Center for Renewable Energy and Power Systems (IRC-REPS) at King Fahd University of Petroleum & Minerals (KFUPM). He received a Ph.D. in electrical engineering (electrical power & energy systems) from KFUPM in 2018. Prior to that, he received the B.Sc. and M.Sc. degrees in electrical & electronic engineering (EEE) from Bangladesh University of Engineering & Technology (BUET) in 2009 and 2013, respectively. He demonstrated his research contributions in 70+ scientific articles (peer-reviewed journals, international conference proceedings, and book chapters). His research interest includes power system fault diagnosis, grid integration of renewable energy resources, power system stability and quality analysis, and machine learning techniques. He received the best research paper awards in two different IEEE flagship conferences (ICEEICT 2014 in Bangladesh and CAIDA 2021 in Saudi Arabia).

Dr. M. A. Abido received his B.Sc. and M.Sc. degrees in electrical engineering (EE) from Menoufiya University, Egypt, in 1985 and 1989, respectively, and Ph.D. from King Fahd University of Petroleum and Minerals (KFUPM), Saudi Arabia, in 1997. He is currently serving at the EE department of KFUPM as university distinguished professor. He's also a senior researcher at K·A·CARE Energy Research & Innovation Center and Interdisciplinary Research Center for Renewable Energy and Power Systems, Dhahran, Saudi Arabia. His research interests are power system control and renewable energy resources integration. Dr. Abido is the recipient of KFUPM Excellence in Research Award, 2002, 2007, and 2012, KFUPM Best Project Award, 2007 and 2010, First Prize Paper Award of the IEEE Industry Applications Society, 2003, Abdel-Hamid Shoman Prize, 2005, Almarai Prize for Scientific Innovation 2017–18, Saudi Arabia, 2018, and Khalifa Award for Higher Education 2017–18, Abu Dhabi, UAE, 2018.

Dr. Ali H. Al-Mohammed received his B.Sc. degree (honors with first class) in electrical engineering from King Fahd University of Petroleum & Minerals (KFUPM), Dhahran, Saudi Arabia, in 1994 and the M.Sc. and Ph.D. degrees from the same university in 1999 and 2013, respectively. Dr. Al-Mohammed has been serving the Saudi Electricity Company (SEC) for more than 27 years in engineering, design, and management of various HV and EHV transmission projects, including substations, overhead transmission lines, underground cables, and smart grid projects. His research interests include power system planning, fault location, asset optimization, substation engineering, phasor measurement units (PMU) applications, and power system protection.

Acknowledgment

All the praises and thanks to the almighty for his countless blessings and everlasting love throughout this endeavor. The authors would like to acknowledge the supports and facilities provided by the Deanship of Research Oversight and Coordination, Interdisciplinary Research Center for Renewable Energy and Power Systems, K·A·CARE Energy Research & Innovation Center, and Electrical Engineering Department at King Fahd University of Petroleum & Minerals (KFUPM), Dhahran, Saudi Arabia. They also acknowledge the support provided by the Saudi Electricity Company (SEC), Dammam, Saudi Arabia.

Furthermore, the authors wish to thank their colleagues and good friends from KFUPM and SEC, including Dr. Fahad A. Al-Sulaiman, Dr. Zakariya Al-Hamouz, Dr. Ibrahim M. El-Amin, Dr. Ibrahim O. Habiballah, Dr. Mahmoud Kassas, Dr. Syed M. Rahman, Dr. Adeyemi C. Adewole, Dr. Md Ismail Hossain, Engr. Mohammed A. Al-Mohammed Saleh, Engr. Abdulaziz J. Al-Nasser, Engr. M. Maaruf, and Mr. Khalid Al-Oudah for their kind supports. We would also like to thank our better halves and the family members. Finally, we are grateful to the Elsevier Team for their excellent and consistent guidance and cooperation throughout the project period.

Introduction

1.1 Introduction

Electric power system (EPS) networks are the most complex and gigantic structures ever devised by human beings to transport energy in the form of electricity. Energy transportation through the EPS networks has several benefits: reliability, flexibility, quick and easy access to power, cost-effectiveness, loss reduction, job creation, better coordination, etc. [1–3]. The EPS networks mainly comprise of three parts: generation, transmission, and distribution. On the generation side, the power plants convert fuel (gas, coal, oil, water, nuclear, renewable, etc.) into electricity at different voltage levels and send them to the transmission system through the transformers. Based on voltage levels, the transmission systems are named as high voltage (HV), extra-high voltage, ultra-high voltage networks. Then, the electricity is transported through transmission lines to various substations. Finally, the consumers receive the electricity at different voltage levels based on their requirements from the substations. Fig. 1.1 presents a simplified structure of a typical EPS network [4–6].

In the past, the distribution networks were mainly responsible for the distribution of energy received from the transmission networks to the customers. Thus, they were passive in nature. Recently, the widespread integration of the distributed energy resources (diesel generators, renewable energy farms, and energy storage systems) into the distribution grids has been conducted for various technical, economic, and environmental benefits. The significant advantages of such integration include reducing greenhouse gas emissions, transmission power losses, primary grid peak demand, improvement of voltage profile, phase imbalance, and reliability, and supplying the reactive power [7–10]. However, integration of such resources introduces several challenges in the EPS networks, for example, power quality, protection, and stability [11–13].

The EPS networks require advanced and sophisticated monitoring, control, and operation considering their gigantic structures, vast covering of landscape, various types of resources integration at different levels (generation, transmission, and distribution), and emerging challenges. Conventional technology, namely the supervisory control and data acquisition (SCADA) systems, plays a significant role in networks monitoring and management due to their efficiency and reliability. Recently, the phasor measurement units (PMUs), hybridized with the global positioning system (GPS), are becoming popular and populating the EPS networks rapidly as they offer faster and time-synchronized data acquisition over the traditional measurement systems with higher accuracy and lower uncertainty. PMU recorded data can be utilized for many applications such as state, harmonic, and parameter estimation; instability and stress point prediction; and fault diagnosis. Besides, as part of energy generation and

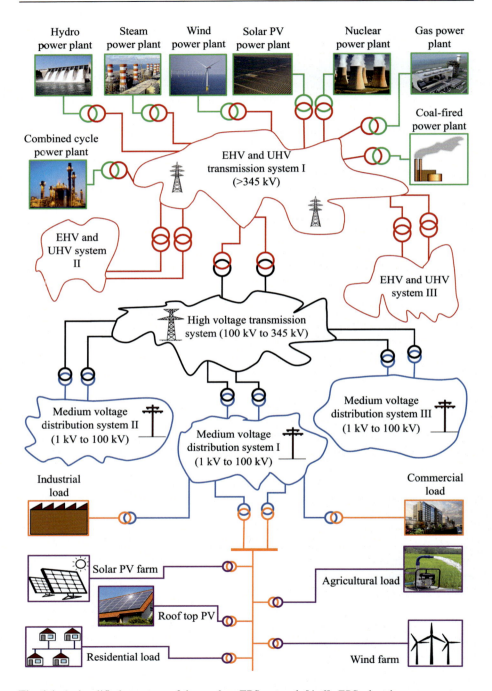

Fig. 1.1 A simplified structure of the modern EPS network [4–6]. *EPS*, electric power system.

distribution systems modernization, they are being switched into the concept of smart grids where data collected from the networks using advanced metering infrastructure (AMI) are appropriately utilized for maximizing efficiency and reliability. The AMI combines multiple technologies, including smart meters, micro phasor measurement unit (μPMU), hierarchical communication networks, data management systems, integration of data into a software application, etc. [14–19].

However, electricity is a commodity in the economic term that should be served instantly where the prime concern of the EPS operators is to deliver safe, secure, reliable, and quality energy to their customers at different voltage levels (transmission and distribution). In modern society, customers are more sensitive to interruptions resulting from network faults. Therefore, there is a growing interest in fault diagnosing in both transmission and distribution networks to reduce outage duration and revenue losses by expediting the restoration process [20–22]. Recent EPS blackouts and outages throughout the world reveal the inevitable upgradation of the traditional protection and fault diagnosis schemes. Besides, the integration of distributed energy resources into the existing grids and their associated uncertainties also emphasizes developing more sophisticated fault diagnosis schemes using data recorded from the advanced measurement devices [23–25].

The remaining parts of this chapter are organized as follows: Section 1.2 presents the fault classifications, their causes and consequences, and the importance of EPS fault diagnosis. Section 1.3 presents the prominent fault diagnosis schemes for the EPS networks and Section 1.4 briefly introduces wide area and phasor measurement technologies by highlighting their potential applications. Sections 1.5 provides the book organization with brief introductions of the remaining chapters. Finally, an overall summary of this chapter is presented in Section 1.6.

1.2 Electric power system fault diagnosis importance

Electrical faults are the abnormal conditions of the EPS networks that deviate the voltages and currents from their nominal values. The EPS networks are susceptible to a wide range of transient and permanent faults as they are primarily overhead type and exposed to trees, vehicles, supporting structures, birds, etc. Moreover, adverse weather conditions like lightning strikes, heavy rains and winds, salt deposition, and ice accumulation also trigger various disturbances, including faults in the EPS networks. In addition, aging and insulation failures of the EPS components are also responsible for fault occurrence. Besides, fire smokes and different kinds of human errors, for example, air ionization, selecting improper devices, forgetting metallic parts after servicing, switching the under servicing circuits, also cause faults in the EPS networks [26–28]. The promising benefits of the underground cables (UCs) over the overhead lines, for example, reliability during bad weather conditions, less space occupation, environmental concerns, lower maintenance requirement, higher efficiency, and cost competitiveness for short distances, lead toward their widespread adoption in the EPS networks [29–31]. However, like overhead transmission lines, the UC is also prone to different types of incipient and permanent faults. One primary reason for the UC faults is the insulation breakdown due to electrical stress, mechanical deficiency, and chemical pollution. In addition, cable aging, environmental condition, moisture, and flashover are also responsible for faults in the UCs [31–35].

The EPS faults can be mainly categorized into two types: open and short circuit faults. The open-circuit faults, known as series faults, occur due to the failure of one or multiple conductors and take place in series with the line. On the other hand, short circuit faults occur when the conductor of different phases encounter each other or the ground. Such faults can be further categorized as symmetrical and unsymmetrical faults. All three phases are involved during the symmetrical faults. Such faults keep the system balanced and are sub-categorized into line-to-line-to-line (LLL) and three-phase-to-ground (LLLG) faults. On the other hand, the unsymmetrical faults give rise to unsymmetrical currents as either one or two phases are involved during such faults. Therefore, their analysis is more challenging than analyzing symmetrical faults as the system becomes unbalanced. However, these faults are further sub-categorized as single-line-to-ground (SLG), line-to-line (LL), and line-to-line-to-ground (LLG). However, amongst different types of short circuit faults, the SLG faults are accounted for around 70% of them. The percentage of LL and LLG faults is approximately 20% and 10%, respectively. The probability of LLLG fault occurrence is about 2% to 3%. Finally, LLL faults are the most severe for the EPS networks that occur rarely. Fig. 1.2 illustrates different types of EPS faults (A, B, C, and G indicate phase A, phase B, phase C, and ground, respectively) [26–28].

Faults in the EPS networks have various adverse effects on system and component levels, including malfunctioning, life-time reduction, and damaging of the components; the inception of electrical fire from the short circuit flashovers and sparks; tripping of the relays that lead towards power outages and revenue losses; and sometimes deaths of birds, animals, and even humans. Therefore, rapid and accurate fault diagnosis (detection,

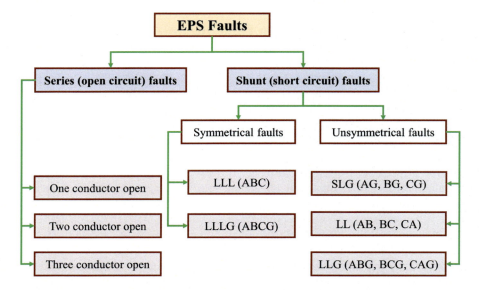

Fig. 1.2 Different types of faults in EPS networks [26–28]. *EPS*, electric power system.

classification, and location) schemes are required to reduce asset damage, minimize financial losses and repair expense, accelerate system restoration for reduction of outage duration and consumer dissatisfaction, and improve system reliability [36–42].

1.3 Electric power system fault diagnosis techniques

Fault diagnosis in EPS networks consists of three parts: detection, classification, and location of the faults. An additional part, namely the fault section or zone identification, is also included in a few cases. Fault diagnoses in transmission systems are more mature; however, they cannot be applied to the distribution networks immediately due to their inherent complexities. Thus, the EPS researchers explored and reported a wide range of fault diagnosis methods for the distribution systems as well, where most of them are developed based on the transmission systems schemes. Therefore, both transmission and distribution networks share similar strategies for fault diagnosis. This section briefly presents the prominent fault detection, classification, and location techniques for the EPS networks.

1.3.1 Fault detection and classification techniques

Fast fault detection allows the protective relays to isolate the faulty parts from the healthy parts that reduce asset damage and allow the continuous power supply to healthy parts. In addition, precise knowledge of fault class provides essential information regarding the location of the faults for quick starting of the restoration process. Thus, fast, precise, and reliable fault detection and classification are crucial maintenance and operational requirements of modern EPS networks. In response, many fault detection and classification techniques were reported in the literature. Mostly, they are based on sequence impedance [43], interharmonic signature [44], statistical cross-alienation coefficients [45], mathematical morphology and recursive least-square [46], Fortescue approach [47], grid information matrix obtained from PMU data [48], ensemble classifier [49], machine learning [50], and deep learning [51]. In addition to the mentioned techniques, EPS network faults are also classified by combining advanced signal processing and machine learning tools (MLTs) [52–60].

1.3.2 Fault location techniques

Commonly used fault location techniques in the EPS networks can be broadly classified into four main categories: impedance-based, traveling wave (TW)-based, knowledge-based (KB), and high frequency-based methods [13,61–65]. Fig. 1.3 illustrates the simplified flowcharts of the prominent fault location schemes for the EPS networks.

1.3.2.1 Impedance-based techniques

The impedance-based techniques are state-of-the-arts fault location schemes that utilize the fundamental frequency voltages and currents measurements. In general, these

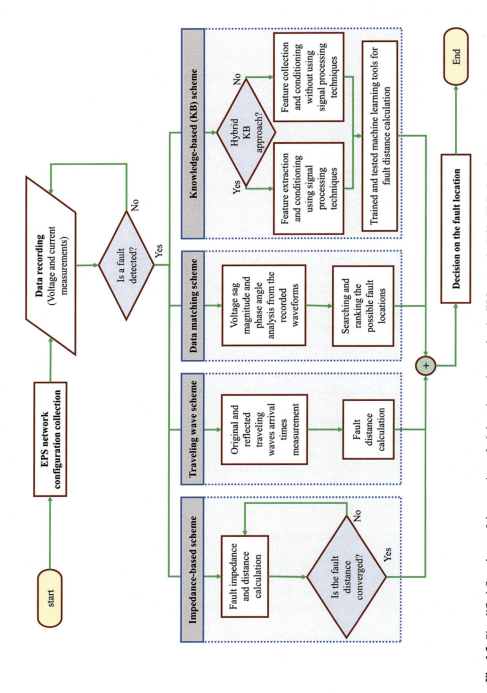

Fig. 1.3 Simplified flowcharts of the prominent fault location schemes for the EPS networks [13]. *EPS*, electric power system.

techniques calculate the impedance seen from the specific buses (nodes) of the networks based on the measured currents and voltages using Ohm's law. Then, the fault distances are determined based on the calculated impedance and available network information [62,63,66,67]. The main advantages of these schemes are their simplicity, ease of implementation, and less computational expense. Besides, they do not require either any sophisticated communication channels or synchronized measurements. The mentioned features made them economically more viable; thus, more popular to the manufacturers and the users. The impedance-based methods can be classified into single-ended or double-ended schemes depending on whether data measurements are taken from one end or both ends of the power transmission lines. The single-ended schemes require voltage and current measurements, knowledge of fault type, and positive and zero-sequence impedances of the transmission lines. A few of them also need prefault current and source impedance data in addition to the previously mentioned items. They are extensively and popularly used in many commercial distance relays. On the other hand, the two-terminal impedance-based schemes use voltage and current measurements from both ends of the power transmission lines. Therefore, communication channels are required to transfer data from one end to another, or data from both-end relays can be processed at a centralized location. They use either the positive-sequence or negative-sequence impedance in their fault location calculation to improve the accuracy. They are independent of the adverse impact of zero-sequence mutual coupling and uncertainty as the zero-sequence components are not used in fault distance calculation. Besides, they do not need information regarding the fault type, fault resistance, and source impedance. Due to the mentioned features, they are relatively more accurate compared to the single-ended schemes. However, the impedance-based schemes require precise line parameter and sequence impedance data that vary with the network operation and ambient conditions; thus, their accuracy is affected for particular implementations [68,69]. Besides, other major drawbacks of these schemes include their hectic iterative processes, frequent offerings of multiple estimations, and sensitivity against prefault loading conditions [13].

1.3.2.2 Traveling wave-based techniques

TWs occur in power systems after lightning strikes, switching operations, and faults. Voltage and current transients travel toward the faulted line terminals after being subjected to any faults. According to the wave reflection theory, such transients continue to bounce back and forth between the fault points and the line terminals until the postfault steady state is reached. Therefore, fault location can be effectively determined by calculating the transient propagation times as their propagation speeds are close to the light speed. The TW-based fault location schemes offer promising solutions in overcoming most of the challenges of the impedance-based methods. Besides, they are less sensitive to fault information uncertainty, operating mode, grounding resistance, and transformer saturation characteristics. The TW schemes, however, are considered as complex and costly as they require high sampling frequency and sophisticated communication channels. Besides, they are primarily applicable for long transmission lines and their accuracy deteriorated significantly for the short overhead lines. Moreover, they cannot be employed on transmission corridors with overhead lines and UCs as the surge impedance drastically changes in such cases. Furthermore, the presence of measurement noises and complex behavior of the fault-originated waves often disturb the effectiveness of the TW schemes [13,69–74].

Like impedance-based schemes, the TW schemes are also classified as single-ended and double-ended schemes. The single-ended schemes use wave sensors at one terminal and do not require communication between the line terminals. On the other hand, double-ended schemes are based on the exact time taken by the TWs to reach the line terminals. GPS is used to ensure the recording of the exact timings. As a result, the double-ended schemes are more accurate than the single-ended schemes. In addition, they do not require much signal processing at the sensors. However, double-ended schemes are more expensive than their single-ended counterparts due to their communication link and time-synchronization requirements. Such requirements also make them less reliable and less robust [69–72]. The TW schemes can also be ramified into three categories as A-, B-, and C-type. The A-type and B-type methods detect the returned TW generated by the faults to determine the fault location using online measurements. Additionally, the A-type methods use single-ended measurements, whereas the B-type methods need double-ended measurements. On the other hand, the C-type methods detect fault location offline using the manually injected TW signals [13].

1.3.2.3 Knowledge-based techniques

Considering the drawbacks of the impedance and TW-based techniques, the EPS researchers explored the third category fault location schemes, namely the KB methods that are comparatively more accurate and less costly. In addition, they are independent of parameter uncertainty (fault information, line parameter, and others). The expert system techniques are the prominent KB techniques used in power-system automation and control to decipher offline tasks such as post-fault analysis, settings coordination, and fault location. Besides, artificial neural networks (ANNs), support vector machines, fuzzy logic systems, and other machine learning techniques are popular in research for locating faults in the EPS networks. The combinations of advanced signal processing tools with a wide range of machine learning approaches have also drawn significant attention in finding faults and other power system transients [75–82]. Another KB fault location scheme, namely the data matching approach, is based on measured and simulated network data. It frequently employs voltage and current measurements gathered from single or numerous network points. They can effectively locate the faults based on "voltage sag" as the measured voltages of the nodes closer to the faults will have severe sags than other nodes [83–85]. Although the KB techniques are simple and do not require elaborate mathematical representation, their effectiveness depends on the quality and quantity of the available training and testing data. Besides, their accuracy is significantly affected by the limited or inaccurate information collected from the low quality and insufficient measurement devices [13].

1.3.2.4 High frequency-based techniques

These schemes measure the high-frequency components of fault-generated current and voltage waveforms that travel between the fault point and the line terminals. Based on the fault conditions, the frequency component of such waveforms varies from a few Hz to kHz. Therefore, they are independent of the power frequency phenomena, including the power swings and current transformer saturation. Besides, they are immune to fault inception angle as the frequency components of the fault generated waveforms do not vary with fault inception angle. These techniques use modal transformation to decompose the multiphase transient signals into modal components. Then, they further decompose the modal components into their wavelet components, hence, the wavelet coefficients. Next, they extract the useful features from the obtained wavelet coefficients to identify the fault branch or path. Finally, the information obtained from the power-frequency signals is employed to compute the fault distance from the primary substation. However, these techniques are not widely adopted due to their complexity and higher costs. They also require high-speed sampling infrastructure and specially tuned filters to measure the high-frequency components [69,70].

1.3.2.5 Other techniques

Apart from the mentioned popular and widely adopted techniques, other efficient and robust schemes are employed to locate temporary and permanent faults in the EPS networks. For instance, the fault indicators-based fault location methods are gaining attention as valuable information on fault location can be obtained from fault indicators installed either in the substations or on towers along the transmission or distribution lines. Besides, another unconventional scheme is based on both very low frequency and very high frequency reception [70]. In addition to the mentioned names, sparse measurements of voltage sag magnitudes [86], voltage sag duration table [87], compressive sensing [88], intelligent multiagent scheme [89], combination of voltage sag and impedance-based [90], mathematical morphology and recursive least-square [46], and minimum entropy theory and Fibonacci search algorithm [91], etc., are also employed to locate faults in the EPS networks.

Unlike overhead lines, UCs have higher capacitances and lower inductances [92]. Besides, they are buried under the ground. Therefore, fault diagnosis in the UCs is much tricky than overhead lines and requires careful investigation [31–34]. Thus, different online and offline fault diagnosis schemes were employed to diagnose UC faults. Such methods include impedance-based [31], TW-based [93], KB (data matching) [94], signal-processing [95], combination of signal processing based machine learning [96], random forest algorithm [30], Bayesian inference [97], and Murray and Varley loop tests [33] techniques.

1.3.3 Factors affecting the accuracy of fault diagnosis techniques

The most prominent fault diagnosis schemes have been discussed in the previous subsections. However, several factors affect the accuracy of the available fault diagnosis schemes, for example, network parameter uncertainty significantly affects the accuracy and reliability of most of the fault diagnosis schemes as many of the network parameters deviated from their initial values due to the ambient condition and operation history. In addition, the presence of *"bad data,"* measurement noises, and loss of data exhibit negative impacts on the accuracy of many diagnosis schemes. Waveform distortion (insufficient sampling frequency, low-resolution measuring devices, and transformer saturation) and bandwidth limitation of the communication infrastructure also heavily affect the effectiveness of the fault diagnosis schemes. The presence of compensating devices (shunt reactors and capacitors or series capacitors), their inaccurate compensation, and mutual effects on the zero-sequence components are also considered as the prime reasons for the lower accuracy of several diagnosis schemes. Besides, fault information (resistance and inception angle) and prefault loading condition uncertainty, dynamic and unbalanced loading conditions, inaccurate system modeling (untransposed lines as the transposed lines and nonconsideration of the presence of capacitors), and oversimplified modeling deteriorate the efficacy of the diagnosis schemes. In a few cases, inaccurate fault types make the fault location tasks challenging. Moreover, lack of practical fault data and incorrect and insufficient (training and testing) data reduce the credibility of many machine learning-based fault diagnosis schemes.

Accuracy is also significantly affected due to the immediate implementation of the schemes developed for the transmission networks on the distribution networks without considering their inherent characteristics, including nonhomogeneity, short distribution lines and cables, multiphase unbalanced loading conditions, intermediate load taps, and laterals. In addition, the recent proliferation of the distributed generators in the distribution networks and frequent network topology changes should also be considered while developing fault diagnosis schemes to achieve better accuracy and robustness. Therefore, it is crucial to eliminate or at least reduce possible factors affecting the accuracy, reliability, trustworthiness, and robustness of the EPS fault diagnosis schemes [98,99].

1.4 Wide area measurement system and phasor measurement units

The introduction of wide-area measurement systems (WAMS) and PMUs in the EPS networks has significantly enhanced the monitoring, dynamic analysis, fault diagnosis, and remedial actions capabilities of the networks. In critical situations, synchronized measurements obtained from the PMU allow fast and reliable emergency actions. Furthermore, in comparison with the traditional measurement approaches, synchronized measurements offer simpler, cheaper, efficient, and reliable solutions. Therefore, power system operators can utilize the existing networks more efficiently with the aid of synchronized measurements. This section presents the WAMS and PMU technologies and highlights their potential applications of the synchronized phasor measurements in the EPS networks.

1.4.1 Historical overview

As a measurement device, the PMU can measure current and voltage, thus, calculate the angle between these measured quantities. Due to time stamping and synchronization features over traditional meters, phase angles from buses at different system locations can be calculated in real-time. This makes the PMU as one of the revolutionizing devices for EPS network monitoring, operation, and control. The development of symmetrical component distance relays (SCDRs) in the 1970s allows early development of the phasor measurement algorithms due to the capability of symmetric positive sequence voltage and current calculation using the recursive discrete Fourier transform (DFT). The recursive algorithm continually updates the data array by removing the oldest and adding the new data to produce a constant phasor. The inception of the GPS in the 1980s added significant features that enabled the modern PMU. Researchers at the Power Systems Laboratory in Virginia Tech used GPS satellite pulses to time stamp and synchronized the phasor data with high accuracy in the mid-1980s. A few years later, the prototype PMU produced by Virginia Tech was supplied to the Bonneville Power Administration (BPA) and the American Electric Power (AEP). The BPA and AEP produced, the Macrodyne 1690, the first commercial PMU unit in 1991 that provided recorded data analysis with essential plotting tools. In 1997, the BPA redesigned the measurement system into a real-time wide area measurement system using a phasor data concentrator (PDC). Now, all major intelligent electronic device (IED) providers in the power system industry manufacture PMU commercially with various features. Besides, several variations of the PDC are also produced to date. To ensure safe and reliable operations of the PMU and PDC, different versions of the IEEE Standards were developed and updated, including IEEE Std 1344-1995, IEEE Std C37.118-2005, IEEE Std C37.118.1-2011, IEEE Std C37.118.2-2011, IEEE Std C37.242-2013, IEEE Std C37.247-2019, IEEE Std PC37.242/D4, Sep 2020, IEEE/IEC International Std 60255-118-1-2018 [99–106].

1.4.2 Phasor definition

The phasors are complex representations of the pure sinusoidal waveforms. The phasor representation of the sinusoidal signal of Eq. (1.1) is presented in Eq. (1.2):

$$x(t) = X_n \cos(\omega t + \theta) \tag{1.1}$$

$$X = \frac{X_n}{\sqrt{2}}(\cos\theta + j\sin\theta) \tag{1.2}$$

Where, X_n, ω, and θ are the magnitude, angular frequency, and phase angle of the sinusoidal signal, respectively. The positive phase angle is measured in a counterclockwise direction from the real axis. All phasors of a single phasor diagram should have the same frequency as the sinusoidal signal frequency is implicit in the phasor definition. Therefore, the sinusoidal signal in a phasor representation is always stationary, resulting in a constant phasor representation [99].

1.4.3 Phasor measurement concept

The phasor measurements are dealt with considering the input signal over a finite data window in practice. Most PMU manufacturers use one cycle of the input signals as the data window. Fig. 1.4A shows a nominal steady-state power frequency signal waveform of Eq. (1.1). If the waveform is started to be observed at time instant $t = 0$, then the waveform can be represented in the complex plane with a magnitude equal to the root mean squared (RMS) value of the signal and a phase angle equal to θ as shown in Fig. 1.4B.

In a digital measurement system, waveform samples are recorded for a nominal period, starting at $t = 0$, then, the DFT of the fundamental frequency component is calculated as:

$$X = \left(\sqrt{2}/N_T\right)\sum_{k=1}^{N_T} x[k]e^{-j2\pi k/N_T} \tag{1.3}$$

Where, X and N_T are the phasor and number of samples per cycle, and $x[k]$ is the waveform samples. If a sufficient sampling rate and precise synchronization with coordinated universal time are maintained, the DFT phasor estimation technique produces an accurate and very usable phasor value for most system conditions. Knowing the three phasor quantities (X_a, X_b, and X_c), the positive, negative, and zero sequence phasors (X_1, X_2, and X_0) can be computed using the following [99]:

$$\begin{bmatrix} X_1 \\ X_2 \\ X_0 \end{bmatrix} = \frac{1}{3}\begin{bmatrix} 1 & \alpha & \alpha^2 \\ 1 & \alpha^2 & \alpha \\ 1 & 1 & 1 \end{bmatrix} \cdot \begin{bmatrix} X_a \\ X_b \\ X_c \end{bmatrix} \tag{1.4}$$

with $\alpha = (-1/2 + j\sqrt{3}/2)$.

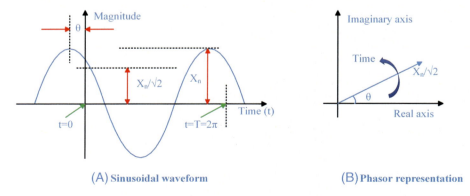

(A) Sinusoidal waveform (B) Phasor representation

Fig. 1.4 Sinusoidal waveform and its phasor representation.

1.4.4 Synchrophasor and the generic phasor measurement unit

The term "synchrophasor" describes the time-synchronized numbers representing both the magnitudes and phase angles of the sinusoidal waveforms. The feature "time-synchronized" enhances the accuracy. Besides, it is crucial to have time-synchronized measurements for effective monitoring and control of the EPS networks spreaded over the vast terrains and geographic regions. However, the hardware configurations of the PMUs vary from manufacturers to manufacturers, and they differ from each other in many aspects. Fig. 1.5 shows the block diagram of a generic PMU having the prime components. The structure is parallel to the relay structure as the PMU was evolved from the SCDR foundations.

The three-phase currents and voltages obtained from the secondary windings of the current and voltage transformers are the analog inputs to the PMU. To match with the analog-to-digital converters (ADCs) requirement, they are converted within the range of ±10 volts. In addition, the application of the antialiasing filters is essential before data sampling to produce a phase delay as a function of the signal frequency. The PMU is intended to compensate for this delay since the sampled data are taken after the antialiasing delay. For synchronization purposes, the used sampling clock is phase-locked to the one-pulse-per-second signal provided by a GPS receiver. The receiver could either be an integral part of the PMU or be installed in the substation to distribute the synchronized pulse to the PMU and other devices requiring it. The digital data of the ADC are sent to the phasor microprocessor for computation of the voltage and current phasors, frequency, rate of change of frequency, and other relevant information. Finally, these timestamped data are transferred through the suitable modems to the PDC for offline or online assessment and monitoring. In general, the PMU provides positive sequence voltage and current phasors. It can also offer phasors for individual phase voltages and currents. With the

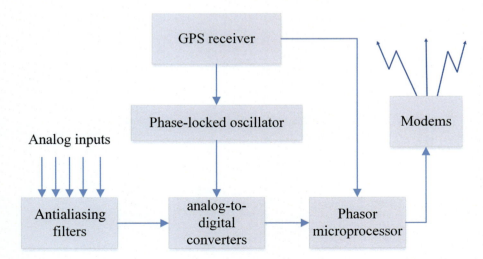

Fig. 1.5 Block diagram of a modern PMU. *PMU*, phasor measurement unit; *GPS*, global positioning system.

availability of the faster ADC and microprocessors devices, the PMU can sample up to 1024 samples/cycle [99,107–109].

1.4.5 Phasor measurement systems

The EPS networks can be monitored and controlled by placing PMUs on each bus of the networks; however, such initiatives are costly due to high capital and operational costs. In addition, such placement cannot be achieved due to the absence of communication infrastructure at a few buses as the power system networks are spreaded over the vast terrains and geographic regions. At the same time, PMU placement on each bus is not even necessary as a single can measure current phasors of all adjacent branches and voltage phasor of the PMU installed buses. Based on the available measurements, the voltage phasors of the adjacent buses can be calculated utilizing Kirchhoff's laws and branch parameters [110–113]. Therefore, PMUs are installed at selected buses (substations) of the EPS networks, and their recorded data can be used locally or sent to remote locations in real-time. However, the communication infrastructure involving the PMU, communication links, and PDCs must exist to realize the full benefits of the time-synchronized measurements. In response, the simplest solution can be the deployment of the PMUs at substations and sending the recorded data to the concentrators at the control centers as the communication infrastructures are developed around those control centers.

Fig. 1.6 depicts a generally accepted architecture of the phasor measurement system. The PMUs installed at EPS substations provide the recorded timestamped measurements, for example, positive-sequence voltages and currents of the monitored buses and feeders. The recorded data are stored in the local data storage devices that can be accessed remotely for diagnostic or post-mortem purposes. The recorded data are made available

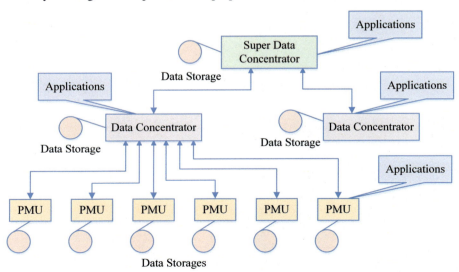

Fig. 1.6 Phasor measurement system architecture. *PMU*, phasor measurement unit.

locally for a few selected local applications. The real-time data are mainly utilized at the higher levels where data from several PMUs are gathered. However, recorded data from the PMUs installed across the EPS networks do not arrive at the mentioned locations simultaneously, instead of with time delays due to their distances and limitations of the communication infrastructures. As many applications are highly time-sensitive, the timestamp on the gathered data is very useful for their effective utilization. In general, the data from several PMUs are collected in the PDCs (first hierarchical level) for regional applications where the bad data is rejected, and others are aligned based on their timestamps; thus, a coherent record of the simultaneously recorded data from a wider part of the EPS networks is created. Likewise, data from the local storages and several PDCs are sent to the super data concentrator (second hierarchical level) for system-level applications. Although most of the data flow is toward the upward hierarchical directions, the communication links are bidirectional as in a few cases, information flow in the reverse direction is also required [99].

1.4.6 Phasor measurement systems application

The PMUs are becoming popular and populating the EPS networks rapidly due to their wide range of benefits and applications to support and maintain power system stability, reliability, and resilience [114]. For instance, North American Grid had only 200 PMUs in 2009 and increased the number to 2500 in 2017 [115–117]. They are considered the most important measuring devices for future electricity grids to deal with the evolving challenges. They offer time-synchronized data acquisition over traditional measurement systems with a faster rate, higher accuracy, and lower uncertainty [15]. According to North American SynchroPhasor Initiative (NASPI), the actual and potential PMU applications can be classified as automation, reliability and market operations, planning, and others. The automation applications consist of automated asset management, control of frequency, voltage, and load, etc. The reliability operation applications include wide-area monitoring and visualization, situational awareness, state estimation, inter-area oscillation analysis and control, asset management, system reclosing and restoration, fault diagnosis, etc. Besides, the market operation applications cover congestion analysis and day and hour ahead operation planning. Different kinds of model benchmarking, development, and validation are the parts of the PMU planning operation. Finally, forensic event analysis and standard compliance can be considered as the miscellaneous applications of the PMU. The mentioned PMU applications can also be classified as real-time and offline applications. Therefore, different monitoring, protection, and control applications can be developed using the data collected from various substations using the PMUs. Faster sampling rate and higher accuracy dramatically enhance power system state estimation; thus, better contingency analysis and other energy management schemes can be developed [99,107,116].

The EPS fault diagnosis is of great significance amongst all PMU applications. PMU recorded time-synchronized voltages and currents data paves the way for the development of more accurate and cost-effective fault diagnosis schemes [98,118–125]. Therefore, PMU recorded data will be employed for diagnosing both transmission and distribution networks faults in the subsequent chapters of this book.

1.5 Book organization

The materials of this book are organized into twelve chapters, where the introductory chapter provides a brief introduction on the EPS networks, types of EPS faults and importance of fault diagnosis, fault diagnosis schemes for transmission and distribution networks, wide area measurement systems, and book organization. **Chapter 2** starts with a general discussion on the classical and metaheuristic optimization techniques. Then, it provides the backgrounds, working principles, flowcharts, and pseudocodes of several popular and widely adopted metaheuristic algorithms. The discussion of the chapter assists the readers in understanding the fundamental features and differences of the explored algorithms that eventually guide them to choose the most appropriate algorithms for their problems.

    Chapter 3 provides a brief history of artificial intelligence (AI) and its foundation and essential components. Then, it discusses the basics of the popular machine learning techniques (a subset of AI), including ANNs, support vector machines, extreme learning machines, fuzzy logic models, genetic programming, and deep learning techniques. The chapter also briefly sheds light on the hybrid, ensemble, and other AI techniques. **Chapter 4** introduces the importance of signal processing techniques (SPTs) in analyzing power system transients. It also discusses the evolution of different advanced SPT along with their advantages and disadvantages. Then, it illustrates two advanced SPTs: the discrete wavelet and Stockwell transforms (STs). After describing the mentioned SPT, the chapter presents the step-by-step feature extraction processes from the recorded three-phase faulty current signals. Besides, it demonstrates how the extracted features are fetched into the AI tools to develop the intelligent fault diagnosis (IFD) scheme for the EPS networks. **Chapter 5** presents an improved optimal PMU placement (OPP) formulation that reduces the number of PMU requirements and increases the number of measurements by ensuring complete network observability. The presented formulation considers the existence of zero injection buses in the networks. It also incorporates the channel limitations of the available PMU. Moreover, the OPP formulation is extended to encompass power system contingencies, for example, single line outage and single PMU loss cases. Then, the chapter employs the grey wolf optimization (GWO) algorithm, a recently developed metaheuristic algorithm, to solve the developed OPP formulation on several IEEE benchmarked transmission and distribution networks. Besides, it compares the results obtained from the GWO and mixed-integer linear programming to confirm the superiority of the metaheuristic technique.

    Chapter 6 presents the use of synchronized phasor measurements of voltage and current acquired by the PMU for online identification of the transmission line parameters and system Thevenin equivalent at a particular node. It illustrates three different methods for identifying the transmission line parameters depending on the number of PMU measurement sets available at the two ends of the line. The presented approaches are tested on a 115 kV transmission system chosen from the Saudi Electricity Company (SEC) network and verified with PSCAD/EMTDC and MATLAB/SIMULINK simulations. **Chapter 7** illustrates two adaptive fault location algorithms based on the PMU measurements for determining fault location on a two-terminal transmission system. Both algorithms recorded three sets of pre-fault current and voltage phasors at the line terminals for online computation of the Thevenin equivalent and line parameters. The adaptive fault location algorithms are tested on a 115 kV transmission network of the SEC and verified with PSCAD/EMTDC and MATLAB/SIMULINK simulations. **Chapter 8** extends the PMU measurement-based adaptive fault location algorithm for a three-terminal power transmission line using three sets of pre-fault current and voltage phasors at the line terminals. The system Thevenin equivalent and the transmission line parameters are computed online through local PMU measurements at each terminal. The adaptive fault location algorithm is applied on a 500 kV system, and the simulation results are achieved using MATLAB/SIMULINK and PSCAD/EMTDC simulations. **Chapter 9** presents an adaptive fault location algorithm for the series compensated lines (SCLs) using the PMU recorded synchronized phasor measurements. The algorithm utilizes the PMU measurements for online computation of the system Thevenin equivalent and line parameters to tackle the parameter uncertainty due to the operating history and environmental conditions. The adaptive fault location algorithm is implemented on a 400 kV transmission network with an SCL, and the study is carried out on PSCAD/EMTDC and MATLAB/SIMULINK platforms.

Chapter 10 illustrates an IFD scheme for a four-node test distribution feeder combining advanced SPT and MLTs. It starts with modeling the mentioned feeder and faults by incorporating the prefault loading conditions and fault information (resistance and inception angle) uncertainty. Then, it extracts valuable features from the recorded current signals employing the SPT, as discussed in Chapter 4. Finally, the extracted features are fetched into three different MLTs namely, the ANNs, support vector machines, and extreme learning machines to develop intelligent fault detection, classification, location schemes. Moreover, the MLT control parameters are also tuned using metaheuristic optimization algorithms as discussed in Chapter 2 for better generalization performance. **Chapter 11** extends the applicability of the developed IFD scheme in Chapter 10 on the IEEE 13-node test distribution feeder under load and renewable energy generation uncertainty. It shows the step-by-step modeling of the selected test feeder incorporated with intermittent renewable energy resources in the RSCAD software and simulation of the modeled feeder in the real-time digital simulator rack. The chapter also presents the load demand and renewable energy generation uncertainty modeling using the appropriate probability density function. Then, it illustrates the faulty data generation and recording processes using the PMU. Moreover, the chapter investigates the IFD scheme efficacy in the presence of measurement noise and under contingencies. **Chapter 12** highlights various utility practices for fault location in transmission networks, distribution systems, and UCs. It discusses different fault location methods along with their deployment in various commercially available fault location solutions. In addition to traditional fault location methods, some advanced fault location approaches in distribution systems are also discussed. Besides, a list of commonly used and less commonly used fault location methods for UCs is also presented. Moreover, the chapter ends with a brief description of the prominent commercially available cable fault location solutions. Finally, the employed software and hardware components for developing the IFD scheme are presented in the **Appendix Section**. It also offers the mathematical definitions of the used statistical indices to analyze the performance of the regression model and extract features from the Discrete Wavelet Transform and ST decomposed signals. Detailed specifications of the IEEE 13-node test distribution feeder are also appended.

1.6 Summary

This chapter briefly introduces EPS networks, types of faults and their consequences, and the importance of precise fault diagnosis. Then, it reviewed and scrutinized the available fault diagnosis (detection, classification, and location) techniques for the transmission and distribution networks. The discussions revealed that fault diagnosing in the EPS networks is still a promising field as most available methods face different challenges. The significant challenges of the EPS fault diagnosis schemes are system parameter uncertainty due to the environmental and operating conditions. Besides, a few techniques suffer from multiple estimations and hectic iterative processes, while others need sophisticated and expensive communications devices and channels with wide bandwidth requirements. Furthermore, the effect of noise in the recorded data or SPTs, lack of practical fault data, and inaccurate and insufficient training and testing data also deteriorate the accuracy of the available fault diagnosis schemes. Moreover, proper modeling of the uncertainty involved in the load demand and renewable energy generation is essential for developing a universal fault diagnosis model. However, the recent proliferation of the AMIs in the EPS networks and the development of high-speed data processing devices might pave the way for developing the universal and IFD schemes by overcoming most of the mentioned challenges. This chapter also sheds light on the wide area measurement systems and PMU technologies by highlighting their potential applications. Finally, the organization of the book with brief introductions of the remaining chapters was illustrated.

References

[1] Southwest Power Pool, A robust transmission grid benefits everyone, 2021. https://www.spp.org/documents/10047/benefits_of_robust_transmission_grid.pdf. (Accessed 4 May 2021).

[2] Union of Concerned Scientists, How the electricity grid works, union of concerned scientists, 2015. https://www.ucsusa.org/resources/how-electricity-grid-works. (Accessed 4 May 2021).

[3] U.S. Energy Information Administration (EIA), Delivery to consumers, U.S. Energy Information Administration (EIA). The U.S. Energy Information Administration, Washington, D.C., USA, 2020. https://www.eia.gov/energyexplained/electricity/delivery-to-consumers.php. (Accessed 21 May 2021).

[4] N. Voropai, Electric power system transformations: a review of main prospects and challenges, Energies 13 (21) (2020) 5639. doi:10.3390/en13215639.

[5] M. Sarwar, Power system: basic structure and functioning, 2019. https://www.eepowerschool.com/energy/power-system-basic-structure-functioning/. (Accessed 4 May 2021).

[6] E. Csanyi, The structure of electric power systems (generation, distribution and transmission of energy), 2017. https://electrical-engineering-portal.com/electric-power-systems. (Accessed 4 May 2021).

[7] A. Ehsan, Q. Yang, Optimal integration and planning of renewable distributed generation in the power distribution networks: A review of analytical techniques, Appl. Energy 210 (Jan) (2018) 44–59, doi:10.1016/J.APENERGY.2017.10.106.

[8] S. Kakran, S. Chanana, Smart operations of smart grids integrated with distributed generation: a review, Renew. Sustain. Energy Rev. 81 (Jan) (2018) 524–535, doi:10.1016/J.RSER.2017.07.045.

[9] A. Rastgou, J. Moshtagh, S. Bahramara, Improved harmony search algorithm for electrical distribution network expansion planning in the presence of distributed generators, Energy 151 (May) (2018) 178–202, doi:10.1016/J.ENERGY.2018.03.030.

[10] M.R. Dorostkar-Ghamsari, M. Fotuhi-Firuzabad, M. Lehtonen, A. Safdarian, Value of distribution network reconfiguration in presence of renewable energy resources, IEEE Trans. Power Syst. 31 (99) (2015) 1–10, doi:10.1109/TPWRS.2015.2457954.

[11] G. Pepermans, J. Driesen, D. Haeseldonckx, R. Belmans, W. D'haeseleer, Distributed generation: definition, benefits and issues, Energy Policy 33 (6) (2005) 787–798, doi:10.1016/j.enpol.2003.10.004.

[12] H. Kuang, S. Li, Z. Wu, Discussion on advantages and disadvantages of distributed generation connected to the grid, 2011 International Conference on Electrical and Control Engineering, ICECE 2011 - Proceedings, (2011) 170–173, doi:10.1109/ICECENG.2011.6057500.

[13] M. Shafiullah, M.A. Abido, A review on distribution grid fault location techniques, Electr. Power Components Syst. 45 (8) (2017) 807–824, doi:10.1080/15325008.2017.1310772.

[14] R. Rashed Mohassel, A. Fung, F. Mohammadi, K. Raahemifar, A survey on advanced metering infrastructure, Int. J. Electr. Power Energy Syst. 63 (Dec) (2014) 473–484, doi:10.1016/j.ijepes.2014.06.025.

[15] F. Ding, C.D. Booth, Protection and stability assessment in future distribution networks using PMUs, 11th IET International Conference on Developments in Power Systems Protection (DPSP 2012). IET, (2012) P34–P34, doi:10.1049/cp.2012.0094.

[16] M. Shafiullah, S.M. Rahman, M.G. Mortoja, B. Al-Ramadan, Role of spatial analysis technology in power system industry: an overview, Renew. Sustain. Energy Rev. 66 (Dec) (2016) 584–595, doi:10.1016/j.rser.2016.08.017.

[17] M.M. Devi, M. Geethanjali, A.R. Devi, Fault localization for transmission lines with optimal phasor measurement units, Comput. Electr. Eng. 70 (Aug) (2018), 163–178, doi:10.1016/J.COMPELECENG.2018.01.043.

[18] S.T. Mak, E. So, Integration of PMU, SCADA, AMI to accomplish expanded functional capabilities of Smart Grid, CPEM Digest (Conference on Precision Electromagnetic Measurements), (2014) 68–69, doi:10.1109/CPEM.2014.6898262.

[19] M. Liu, State estimation in a smart distribution system, Hkie Trans. 24 (1) (2017) 1–8, doi:10.1080/1023697X.2016.1231015.

[20] C.M. Furse, M. Kafal, R. Razzaghi, Y.J. Shin, Fault diagnosis for electrical systems and power networks: a review, IEEE Sens. J. 21 (2) (2021) 888–906, doi:10.1109/JSEN.2020.2987321.

[21] V.H. Ferreira, et al., A survey on intelligent system application to fault diagnosis in electric power system transmission lines, Electr. Power Syst. Res. 136 (Jul) (2016) 135–153, doi:10.1016/j.epsr.2016.02.002.

[22] H. Hwang Goh, et al., Fault Location Techniques in Electrical Power System: A Review, Indones. J. Electr. Eng. Comput. Sci. 8 (1) (2017) 206–212, doi:10.11591/ijeecs.v8.i1. pp206-212.

[23] S.S. Gururajapathy, H. Mokhlis, H.A. Illias, Fault location and detection techniques in power distribution systems with distributed generation: A review, Renew. Sustain. Energy Rev. 74 (Jul) (2017) 949–958, doi:10.1016/j.rser.2017.03.021.

[24] R. Kumar, D. Saxena, A literature review on methodologies of fault location in the distribution system with distributed generation, Energy Technol. 8 (3) (2020) 1901093, doi:10.1002/ente.201901093.

[25] S. Jadidi, H. Badihi, Y. Zhang, Fault diagnosis in microgrids with integration of solar photovoltaic systems: a review, IFAC-PapersOnLine 53 (2) (2020) 12091–12096, doi:10.1016/j.ifacol.2020.12.763.

[26] D. Prajapat, Faults and effects in electrical power system, 2021. https://madhavuniversity.edu.in/faults-and-effects-in-electrical-power-system.html. (Accessed 5 May 2021).

[27] H.S.H. Al-Suraihi, What are the different types of faults in power system? 2019. https://circuitglobe.com/types-of-faults-in-power-system.html. (Accessed 5 May 2021).

[28] A. Al-Mohammed, M. Abido, Fault location based on synchronized measurements: a comprehensive survey, Sci. World J. 2014 (2014). http://www.hindawi.com/journals/tswj/2014/845307/abs/ (Accessed 23 May 2016).

[29] X. qin, et al., A cable fault recognition method based on a deep belief network, Comput. Electr. Eng. 71 (Oct) (2018) 452–464, doi:10.1016/j.compeleceng.2018.07.043.

[30] J. Kaewmanee, T. Indrasindhu, T. Menaneatra, T. Tosukolvan, Underground cable fault location via random forest algorithm, 2019 IEEE PES GTD Grand International Conference and Exposition Asia, GTD Asia 2019, (2019) 270–273, doi:10.1109/GTDAsia.2019.8715921.

[31] O. Naidu, N. George, and D. Pradhan, A new fault location method for underground cables in distribution systems, 2017, doi:10.1109/SGBC.2016.7936076.

[32] M. Jannati, B. Vahidi, S.H. Hosseinian, Incipient faults monitoring in underground medium voltage cables of distribution systems based on a two-step strategy, IEEE Trans. Power Deliv. 34 (4) (2019) 1647–1655, doi:10.1109/TPWRD.2019.2917268.

[33] A. Nag, A. Yadav, A.Y. Abdelaziz, M. Pazoki, Fault location in underground cable system using optimization technique, 2020 1st International Conference on Power, Control and Computing Technologies, ICPC2T 2020, (2020) 261–266, doi:10.1109/ICPC2T48082.2020.9071462.

[34] O.A. Gashteroodkhani, M. Majidi, M. Etezadi-Amoli, A.F. Nematollahi, B. Vahidi, A hybrid SVM-TT transform-based method for fault location in hybrid transmission lines with underground cables, Electr. Power Syst. Res. 170 (May) (2019) 205–214, doi:10.1016/j.epsr.2019.01.023.

[35] X. Yang, M.-S. Choi, S.-J. Lee, C.-W. Ten, S.-I. Lim, Fault location for underground power cable using distributed parameter approach, IEEE Trans. Power Syst. 23 (4) (2008) 1809–1816, doi:10.1109/TPWRS.2008.2002289.

[36] J. Doria-García, C. Orozco-Henao, R. Leborgne, O.D. Montoya, W. Gil-González, High impedance fault modeling and location for transmission line, Electr. Power Syst. Res. 196 (Jul) (2021) 107202, doi:10.1016/j.epsr.2021.107202.

[37] A. Mukherjee, P.K. Kundu, A. Das, A supervised principal component analysis-based approach of fault localization in transmission lines for single line to ground faults, Electr. Eng. 1 (4) (2021) 3, doi:10.1007/s00202-021-01221-9.

[38] J. Sadeh, A. Adinehzadeh, Accurate fault location algorithm for transmission line in the presence of series connected FACTS devices, Int. J. Electr. Power Energy Syst. 32 (4) (2010) 323–328, doi:10.1016/j.ijepes.2009.09.001.
[39] W. Bo, Q. Jiang, and Y. Cao, Transmission network fault location using sparse PMU measurements, 2009, doi:10.1109/SUPERGEN.2009.5348286.
[40] C.A. Apostolopoulos, G.N. Korres, A novel algorithm for locating faults on transposed/untransposed transmission lines without utilizing line parameters, IEEE Trans. Power Deliv. 25 (4) (2010) 2328–2338, doi:10.1109/TPWRD.2010.2053223.
[41] M. Shafiullah, M. Ijaz, M.A. Abido, Z. Al-Hamouz, Optimized support vector machine & wavelet transform for distribution grid fault location, 11th IEEE International Conference on Compatibility, Power Electronics and Power Engineering (CPE-POWERENG), IEEE, 2017, pp. 77–82, doi:10.1109/CPE.2017.7915148.
[42] A. Aljohani, T. Sheikhoon, A. Fataa, M. Shafiullah, and M. A. Abido, Design and implementation of an intelligent single line to ground fault locator for distribution feeders, 2019, doi:10.1109/ICCAD46983.2019.9037950.
[43] M. Abdel-Akher, K.M. Nor, Fault Analysis of Multiphase Distribution Systems Using Symmetrical Components, IEEE Trans. Power Deliv. 25 (4) (2010) 2931–2939, doi:10.1109/TPWRD.2010.2046682.
[44] J.R. Macedo, J.W. Resende, C.A. Bissochi, D. Carvalho, F.C. Castro, Proposition of an interharmonic-based methodology for high-impedance fault detection in distribution systems, IET Gener. Transm. Distrib. 9 (16) (2015) 2593–2601, doi:10.1049/iet-gtd.2015.0407.
[45] M.M.A. Mahfouz, M.A.H. El-Sayed, Smart grid fault detection and classification with multi-distributed generation based on current signals approach, IET Gener. Transm. Distrib. 10 (16) (2016) 4040–4047, doi:10.1049/iet-gtd.2016.0364.
[46] T. Gush, et al., Fault detection and location in a microgrid using mathematical morphology and recursive least square methods, Int. J. Electr. Power Energy Syst. 102 (2018) 324–331, doi:10.1016/j.ijepes.2018.04.009.
[47] C. Zhang, J. Wang, J. Huang, P. Cao, Detection and classification of short-circuit faults in distribution networks based on Fortescue approach and Softmax regression, Int. J. Electr. Power Energy Syst. 118 (Jun) (2020) 105812, doi:10.1016/j.ijepes.2019.105812.
[48] Z. Wang, F. Wang, Earth fault detection in distribution network based on wide-area measurement information, 2011 International Conference on Electrical and Control Engineering, IEEE, 2011, pp. 5855–5859, doi:10.1109/ICECENG.2011.6057020.
[49] T. Gush, S.B.A. Bukhari, S. Admasie, and C.H. Kim, An intelligent fault classification method for microgrids based on discrete orthonormal s-transform and ensemble classifier, 2019, doi:10.1109/ITEC-AP.2019.8903837.
[50] L.C. Acacio, P.A. Guaracy, T.O. Diniz, D.R.R.P. Araujo, L.R. Araujo, Evaluation of the impact of different neural network structure and data input on fault detection, 2017 IEEE PES Innovative Smart Grid Technologies Conference - Latin America (ISGT Latin America), IEEE, 2017, pp. 1–5, doi:10.1109/ISGT-LA.2017.8126699.
[51] D. Mnyanghwalo, H. Kundaeli, E. Kalinga, N. Hamisi, Deep learning approaches for fault detection and classifications in the electrical secondary distribution network: Methods comparison and recurrent neural network accuracy comparison, Cogent Eng 7 (1) (2020) 1857500, doi:10.1080/23311916.2020.1857500.
[52] F. Lucas, P. Costa, R. Batalha, D. Leite, I. Škrjanc, Fault detection in smart grids with time-varying distributed generation using wavelet energy and evolving neural networks, Evol. Syst. 11 (2) (2020) 165–180, doi:10.1007/s12530-020-09328-3.

[53] Y.D. Mamuya, Y.-D. Lee, J.-W. Shen, M. Shafiullah, C.-C. Kuo, Application of machine learning for fault classification and location in a radial distribution grid, Appl. Sci. 10 (14) (2020) 4965, doi:10.3390/app10144965.
[54] S. Jana, G. Dutta, Wavelet entropy and neural network based fault detection on a non radial power system network, IOSR J. Electr. Electron. Eng. 2 (3) (2012) 26–31.
[55] I. Nikoofekr, M. Sarlak, S.M. Shahrtash, Detection and classification of high impedance faults in power distribution networks using ART neural networks, 21st Iranian Conference on Electrical Engineering (ICEE). IEEE, (2013) 1–6, doi:10.1109/IranianCEE.2013.6599760 2013.
[56] J. Klomjit, A. Ngaopitakkul, Selection of proper input pattern in fuzzy logic algorithm for classifying the fault type in underground distribution system, 2016 IEEE Region 10 Conference (TENCON). IEEE, (2016) 2650–2655, doi:10.1109/TENCON.2016.7848519.
[57] W. Li, X. Miao, X. Zeng, Short Circuit Fault Type Identification of Low Voltage AC System Based on Black Hole Particle Swarm and Multi-level SVM, Proceedings - 2020 Chinese Automation Congress, CAC 2020, (2020) 208–213, doi:10.1109/CAC51589.2020.9327638.
[58] M. Mishra, P.K. Rout, Detection and classification of micro-grid faults based on HHT and machine learning techniques, IET Gener. Transm. Distrib. 12 (2) (2018) 388–397, doi:10.1049/iet-gtd.2017.0502.
[59] M. Shafiullah, M.A. Abido, S-transform based FFNN approach for distribution grids fault detection and classification, IEEE Access 6 (1) (2018) 8080–8088, doi:10.1109/ACCESS.2018.2809045.
[60] A. Aljohani, A. Aljurbua, M. Shafiullah, M.A. Abido, Smart fault detection and classification for distribution grid hybridizing ST and MLP-NN, 2018 15th International Multi-Conference on Systems, Signals & Devices (SSD), IEEE, Hammamet, Tunisia, 2018, pp. 1–5.
[61] A. Bahmanyar, S. Jamali, A. Estebsari, E. Bompard, A comparison framework for distribution system outage and fault location methods, Electr. Power Syst. Res. 145 (Apr) (2017) 19–34, doi:10.1016/J.EPSR.2016.12.018.
[62] L. Awalin, H. Mokhlis, A. Bakar, Recent developments in fault location methods for distribution networks, Prz. Elektrotechniczny R88 (12a) (2012) 206–212. http://eprints.um.edu.my/7868/. (Accessed 26 November 2015).
[63] M. Shafiullah and M. A. Abido, Distribution grid fault analysis under load and renewable energy uncertainties (Pending), US20200403406A1, 2020.
[64] A. Mukherjee, P.K. Kundu, A. Das, Transmission line faults in power system and the different algorithms for identification, classification and localization: a brief review of methods, J. Inst. Eng. (India): B 102 (4) (2021) 1–23, doi:10.1007/s40031-020-00530-0.
[65] A. R. Jonnalagadda and G. Hagos, Review of performance of impedance based and travelling wave based fault location algorithms in double circuit transmission lines, 3 (4) (2015) 65, doi:10.11648/J.JEEE.20150304.11.
[66] E.C. Senger, G. Manassero, C. Goldemberg, E.L. Pellini, Automated fault location system for primary distribution networks, IEEE Trans. Power Deliv. 20 (2) (2005) 1332–1340, doi:10.1109/TPWRD.2004.834871.
[67] M.A. Gabr, D.K. Ibrahim, E.S. Ahmed, M.I. Gilany, A new impedance-based fault location scheme for overhead unbalanced radial distribution networks, Electr. Power Syst. Res. 142 (Jan) (2017) 153–162, doi:10.1016/j.epsr.2016.09.015.
[68] A. Gaikwad, Transmission line protection support tools: fault location algorithms and the potential of using intelligent electronic device data for protection applications, Palo Alto, CA, 2013. https://www.epri.com/research/products/3002002381. (Accessed 29 June 2021).

[69] Shreya Parmar, Fault location algorithms for electrical power transmission lines (methodology, design and testing), Delft University of Technology, 2015.
[70] J. Izykowski, Fault location on power transmission lines, Oficyna Wydawnicza Politechniki Wrocławskiej, Wrocław, 2008.
[71] A.M. Elhaffar, Power transmission line fault location based on current travelling waves, Helsinki University of Technology, 2008.
[72] MWFTR, MWFTR - learning environment, MWFTR, 2021. http://www.mwftr.com/ (Accessed 29 June 2021).
[73] J. Tang, X. Yin, Z. Zhang, Traveling-wave-based fault location in electrical distribution systems with digital simulations, TELKOMNIKA (Telecommunication Comput. Electron. Control. 12 (2) (2014) 297, doi:10.12928/telkomnika.v12i2.67.
[74] H. Livani, C.Y. Evrenosoglu, A machine learning and wavelet-based fault location method for hybrid transmission lines, IEEE Trans. Smart Grid 5 (1) (2014) 51–59, doi:10.1109/TSG.2013.2260421.
[75] M. Shafiullah, M.A. Abido, Z. Al-Hamouz, Wavelet-based extreme learning machine for distribution grid fault location, IET Gener. Transm. Distrib. 11 (17) (2017) 4256–4263, doi:10.1049/iet-gtd.2017.0656.
[76] J. Von Euler-chelpin, Distribution grid fault location an analysis of methods for fault location in LV and MV power distribution grids, Uppsala University, Disciplinary Domain of Science and Technology, Technology, Department of Engineering Sciences (2018). http://www.diva-portal.org/smash/record.jsf?pid=diva2%3A1218803&dswid=3388.
[77] M. Shafiullah, et al., An Intelligent Approach for Power Quality Events Detection and Classification, First International Conference on Artificial Intelligence & Data Analytics (CAIDA 2021), IEEE, Riyadh, Saudi Arabia, 2021, pp. 1–6.
[78] R. Liang, G. Fu, X. Zhu, X. Xue, Fault location based on single terminal travelling wave analysis in radial distribution network, Int. J. Electr. Power Energy Syst. 66 (Mar) (2015) 160–165, doi:10.1016/j.ijepes.2014.10.026.
[79] M. Ijaz, M. Shafiullah, M.A. Abido, Classification of power quality disturbances using Wavelet Transform and Optimized ANN, 2015 18th International Conference on Intelligent System Application to Power Systems (ISAP), Proceedings of the Conference on, 2015, pp. 1–6, doi:10.1109/ISAP.2015.7325522.
[80] A.C. Adewole, R. Tzoneva, S. Behardien, Distribution network fault section identification and fault location using wavelet entropy and neural networks, Appl. Soft Comput. 46 (Sept) (2016) 296–306, doi:10.1016/j.asoc.2016.05.013.
[81] M. Shafiullah, M. Abido, T. Abdel-Fattah, Distribution grids fault location employing st based optimized machine learning approach, Energies 11 (9) (2018) 2328, doi:10.3390/en11092328.
[82] M. Shafiullah, M.A.M. Khan, S.D. Ahmed, PQ disturbance detection and classification combining advanced signal processing and machine learning tools, in: P. Sanjeevikumar, C. Sharmeela, J.B. Holm-Nielsen, P. Sivaraman (Eds.), Power Quality in Modern Power Systems, Academic Press, Massachusetts, USA, 2021, pp. 311–335.
[83] H. Mokhlis, H.Y. Li, A.R. Khalid, The application of voltage sags pattern to locate a faulted section in distribution network, Int. Rev. Electr. Eng. 5 (1) (2010) 173–179.
[84] H. Mokhlis, H. Li, Non-linear representation of voltage sag profiles for fault location in distribution networks, Int. J. Electr. Power Energy Syst. 33 (1) (2011) 124–130, doi:10.1016/j.ijepes.2010.06.020.
[85] T. Zheng, X. Xiao, Y. Wang, W. Zhang, Distribution System Fault Location Considering Voltage Sag Characteristics, 2011 Asia-Pacific Power and Energy Engineering Conference, Proccedings of the Conference on, 2011 1–4, doi:10.1109/APPEEC.2011.5748505.

[86] S. Lotfifard, M. Kezunovic, M.J. Mousavi, Voltage sag data utilization for distribution fault location, IEEE Trans. Power Deliv. 26 (2) (2011) 1239–1246, doi:10.1109/TPWRD.2010.2098891.

[87] S.R. Naidu, E. Guedes da Costa, G.V. Andrade, Fault location in distribution systems using the voltage sag-duration table, 2014 11th IEEE/IAS International Conference on Industry Applications, 2014, pp. 1–7, doi:10.1109/INDUSCON.2014.7059401.

[88] M. Majidi, A. Arabali, M. Etezadi-Amoli, Fault location in distribution networks by compressive sensing, IEEE Trans. Power Deliv. 30 (4) (2015) 1761–1769, doi:10.1109/TPWRD.2014.2357780.

[89] M.S. Rahman, N. Isherwood, A.M.T. Oo, Multi-agent based coordinated protection systems for distribution feeder fault diagnosis and reconfiguration, Int. J. Electr. Power Energy Syst. 97 (Apr) (2018) 106–119, doi:10.1016/J.IJEPES.2017.10.031.

[90] M. Daisy, R. Dashti, Single phase fault location in power distribution network using combination of impedance based method and voltage sage matching algorithm, Electrical Power Distribution Networks Conference (EPDC), 2015 20th Conference on, 2015 166–172, doi:10.1109/EPDC.2015.7330490.

[91] T. Zhang, H. Yu, P. Zeng, L. Sun, C. Song, J. Liu, Single phase fault diagnosis and location in active distribution network using synchronized voltage measurement, Int. J. Electr. Power Energy Syst. 117 (May) (2020) 105572, doi:10.1016/j.ijepes.2019.105572.

[92] D.S. Gastaldello, A.N. Souza, C.C.O. Ramos, P. Da Costa Junior, M.G. Zago, Fault location in underground systems using artificial neural networks and PSCAD/EMTDC, INES 2012 - IEEE 16th International Conference on Intelligent Engineering Systems, Proceedings, 2012, pp. 423–427, doi:10.1109/INES.2012.6249871.

[93] P.F. Gale, A. Wang, Type D travelling wave fault location on branched underground low-voltage networks, J. Eng. 2018 (15) (2018) 1229–1233, doi:10.1049/joe.2018.0215.

[94] H. Mokhlis, A.H.A. Bakar, H. Mohamad, H.Y. Li, Voltage sags matching to locate faults for underground distribution networks, Adv. Electr. Comput. Eng. 11 (2) (2011) 43–48.

[95] A.H.A. Bakar, M.S. Ali, C. Tan, H. Mokhlis, H. Arof, H.A. Illias, High impedance fault location in 11kV underground distribution systems using wavelet transforms, Int. J. Electr. Power Energy Syst. 55 (Feb) (2014) 723–730, doi:10.1016/j.ijepes.2013.10.003.

[96] S. Barakat, M.B. Eteiba, W.I. Wahba, Fault location in underground cables using ANFIS nets and discrete wavelet transform, J. Electr. Syst. Inf. Technol. 1 (3) (2014) 198–211, doi:10.1016/j.jesit.2014.12.003.

[97] Y. Xiang, J.F.G. Cobben, A Bayesian Approach for Fault Location in Medium Voltage Grids With Underground Cables, IEEE Power Energy Technol. Syst. J. 2 (4) (2015) 116–124, doi:10.1109/jpets.2015.2477598.

[98] M. Shafiullah, Fault diagnosis in distribution grids under load and renewable energy uncertainties, King Fahd University of Petroleum & Minerals, Dhahran, Saudi Arabia, 2018.

[99] A.H. Al-Mohammed, Adaptive fault location in power system networks based on synchronized phasor measurements, King Fahd University of Petroleum and Minerals, Dhahran, Saudi Arabia, 2012.

[100] J.R. Altman, A practical comprehensive approach to PMU placement for full observability, Virginia Tech, Virginia, USA, 2008.

[101] A.G. Phadke, et al., The wide world of wide-area measurement, IEEE Power Energy Mag 6 (5) (2008) 52–65, doi:10.1109/MPE.2008.927476.

[102] IEEE Power & Energy Society, IEEE Std C37.118.2-2011 (Revision of IEEE Std C37.118-2005) - IEEE Standard for Synchrophasor Data Transfer for Power Systems. In: IEEE Stand. 2011.

[103] IEEE Power & Energy Society, IEEE Std C37.242-2013 - IEEE guide for synchronization, calibration, testing, and installation of phasor measurement units (PMUs) for power system protection and control. In: IEEE Stand. 2013, p. 95.
[104] IEEE Power & Energy Society, IEEE Std C37.247-2019 - IEEE Standard for Phasor Data Concentrators for Power Systems. In: IEEE Stand. 2019.
[105] IEEE Power & Energy Society, IEEE Std PC37.242/D4, Sep 2020 - IEEE Draft Guide for Synchronization, Calibration, Testing, and Installation of Phasor Measurement Units (PMUs) for Power System Protection and Control. In: IEEE Stand. 2021.
[106] IEEE Power & Energy Society, 60255-118-1-2018 - IEEE/IEC International Standard - Measuring relays and protection equipment - Part 118-1 : Synchrophasor for power systems - Measurements. In: IEEE Stand. 2018.
[107] M. Hojabri, U. Dersch, A. Papaemmanouil, P. Bosshart, A comprehensive survey on phasor measurement unit applications in distribution systems, Energies 12 (23) (2019) 4552, doi:10.3390/EN12234552.
[108] Phasor measurement unit, Elspec Ltd., 2021. https://www.elspec-ltd.com/metering-protection/g5-phasor-measurement-unit/. (Accessed 16 July 2021).
[109] Candura Instruments, PQPro power quality analyzer, 2021. https://www.candura.com/products/pqpro.html. (Accessed 16 July 2021).
[110] M. Shafiullah, M.J. Rana, M.S. Alam, M.A. Uddin, Optimal placement of Phasor Measurement Units for transmission grid observability, 2016 International Conference on Innovations in Science, Engineering and Technology (ICISET), Chittagong, Bangladesh, IEEE, 2016, pp. 1–4, doi:10.1109/ICISET.2016.7856492.
[111] E. Abiri, F. Rashidi, T. Niknam, M.R. Salehi, Optimal PMU placement method for complete topological observability of power system under various contingencies, Int. J. Electr. Power Energy Syst. 61 (Oct) (2014) 585–593, doi:10.1016/j.ijepes.2014.03.068.
[112] M. Shafiullah, M. Abido, M. Hossain, A. Mantawy, An improved OPP problem formulation for distribution grid observability, Energies 11 (11) (2018) 3069, doi:10.3390/en11113069.
[113] M. Shafiullah, M.I. Hossain, M.A. Abido, T. Abdel-Fattah, A.H. Mantawy, A modified optimal PMU placement problem formulation considering channel limits under various contingencies, Meas. J. Int. Meas. Confed. 135 (Mar) (2019), 875–885, doi:10.1016/j.measurement.2018.12.039.
[114] S.M. Ashraf, A. Gupta, D.K. Choudhary, S. Chakrabarti, Voltage stability monitoring of power systems using reduced network and artificial neural network, Int. J. Electr. Power Energy Syst. 87 (May) (2017) 43–51, doi:10.1016/J.IJEPES.2016.11.008.
[115] North American SynchroPhasor Initiative, Synchrophasor technology fact sheet, 2014. https://www.naspi.org/sites/default/files/reference_documents/33.pdf?fileID=1326. (Accessed 16 July 2021).
[116] A. Silverstein, Synchrophasors & the grid, 2017. https://www.energy.gov/sites/prod/files/2017/09/f36/2_ModernGrid-networked Measurement and Monitoring Panel - Alison Silverstein%2C NASPI.pdf. (Accessed 16 July 2021).
[117] North American SynchroPhasor Initiative, NASPI PMU map March 2017, 2017. https://www.naspi.org/node/749. (Accessed 16 July 2021).
[118] A. Esmaeilian, T. Popovic, M. Kezunovic, Transmission line relay mis-operation detection based on time-synchronized field data, Electr. Power Syst. Res. 125 (Aug) (2015) 174–183, doi:10.1016/J.EPSR.2015.04.008.
[119] A. Esmaeilian, M. Kezunovic, Fault location using sparse synchrophasor measurement of electromechanical-wave oscillations, IEEE Trans. Power Deliv. 31 (4) (2016) 1787–1796, doi:10.1109/TPWRD.2015.2510585.

[120] N.M. Khoa, D.D. Tung, Locating fault on transmission line with static var compensator based on phasor measurement unit, Energies 11 (9) (2018) 2380, doi:10.3390/EN11092380.
[121] W. Fan, Y. Liao, Wide area measurements based fault detection and location method for transmission lines, Prot. Control Mod. Power Syst. 4 (1) (2019) 1–12, doi:10.1186/S41601-019-0121-9 2019.
[122] A. Saber, PMU-based fault location technique for three-terminal parallel transmission lines with series compensation, Electr. Power Compon. Syst. 48 (4–5) (2020) 410–422, doi:10.1080/15325008.2020.1793836.
[123] S. Belagoune, N. Bali, A. Bakdi, B. Baadji, K. Atif, Deep learning through LSTM classification and regression for transmission line fault detection, diagnosis and location in large-scale multi-machine power systems, Measurement 177 (Jun) (2021) 109330, doi:10.1016/j.measurement.2021.109330.
[124] M. Gilanifar, H. Wang, J. Cordova, E.E. Ozguven, T.I. Strasser, R. Arghandeh, Fault classification in power distribution systems based on limited labeled data using multi-task latent structure learning, Sustain. Cities Soc. 73 (Oct) (2021) 103094, doi:10.1016/J.SCS.2021.103094.
[125] J.J. Chavez, et al., PMU-voltage drop based fault locator for transmission backup protection, Electr. Power Syst. Res. 196 (Jul) (2021) 107188, doi:10.1016/J.EPSR.2021.107188.

Metaheuristic optimization techniques

2.1 Introduction

Optimization processes are an integral part of our daily lives that can be defined as finding effective ways of available resource utilization without violating any operational constraints. They involve several steps, including the mathematical definition of the problems, system properties, identification of the variables and parameters, and conditions to be satisfied that yield the most desirable outcomes. The optimization problems that arise from the practical applications can be either constrained or unconstrained. In addition, they can be either continuous or discrete, linear or nonlinear, differentiable or nondifferentiable, etc. Throughout the years, the researchers explored and proposed many approaches to carry out solutions for the day-to-day optimization problems that can be ramified into two major groups: classical and metaheuristic techniques. Fig. 2.1 presents a generalized and simplified classification of the optimization techniques [1–7].

Among many classical approaches, convex programming, linear programming (LP), non-LP, quadratic programming, integer programming, mixed-integer LP, gradient descent, and Newton-Raphson methods are widely used. Conversely, the metaheuristic techniques are either natural, biological, or physical system inspired that can be ramified as either trajectory or population-based algorithms where the population-based algorithms can be further ramified into evolutionary and swarm-based metaheuristic techniques. Popular metaheuristics techniques include genetic algorithm (GA), differential evolution (DE), tabu search (TS), simulated annealing (SA), particle swarm optimization, backtracking search algorithm (BSA), and grey wolf optimization (GWO) [1–7].

The remaining parts of this chapter are organized as follows: Section 2.2 provides a brief introduction to classical optimization techniques, whereas Section 2.3 provides detailed discussions on different metaheuristic optimization techniques. Finally, an overall summary of this chapter is presented in Section 2.4.

2.2 Classical optimization techniques

The classical optimization techniques are analytical and often deal with the problems involving continuous and differential functions to offer optimal solutions. Such methods handle problems having single and multiple variables with no constraint or with equality and nonequality constraints. The classical techniques often use the Lagrange multiplier for the problems with equality constraints and Karush–Kuhn–Tucker (KKT) conditions for the problems with inequality constraints to achieve the optimum solutions [8]. This section introduces several popular classical optimization techniques such as LP, integer programming, mixed-integer LP, and non-LP.

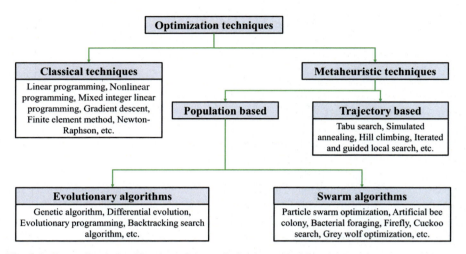

Fig. 2.1 Generalized classification of the optimization techniques [1–7].

2.2.1 Linear programming

LP achieves the optimal outcomes from the mathematical models whose requirements are linear by nature. More specifically, it optimizes linear objective functions subject to linear equality and inequality constraints. The feasible regions of such problems are convex polytopes, where the LP algorithm finds the points on the polytopes where the objective functions have the optimal values. The standard form of the LP problems consists of three parts namely, the objective function, constraints, and nonnegative variables are represented below for a problem with n number of variables [9]:

Part 1: a linear function to be minimized or maximized

$$f(x_1, x_2, \ldots, x_n) = c_1 x_1 + c_2 x_2 + \ldots + c_n x_n \tag{2.1}$$

Part 2: constraints of the problem

$$a_{11} x_1 + a_{12} x_2 + \ldots + a_{1n} x_n \leq b_1 \tag{2.2}$$

$$a_{21} x_1 + a_{22} x_2 + \ldots + a_{2n} x_n \leq b_2 \tag{2.3}$$

$$\ldots\ldots\ldots\ldots\ldots\ldots\ldots\ldots\ldots\ldots\ldots\ldots \tag{2.4}$$

$$a_{n1} x_1 + a_{n2} x_2 + \ldots + a_{nn} x_n \leq b_n \tag{2.5}$$

Part 3: nonnegative variables

$$x_1 \geq 0;\ x_2 \geq 0;\ \ldots x_n \geq 0 \tag{2.6}$$

The matrix representation of the problem becomes:

$$\text{maximize/minimize}: \{c^T x \mid Ax \leq b;\ x \geq 0\} \tag{2.7}$$

Here x represents the vector of variables, c is the vector of variable coefficients, and b is another vector of coefficients. The term A is a matrix of coefficients with a dimension of $n \times n$, and the term $(\bullet)^T$ represents matrix transpose where n is the number of variables. The term $c^T x$ is the objective function to be optimized where the inequalities $Ax \le b$ and $x \ge 0$ specify the convex polytope.

The LP problems can be solved graphically with a very low number of variables. The graphical approach is not practically possible with problems having more than two variables; therefore, the simplex method is employed to solve the LP problems in such cases. However, the representation of the LP problem in Eq. (2.7) is referred to as the *primal* problem that can be converted into a *dual problem* where the minimization problems turn into maximization problems and vice versa. The dual representation of the Eq. (2.7) can be represented as [9]:

$$\text{minimize/maximize} : \left\{ b^T y \mid A^T y \ge c; y \ge 0 \right\} \tag{2.8}$$

The LP is applied to various fields of studies such as mathematics, economics, business, food, agriculture, and engineering problems. A wide range of industries, including energy, telecommunication, transportation, and manufacturing, also use LP widely for optimization purposes [10–12].

2.2.2 Mixed-integer linear programming

The LP problems where the objective functions and the constraints are linear in which all the variables are restricted to be integers are known as the integer linear programming (ILP) problems. The matrix representation of the ILP problem becomes:

$$\text{maximize/minimize} : \left\{ c^T x \mid Ax \le b; x \ge 0; x \in \mathbb{Z}^n \right\} \tag{2.9}$$

Where b and c are the requirement and cost vectors, A is the coefficient matrix, Z refers to integer numbers, and x is unknown vectors and supposed to take only integer entries. However, the slack variable s is introduced to represent the ILP representation in a standard form that changes the ILP representation as:

$$\text{maximize/minimize} : \left\{ c^T x \mid Ax + s = b; x \ge 0; s \ge 0; x \in \mathbb{Z}^n \right\} \tag{2.10}$$

In a few cases, the ILP problems are known as pure ILP if all variables of the problems are restricted to be integers, and they are known as the decision or binary or Boolean or "0–1" ILP if the variables are restricted to take either "0" or "1" values. Conversely, in the mixed-integer linear programming (MILP) problems, some variables are restricted to be integers, and others can be nonintegers. Thus, MILP combines both integer and fractional (discrete and continuous) variables [13]. The MILP is a powerful but flexible technique in solving complex problems. It is often employed for system analysis and optimization. For instance, the MILP technique is used to solve a wide range of engineering and decision-making problems, including flight path planning [14], internet of things devices placement [15], phasor measurement unit placement [16], power flow optimization [17], and unit commitment in power system operation [18]. However, the technique is not new, but the invention of faster computation systems and software packages made it popular for researchers. The method is equally useful for pure-integer, pure-binary, and mixed (discrete and continuous) problems. The MILP problems are usually solved employing the linear-programming-based cutting plane, branch-and-bound, branch-and-cut, relaxation, and decomposition algorithms [19,20].

2.2.3 Nonlinear programming

Like LP problems, the nonlinear programming (NLP) problems also consist of three main parts: objective functions, constraints, and variable bounds. The fundamental difference is that in NLP, either the objective function or at least one of the constraints is nonlinear. As most of the real-world systems are nonlinear in nature, it is expected that most of the optimization problems are nonlinear [21]. Considering the problem complexities, different approaches such as substitution, Lagrange multiplier, KKT, fractional programming, and quadratic programming methods are employed to find the optimal solutions. The nonlinear problems are also solved using the iterative approach, where the users apply different trial and error procedures that converge to a solution in a finite number of steps. The users evaluate either Hessians or gradients or the function values to proceed toward the solutions in such methods. Different techniques like the interior point or Newton's methods evaluate the Hessians, whereas the coordinate descent, gradient descent, or quasi-Newton methods evaluate the gradients. Besides, the interpolation and pattern search approaches are employed to find function values. The NLP optimization techniques are often used to solve real-world problems, including bid evaluation, transportation, manufacturing, vehicle design and costing, parameter estimation, process optimization, telescope design, and structural optimization [22,23].

The classical optimization techniques have several drawbacks, including susceptibility to being trapped into the local optimum, complexity to be implemented in some cases, difficulties in solving discrete, and nondifferentiable optimization problems, the massive computer memory requirement, and susceptibility to numerical noises. Besides, the wrong selection of parameters may lead to a numerical ill-conditioning situation [24]. Despite having limitations, the classical optimization techniques form the basis for developing advanced optimization techniques to deal with complex real-life problems.

2.3 Metaheuristic techniques

As stated in the previous section, most real-world optimization problems comprise a wide range of complexities, including nonlinearity, nonconvexity, discontinuity, mixed nature of variables (discrete and continuous), and higher dimensionality. Besides, in a few cases, appropriate mathematical modeling of the optimization problems cannot be formulated. In such cases, the classical optimization approaches are either ineffective, impractical, computationally burdensome, or inapplicable. In response, metaheuristic search optimization techniques are introduced to deal with the mentioned complex problems. Glover, in 1986 introduced the term "metaheuristic" combining two terms, "meta" (meaning after or beyond) and "heuristic" (meaning find or discover). They are high-level problem-independent algorithmic frameworks to find, generate, or select the guidelines or strategies that provide sufficiently good solutions in a reasonable amount of time to any optimization problem, even to the problems with insufficient or limited information. However, they do not necessarily guarantee the optimal solutions to optimization problems; instead, they offer near-optimal solutions. Such techniques received widespread attention in various application domains due to their ease of implementation along with the mentioned appeal [1,25,26]. As illustrated in Fig. 2.1, the metaheuristic techniques are primarily divided into the trajectory and the population-based approaches. The population-based approaches are further ramified into two groups namely, the evolutionary and the swarm optimization techniques. The following parts of this section will demonstrate several metaheuristic approaches.

2.3.1 Trajectory-based metaheuristic techniques

The trajectory-based metaheuristic techniques iteratively improve a single solution to form a search trajectory in the solution space. They start with a single solution and replace the current one with another (often the best) found in the neighborhood. In general, they discover a locally optimal solution quickly; therefore, they are known as the exploitation-oriented techniques that promote intensification in the search space. The most common and popular trajectory-based search optimization techniques are SA, TS, guided local search, and iterated local search [1,2].

2.3.1.1 Simulated annealing

SA is a trajectory-based optimization method that simulates the physical annealing process in the optimization field by mapping the physical cooling process elements onto the optimization problem elements. Annealing is the physical process of melting a solid by heating it, then cooling it down slowly by decreasing the ambient temperature in steps. It maintains a constant temperature in each step for a while that is sufficient for the solid to reach the thermal equilibrium. Such an annealing process decreases the defects of the materials, thus, minimizes system energy. In 1953, Metropolis et al. [27] proposed a method for fast computing machines by simulating the physical annealing process. Their proposal forms the basis of the SA algorithm that consists of a modified Monte Carlo integration over configuration space. In the early 1980s, Kirkpatrick et al. [28] provided a detailed analogy of the annealing process in optimizing large and complex systems. In the SA algorithm, the physical system configurations are analogous to the solutions of the optimization problems. The cost of a solution is analogous to the energy of the configuration of the system. Besides, the SA algorithm introduces a control parameter that plays the role of physical system temperature. Steps of a generic SA algorithm are illustrated below, assuming, without loss of generality that the problem under investigation is a minimization one [29–31]:

Step 1: Initialization

The SA algorithm starts by generating a random initial solution (x_{initial}) from the search space with a dimension of the optimization problem, D, using Eq. (2.11). The algorithm also sets the initial value of the control parameter (C_p), multiplying factors (α and β), the maximum number of iterations (*ItMax*), and other required parameters.

$$x_{\text{initial}} = U\left(low_j, up_j\right) \qquad \forall j \qquad (2.11)$$

Here, $j = 1, 2, 3, \ldots, D$, and the terms "*up*" and "*low*" refer to the upper and the lower boundaries of the variables to be optimized, respectively. Finally, the term "*U*" stands for uniform distribution.

Step 2: Fitness evaluation and updating the current and best solutions

After initialization, the SA algorithm evaluates the fitness of the initial solution (J_{initial}) based on the specified objective function of the optimization problem. It then stores the initial solution as the best solution (x_{best}) and as the current solution (x_{current}). It also saves the fitness of the initial solution as the best fitness (J_{best}) and current fitness (J_{current}). This yields,

$$x_{\text{best}} = x_{\text{current}} = x_{\text{initial}} \qquad (2.12)$$

$$J_{\text{best}} = J_{\text{current}} = J_{\text{initial}} \qquad (2.13)$$

Step 3: Generation of the trial solution and fitness evaluation

This step generates a trial solution in the neighborhood of the current solution (x_{current}) and evaluates the fitness of the trial solutions (J_{trial}).

Step 4: Condition-based selection/rejection of the trial population

The algorithm accepts the trial solution as the current solution if anyone of the following conditions (C_1 or C_2) is satisfied:

$$C_1 : J_{\text{trial}} < J_{\text{current}} \qquad (2.14)$$

$$C_2 : \text{if } J_{\text{trial}} > J_{\text{current}} \text{ and } e^{\left(-\frac{J_{\text{trial}} - J_{\text{current}}}{C_p}\right)} \geq r \qquad (2.15)$$

Here, r is a uniform random number generated in the range of (0, 1). The second condition provides a chance to a particular trial solution even if it is lower fitted than the current solution with the hope of reaching a better solution through it in a future iteration. If the trial solution is accepted based on the second condition (C_2), the SA updates x_{current} and J_{current}. However, if the trial solution is accepted based on the first condition (C_1), the SA updates x_{current}, x_{best}, J_{current}, and J_{best}. Conversely, if the trial population is not accepted, the SA algorithm multiplies the control parameter (C_p) with a factor (α) larger than one and returns to the previous step. It continues the mentioned process until a trial population is selected. The method of increasing the C_p value is equivalent to the heating process in the physical systems.

Step 5: Checking the termination criteria

The SA algorithm checks the termination criteria if the trial solution is accepted based on the first condition (C_1) and skips this step if it is accepted based on the second condition (C_2). It is worthy of mentioning that the algorithm starts checking the termination criteria after a prespecified number of iterations to avoid premature convergence. It terminates the optimization process if the values of the best solution do not change for a pr-specified number of iterations or it reaches the maximum number of iterations. Then, the technique displays x_{best} as the optimal solution to the optimization problem under investigation. Otherwise, it proceeds to the next step.

Step 6: Checking thermal equilibrium condition and updating C_p value

This step first checks the thermal equilibrium condition for example, no change in the trial population for a certain number of iterations. If the condition is satisfied, the SA updates the control parameter value by multiplying it with a factor (β) smaller than 1 (e.g., 0.80 to 0.99) that is equivalent to the cooling schedule of the physical systems. Then, the SA algorithm returns to Step 3. However, if the thermal equilibrium condition is not satisfied, it returns to Step 3 directly without updating the C_p value. Fig. 2.2 shows a generic flowchart, and Alg. 2.A presents the pseudocode of the SA technique.

Alg. 2.A: Pseudocode of the simulated annealing algorithm

Input: SA parameters and optimization problem information
1: Initialization of the initial solution
2: Fitness evaluation of the solution
3: Storing the initial solution as the current and best solutions and fitness
4: **for** *iteration = 2: ItMax*, **do**
5:     Generating a trial solution in the neighborhood of the current solution inside the search space
6:     Evaluating the fitness of the trial solution
7:     **if** the conditions of Eqs. (2.14) and (2.15) are satisfied, **then**.
8:         **if** the condition of Eq. (2.14) is satisfied, **then**.
9:             Updating current and best solutions and associated fitness values
10:            **if** termination criteria are met, **then**
11:                 Stopping the SA algorithm
12:            **else if**
13:                 Going to Line # 17
14:            **end if**
15:         **else if** the condition of Eq (2.15) satisfied, **then**.
16:            Updating current solution and associated fitness value
17:     **end if**
18:     **if** thermal equilibrium condition met, **then**
19:         Updating C_p value multiplying with a factor (β) smaller than 1 (**cooling down process**)
20:     **else if**
21:         Returning to Line # 5
22:     **end if**
23:     **else if** the trial population is not selected, **then**
24:         Updating C_p value multiplying with a factor (α) larger than 1 (**heating up the process**)
25:     **end if**
26: **end for**
27: Assigning the best solution as the global best solution.
Output: The best solution to the optimization problem under investigation.

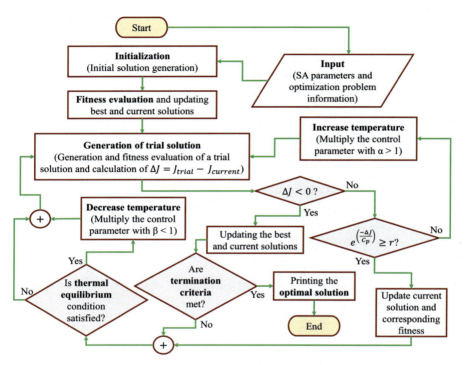

Fig. 2.2 Generic flowchart of the SA algorithm. *SA*, simulated annealing.

2.3.1.2 Tabu search algorithm

TS is another trajectory-based metaheuristic technique that guides the local (neighborhood) search procedure by exploring the solution space beyond local optimality. Like SA, TS starts with a single solution and searches for better solutions, applying actions and moving between the neighboring solutions. It manages the acceptance, applicability, and availability of actions using a set of rules. Unlike the descent algorithms that are unidirectional in nature and end at the local optimal solutions in general, the TS improves local search methods performance through exploration and avoids entrapment in local optimal solutions using flexible memory of the search history. The main feature of the algorithm is the "taboo (tabu being a different spelling of the same word) list" or "forbidden list" of solutions that are prepared from the visited potential solutions or the rule violating solutions. The TS does not revisit the tabu solutions in a certain period. Firstly, Fred Glover introduced the TS algorithm in 1986 [32] and gave it a formal shape in 1990 [33,34]. With time, many variations of the TS algorithm have been proposed, including directed TS [35], robust TS [36], and reactive TS [37]. Due to its simplicity and ease of implementation, the TS has been adopted in many fields such as vehicle routing, manufacturing, supply chain, power system, and telecommunication. The steps of a generic TS algorithm are illustrated below, assuming, without loss of generality that the problem under investigation is a minimization one [38–40]:

Step 1: Initialization

The TS algorithm starts by initializing the initial solution ($x_{initial}$) from the search space with a dimension, D, of the optimization problem under investigation. It also sets the number of trial solutions (n_{trial}), tabu list size (L), the maximum number of iterations (*ItMax*), and other required parameters along with the cost/objective function of the problem. Similar to SA, TS employs the following equation to generate an initial solution:

$$x_{initial} = U(low_j, up_j) \quad \forall j \quad (2.16)$$

Here, $j = 1, 2, 3, ..., D$, and the terms "*up*" and "*low*" refer to the upper and the lower boundaries of the variables to be optimized, respectively. Finally, the term "*U*" stands for uniform distribution.

Step 2: Fitness evaluation and updating the current and best solution

After initialization, the TS algorithm evaluates the initial solution fitness ($J_{initial}$) based on the objective function of the optimization problem. Then, it stores the initial solution as both the best solution (x_{best}) and the current solution ($x_{current}$). It also saves $J_{initial}$ as the best fitness (J_{best}) and the current fitness ($J_{current}$) of the problem. This yields,

$$x_{best} = x_{current} = x_{initial} \quad (2.17)$$

$$J_{best} = J_{current} = J_{initial} \quad (2.18)$$

Step 3: Generation of trial solutions and updating the best solution

This step generates a prespecified number of trial solutions in the current solution neighborhood and evaluates their fitnesses. Then, it sorts the solutions and their associated fitnesses in ascending order. If i^{th} trial solution and its associated fitness are denoted by x_{trial}^i and J_{trial}^i, where $i = 1, 2, 3, ..., n_{trial}$, then, x_{trial}^1 and J_{trial}^1 stand for the best trial solution and its associated fitness. If $J_{trial}^1 < J_{best}$, the algorithm updates x_{best} and J_{best}, then it proceeds to the next step.

Step 4: Checking the termination criteria

This step checks the termination criteria of the TS algorithm after a prespecified number of iterations to avoid premature convergence. It terminates the optimization process if the values of the best solution do not change for a prespecified number of iterations or it reaches the maximum number of iterations. Then, the technique displays x_{best} as the optimal solution to the optimization problem. Otherwise, it proceeds to the next step.

Step 5: Updating the tabu list and current solution

This step checks the tabu status of the x_{trial}^i starting from the best one (e.g., $i = 1$). If the best trial solution is not in the tabu list, it will be considered as the current solution and added to the tabu list, and the algorithm will move to Step 3. Otherwise, the algorithm checks the tabu status of the second-best trial solution and continues until it checks all trial solutions as per their ascending fitness values. The first nontabu-listed solution encountered will be considered as the current solution and added to the tabu list, and the algorithm will move to Step 3. If all trial solutions are listed in the tabu list, the TS algorithm moves to Step 3 without updating the tabu list and current solution that is unlikely to occur if $L < n_{\text{trial}}$.

It is worth noting that the tabu list size plays a vital role in the high-quality solution search as the algorithm observes occurring of cycling if the size is too small and the deterioration of the solution quality by forbidding many moves if the size is too large. According to Ref. [32,41], a tabu list of sizes between 7 and 15 worked well for many of the optimization problems. Additional elements like intensification, diversification, allowing infeasible solutions, aspiration criteria, and surrogation are incorporated in a few cases to make the TS algorithm more efficacious [42]. Fig. 2.3 shows a generic flowchart, and Alg. 2.B presents the pseudocode of the TS technique.

Alg. 2.B: Pseudocode of the tabu search algorithm

Input: TS parameters and optimization problem information
1: Initialization of the initial solution
2: Fitness evaluation of the solution
3: Storing the initial solution as the current and best solutions and fitness
4: **for** *iteration=2: ItMax,* **do**
5:     Generating the trial solutions in the neighborhood of the current solution inside the search space
6:     Evaluating the fitness of the trial solutions
7:     Sorting the trial solutions in ascending order as per their fitness values
7:     **if** the fitness of the best trial solution is better than the previously-stored best solution, **then**
8:         Updating current and best solutions and associated fitness values
9:     **end if**
10:    **for** *i=1: n*$_{\text{trial}}$, **do**
11:        **if** the trial solution (x^i_{trial}) not found in the tabu list, **then**
12:            Adding the trial solution (x^i_{trial}) in the tabu list and setting $x_{\text{current}} = x^i_{\text{trial}}$
13:            Stopping the for loop
14:        **end if**
15:    **end for**
16:    **if** termination criteria met, **then**
17:        Stopping the TS algorithm
18:    **end if**
19: **end for**
20: Assigning the best solution as the global best solution.
Output: The best solution to the optimization problem under investigation.

Despite having multiple advantages, the trajectory-based optimization algorithms are computationally expensive and can get stuck in a particular area of the search space (local optima) [43,44].

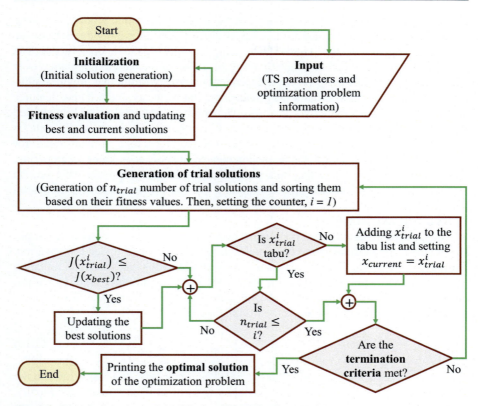

Fig. 2.3 Generic flowchart of the TS algorithm. *TS*, tabu search.

2.3.2 *Evolutionary metaheuristic techniques*

The evolutionary metaheuristic techniques are the generic population-based random search algorithms that consist of a population of individuals where each of them represents a solution in the search space. The individuals are randomly initialized and then exposed to the learning processes of selection, recombination, and mutation from generation to generation that evolve the newly created individuals towards more favorable regions in the search space. The evolutionary algorithms terminate their search based on the prespecified termination criterion, and the best individual of the last generation represents the optimal (near optimal) solution. Though researchers proposed many evolutionary algorithms, their basic structures are somehow like each other [7,45]. This section illustrates three widely adopted evolutionary algorithms: GA, DE, and BSA.

2.3.2.1 *Genetic algorithm*

In the 1970s, J. H. Holland developed the stochastic global search algorithm inspired by Charles Darwin's theory of natural evolution [1]. It reflects the natural selection process where the fittest individuals are selected for reproduction to produce offspring for the next generation. At the same time, it discards the unfit individuals from the race. After selecting the fittest individuals, it applies other biological operators such as recombination (crossover) and mutation to evolve toward the natural population over many generations. GA is a problem-independent search algorithm that provides high-quality solutions even without a deep understanding of the problem. Other features such as parallelism, simplicity, ease of implementation, adaptability, and diversified search capability are the primary enablers of its widespread attention. To date, it is the most popular and widely adopted metaheuristic technique in many fields for example, life science, social science, mathematics, finance, economics, engineering, and management. It can be either binary-coded or real-coded. Major steps of a generic real-coded GA are illustrated below, assuming that the problem under investigation is a minimization one:

Step 1: Initialization

The GA generates its initial population randomly based on the prespecified population size (N) and dimension (D) of the optimization problem under investigation. During this stage, the GA also sets other parameters like the crossover and mutation probabilities (p_c and p_m) and the number of elites (n_{elite}), and other required parameters along with the maximum number of generations (*GenMax*). The GA employs the following equation to generate the initial population:

$$P = U\left(low_j, up_j\right) \qquad \forall\, i\, \&\, j \qquad (2.19)$$

Here, $i = 1, 2, 3, \ldots, N$ and $j = 1, 2, 3, \ldots, D$. The terms "*up*" and "*low*" refer to the upper and the lower boundaries of the variables to be optimized, respectively. Finally, the term "*U*" stands for uniform distribution.

Step 2: Fitness evaluation and updating the global best solution

The GA evaluates the fitness of each individual of initially generated/updated populations based on the objective function of the optimization problem. It stores the best fitness value and associated individual as the best objective function value (J_{best}) and the global best solution (P_{best}), respectively.

Step 3: Checking the termination criteria

This step checks the termination criteria after a prespecified number of generations to avoid premature convergence. It terminates the optimization process if the values of the global best solution do not change for a prespecified number of generations or the algorithm reaches the maximum number of generations. Then, the technique displays the global best individual as the optimal solution to the optimization problem. Otherwise, it proceeds to the next step.

Step 4: Selection

This step selects the fittest individuals following the principle of "survival of the fittest" and forms a mating pool (intermediate population). This step aims at facilitating fitter individuals to mate to produce better offsprings in the subsequent generation. The researchers proposed many selection methods for GA over the years. The most common selection procedures are the roulette wheel selection, rank selection, tournament selection, and Boltzmann selection [46]. In one variation of the tournament selection method, the GA chooses two random individuals, selects the fitter one as the first parent (p_1), and follows the same procedure to select the second parent (p_2). Besides, a certain amount of top fittest individuals, also known as the elites, are selected and transferred to the next generation, generally, without passing through the crossover and mutation operations. Such selection of elitism guarantees the improvement of populations and assists in faster convergence. However, the elite individuals will still be in the pool of the selection process [47].

Step 5: Crossover

After the selection of two parents, the GA passes them through its crossover operator. Like selection methods, many GA crossover variations (e.g., flat, simple, arithmetical, discrete, linear, and BLX-α) have also been evolved throughout the decades. The BLX-α crossover strategy generates two offspring (o_1 and o_2) from the two parents selected in the previous step (p_1 and p_2) using the following equations [48]:

$$c_{max} = \max(p_1, p_2) \quad \forall j \quad (2.20)$$

$$c_{min} = \min(p_1, p_2) \quad \forall j \quad (2.21)$$

$$I = c_{max} - c_{min} \quad \forall j \quad (2.22)$$

$$o_1 = U(c_{max} - I \cdot \alpha, c_{max} + I \cdot \alpha) \quad \forall j \quad (2.23)$$

$$o_2 = U(c_{max} - I \cdot \alpha, c_{max} + I \cdot \alpha) \quad \forall j \quad (2.24)$$

The value of "α" can be any number between [0, 1]. However, the *BLX-α* crossover turns into a flat crossover if "$\alpha = 0$" [48,49]. It is worthy to note that the GA performs the crossover operation if the crossover probability (p_c) is less than a uniform random number generated in the range of [0, 1]; otherwise, it passes the parents as the offspring directly to the next round of generation. The crossover probability is set to a higher number for example [0.60–1.0], which means a significant number of individuals will be going through this operation.

Step 5: Mutation

In general, the mutation operation takes place after the crossover operation. It maintains genetic diversity from generation to another by altering one or more gene values in a chromosome. It creates adaptive individuals that assist the algorithm from not getting trapped into the local optimum. As pointed out earlier, the mutation probability is set during the initialization stage of the GA that dictates the occurrence of mutation on the offspring. However, this probability value is set to a lower number for example [0.01–0.10], which means fewer individuals will be going through this operation. Like selection and crossover methods, the researchers proposed many mutation methods (random mutation, nonuniform mutation, boundary mutation, and power mutation) over the years [48,50]. Mathematical representation of the random mutation method is given in the following equation when a certain offspring is selected for mutation operation based on the mutation probability:

$$\text{if } r_1 < r_2 : o_{\#,j} = U(low_j, up_j) \quad \forall j \quad (2.25)$$

Here, r_1 and r_2 are randomly generated numbers from the range of [0, 1].

Step 6: Checking the variables boundary

After updating the individuals through selection, crossover, and mutation operations, their boundaries (constraints of the objective function) are checked as some of the elements of some of the individuals might go beyond the search space. Among several boundary condition mechanisms, one is to set the elements to the respected upper bounds if they become larger than that of the upper bound values. Conversely, if the values of the elements become lower than that of the lower bound values, the GA sets those specific violating elements of the individual to respected lower bounds. The mentioned boundary checking mechanism can be expressed mathematically as follows [51]:

$$\text{if } P_{i,j} > up_j : P_{i,j} = up_j \text{ ; else if } P_{i,j} < low_j : P_{i,j} = low_j; \qquad \forall \ i \ \& \ j \qquad (2.26)$$

The GA returns to Step 2 after checking the boundary and repairing the individuals in case of any violation. Fig. 2.4 shows a generic flowchart, and Alg. 2.C presents the pseudocode of the GA.

Alg. 2.C: Pseudocode of the genetic algorithm

Input: GA parameters and optimization problem information
1: Initialization of the initial population
2: **for** *generation=1: GenMax*, **do**
3:     Fitness evaluation of each individual of the initial/updated population
4:     Storing/updating the best fitness and associated individual as the global best solution
5:     **if** termination criteria met, **then**
6:         Stopping the GA algorithm
7:     **else**
8:         Selecting the parent pairs and the elites
9:         **if** a randomly generated number between 0 and 1 is less the crossover probability, **then**
10:           Executing crossover operation
11:         **end if**
12:         **if** another randomly generated number between 0 and 1is less the mutation probability, **then**
13:           Executing mutation operation
14:         **end if**
15:         Checking boundaries of the newly generated population and repairing them in case of violation
16::   **end if**
17: **end for**
18: Assigning the best fittest individual as the global best solution.
Output: The best solution to the optimization problem under investigation.

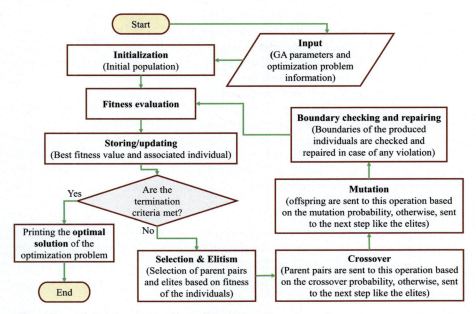

Fig. 2.4 Generic flowchart of the genetic algorithm.

2.3.2.2 Differential evolution

In 1995, R. Storn and K. Price proposed another widely adopted evolutionary metaheuristic optimization technique namely, the DE [52–54]. The DE algorithm demonstrates better preference than its peer algorithms due to its simplicity, robustness, parallelism, diversification, faster convergence, lower space complexity, the requirement of a minimal number of control variables, and adaptability [54–56]. Like GA, the DE algorithm has been adopted in many fields, including science and engineering, single and multiobjective function optimization, market modeling, clustering, neural network training, and feature selection [57,58]. In general, the DE algorithm comprises the following steps assuming that the problem under investigation is a minimization one:

Step 1: Initialization

Like other metaheuristic optimization techniques, the DE algorithm starts with the initialization step, where it generates the initial population randomly based on the pre-specified population size (N) and dimension (D) of the optimization problem. Besides, the DE algorithm sets other parameters like the mutation factors (F_1 and F_2), crossover factor (C_R), and other required parameters along with the maximum number of generations (*GenMax*). It employs the following equation for initial population generation:

$$P = U\left(low_j, up_j\right) \qquad \forall\, i\, \&\, j \qquad (2.27)$$

Here, $i = 1, 2, 3, \ldots, N$ and $j = 1, 2, 3, \ldots, D$. The terms "*up*" and "*low*" refer to the upper and the lower boundaries of the variables to be optimized, respectively. Finally, the term "*U*" stands for uniform distribution.

Step 2: Fitness evaluation and updating the global best solution

In this stage, the DE algorithm evaluates the fitness of each individual of the initially generated population (J_P) based on the objective function of the optimization problem. It stores the best fitness value and associated individual as the best objective function value (J_{best}) and the best solution (P_{best}), respectively.

Step 3: Checking the termination criteria

After fitness evaluation and updating the best solution, the DE algorithm checks the termination criteria after a prespecified number of generations to avoid premature convergence. It terminates the optimization process if the value of the best solution does not change for a prespecified number of generations or the algorithm reaches the maximum number of generations. Then, the technique displays the best individual as the optimal solution to the optimization problem. Otherwise, it proceeds to the next step.

Step 3: Mutation

In the DE algorithm, the mutation operation is the first step toward generating the new pool of individuals. It generates the mutant vector (M) using several ways where one of them is given in Eq. (2.28) [56]:

$$M_{i,j} = P_{i,j} + F_1 \times \left(P_{rand1,j} - P_{rand2,j} \right) + F_2 \times \left(P_{best,j} - P_{i,j} \right) \qquad \forall\, i\, \&\, j \qquad (2.28)$$

Here, P_{best} refers to the best individual as stored in step 2; P_{rand1} and P_{rand2} are randomly selected two different individuals from the population pool; F_1 and F_2 are mutation factors (usually less than 1.00) are set during the initialization stage.

Step 4: Crossover

The DE algorithm further perturbs the mutant vector individuals by applying the crossover operation that assists the algorithm in achieving diversified individuals. In this step, a trial solution is formed as a combination of mutant solution and parent solution. For each dimension (parameter), the DE algorithm generates a uniform random number (r) in the range of (0, 1). If the generated number of a specific individual is greater than the crossover factor (C_R), that plays a role in controlling the smoothness of the convergence; the DE algorithm selects that parameter of the trial vector (T) from the mutant vector (M), otherwise, from the parent vector (P). The C_R is usually set to a small number that facilitates more individuals from the mutant vector. However, the DE algorithm crossover operations can be mathematically illustrated by the following equation:

$$\text{if } r < C_R : T_{i,\#} = M_{i,\#}; \text{ else } T_{i,\#} = P_{i,\#} \qquad \forall\, i \qquad (2.29)$$

Step 5: Checking the variables boundary

After updating the parameters through mutation and crossover operations, their boundaries are checked as some of them might go beyond the search space. Among several boundary condition mechanisms, one is to set the parameters to the respected violated bounds. The mentioned boundary checking mechanism can be expressed mathematically as follows [51]:

$$\text{if } T_{i,j} > up_j : T_{i,j} = up_j\, ; \text{ else if } T_{i,j} < low_j : T_{i,j} = low_j; \qquad \forall\, i\, \&\, j \qquad (2.30)$$

The DE algorithm returns to the next step after checking the boundary and repairing the individuals in case of any violation.

Step 6: Selection

The DE algorithm evaluates the fitness of all trial solutions (J_T). It then compares the fitness of the parent vector with the fitness of the trial vector and the winner will be considered as a parent in the next generation. At the same time, the DE algorithm updates the corresponding fitness of the new parent accordingly. The following equation illustrates the selection process of the DE algorithm:

$$\text{if } J_{T,i} > J_{P,i} : P_{i,\#} = T_{i,\#} \ \& \ J_{P,i} = J_{T,i} \qquad \forall \ i \qquad (2.31)$$

After updating all parent vectors and their corresponding fitness values, the DE algorithm updates the best individual and associated fitness. Then, it returns to Step 3. Fig. 2.5 shows a generic flowchart, and Alg. 2.D presents the pseudocode of the DE algorithm.

Alg. 2.D: Pseudocode of the DE algorithm

Input: DE parameters and optimization problem information
1: Initialization of the population (P)
2: **for** *generation=1: GenMax,* **do**
3:     Fitness evaluation of P individuals
4:     Storing/updating the best fitness and associated individual as the best solution
5:     **if** termination criteria met, **then**
6:         Stopping the DE algorithm
7:     **else**
8:         Generation of the mutant vector (M)
9:         **for** *dimension=1: N,* **do**
10:           **if** a randomly generated number is greater than the crossover factor, **then**
11:               Select trial vector (T) parameter from the M vector
12:           **else**
13:               Update T vector parameter and corresponding fitness from the P vector
14:           **end if**
12:           Checking boundaries of T vector individuals and repairing them in case of violation
13:           Fitness evaluation of T vector individuals
15:           **if** the T vector fitness is better than the corresponding P vector, **then**
18:               Update P vector and corresponding fitness from the T vector
19:           **else**
20::   Keep P vector individual and corresponding fitness as is
21:           **end if**
22:         **end for**
23:     **end if**
24: **end for**
25: Assigning the best fittest individual as the best solution.
Output: The best solution to the optimization problem under investigation.

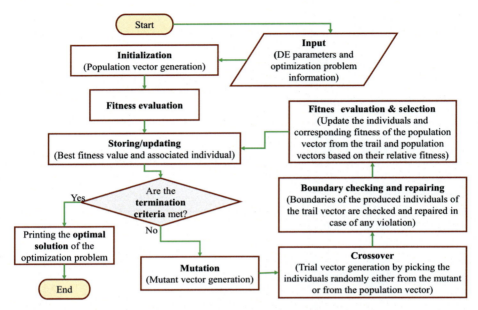

Fig. 2.5 Generic flowchart of the DE algorithm. *DE*, differential evolution.

2.3.2.3 Backtracking search algorithm

In 2013, Pinar Civicioglu developed the BSA, where the leading operators of the algorithm are initialization, selection-I, mutation, crossover, and selection-II. The BSA shares similar names for the operators like other evolutionary algorithms for example, GA and DE. However, the mutation, crossover, and selection operations of the BSA are different from other algorithms. According to the algorithm developer, it produces a diversified and effective population in each generation through its distinct operators. Besides, it controls the search direction and its amplitude in a balanced way as per the global and the local search criteria. Another distinct feature of the algorithm is its historical population that attain diversity even in the advanced generations by playing a dynamic role [59]. Major steps of the BSA algorithm are illustrated below, considering the problem under investigation as a minimization one:

Step 1: Initialization

The BSA uses the following Eq. (2.32) to generate a set of individuals in the initial population (P) randomly based on the prespecified population size (*N*) and dimension (*D*) of the optimization problem. It also generates another set of the historical population (H) of the same size using Eq. (2.33) during the initialization stage. BSA also sets the maximum number of generations (*GenMax*), control parameter (F_c), control parameter multiplying factor (m_f) values, and other required parameters.

$$P = U\left(low_j, up_j\right) \qquad \forall\, i\, \&\, j \qquad (2.32)$$

$$H = U\left(low_j, up_j\right) \qquad \forall\, i\, \&\, j \qquad (2.33)$$

Here, $i = 1, 2, 3, \ldots, N$ and $j = 1, 2, 3, \ldots, D$. The terms "*up*" and "*low*" refer to the upper and the lower boundaries of the parameters/variables to be optimized, respectively. Finally, the term "*U*" stands for uniform distribution.

Step 2: Fitness evaluation

This step evaluates the fitness of each individual of the population (J_P) based on the objective function of the optimization problem. The BSA stores the best fitness value and associated individual as the best objective function value (J_{best}) and the global best solution (P_{best}), respectively.

Step 3: Selection I

This step updates the historical population to find the search track for the BSA by generating two random numbers (r_1, r_2) for each parameter in the range of [0, 1]. Then, it employs the "*if-else*" rule as per Eq. (2.34) and updates the historical population. Then, the BSA shuffles the positions of the parameters of the updated historical population using Eq. (2.35).

$$\text{if } r_1 < r_2 : H_{i,\#} = P_{i,\#}; \text{else } H_{i,\#} = H_{i,\#} \qquad \forall\, i \qquad (2.34)$$

$$H = \# \text{ shuffle } (H) \qquad (2.35)$$

Step 4: Mutation

The BSA produces the primary trial population (T) in this step using Eq. (2.36), where F_c controls the search direction and amplitude of the difference matrix (H-P). Besides, the control parameter value is updated in each generation using Eq. (2.37), where the multiplying factor (m_f) is a fixed number less than one. In general, the F_c and m_f values are set as "2" and "0.99," respectively.

$$T = P + F_c \cdot (H - P) \qquad (2.36)$$

$$F_c = m_f F_c \qquad (2.37)$$

After the mutation operation, it checks the violating elements of the trial population and brings back the violating elements inside the search space. For instance, if the element of the ith row and j^{th} column of the T either become larger or lower than that of the boundary values, then the BSA uses Eq. (2.38) to update them.

$$T_{i,j} = U\left(low_j, up_j\right) \qquad \forall\, i \,\&\, j \qquad (2.38)$$

Step 5: Crossover

This is the most complex operators of the BSA technique, and it produces the final trial population (Q). The BSA starts the process by generating a binary integer-valued matrix (B) having the same size (number of rows and columns) as the trial population. The matrix elements are randomly generated and take either "*0*" or "*1*" values. Then, the BSA selects the elements of the final trial population from either primary trial population (T) or regular population (P) using the following equation:

$$\text{if } B_{i,j} = 0 : Q_{i,j} = T_{i,j}; \text{else } Q_{i,j} = P_{i,j}, \qquad \forall\, i \,\&\, j \qquad (2.39)$$

Step 5: Selection-II

In this step, the BSA evaluates the fitness of the final trial population (J_Q) individuals. Then, it updates the regular population matrix (P) and associated fitness vector (J_P) based on the relative fitness using Eq. (2.40). This step updates the best fitness value and the global best solution for the optimization problem if the best fitness value of the updated fitness vector (J_P) is better than that of the previously stored values.

$$\text{if } J_{Q,i} < J_{P,i} : P_{i,\#} = Q_{i,\#} \text{ \& } J_{P,i} = J_{Q,i} \qquad \forall\, i \qquad (2.40)$$

Step 6: Stopping criteria

Like other metaheuristic algorithms, the BSA checks the termination criteria after a prespecified number of generations to avoid premature convergence after updating the global best solution. It terminates its operation if the best fitness value does not change for a prespecified number of generations or the algorithm reaches a prespecified number of generations. The technique then displays the global best individual as the optimal solution to the optimization problem under investigation. Otherwise, it returns to Step 3. Fig. 2.6 shows a generic flowchart, and Alg. 2.E presents the pseudocode of the BSA technique.

Alg. 2.E: Pseudocode of the BSA

Input: BSA parameters and optimization problem information
1: Initialization of the initial population (P_{pop}) and historical population (P_{his})
2: **for** *generation=1: GenMax*, **do**
3:     Fitness evaluation of the initial population (P_{pop})
4:     Storing/updating the best fitness and associated individual as the best solution
5:     **if** termination criteria met, **then**
6:         Stopping the BSA algorithm
7:     **else**
8:         Updating individuals of the P_{his} from the P_{pop} and P_{his} randomly.
9:         Preparing the initial trial population (T_m) and checking and updating them in case of violation.
10:        Preparing final trial population (T_f) and evaluating the fitness of the individuals.
11:        Updating individuals of the P_{pop} from the P_{pop} and T_f based on individuals relative fitness.
12:    **end if**
13: **end for**
14: Assigning the position associated with the best fittest individual as the global best solution.
Output: The best solution to the optimization problem under investigation.

2.3.3 Swarm-based metaheuristic techniques

The swarm-based metaheuristic techniques adopt the exceptional collective intelligence features of various swarm behaviors in nature that are mainly related to how swarm

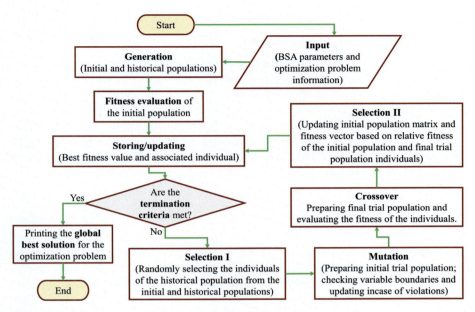

Fig. 2.6 Generic flowchart of the BSA. *BSA*, backtracking search algorithm.

individuals communicate to reach out to the best food source. They communicate in different forms such as chemical messenger (ant pheromone), broadcasting (firefly communication style), and dance (bee waggle dance). The swarm individuals execute the search based on personal (cognitive behavior) and combined (social behavior) experiences. In general, the swarms are composed of decentralized search agents without having any supervisor. They do not have a predefined global plan where the solution emerges based on their experience and communication. However, the swarms are robust; they can research near-optimal solutions even after the failure of several agents. In addition to the mentioned behaviors of the swarm, the researchers blended other mechanisms for example, exploration and exploitation to enhance the capability of the swarm intelligent techniques [60–62]. This section illustrates three widely adopted swarm-based metaheuristic algorithms: particle swarm optimization, constriction factor particle swarm optimization, and GWO.

2.3.3.1 Particle swarm optimization

In 1995, Kennedy and Eberhart [63] proposed the swarm-based metaheuristic technique called the particle swarm optimization (PSO) by combining the cognitive and social behavior of the swarms. The PSO algorithm was motivated by organism behavior such as bird flocking and fish schooling that is flexible and easy to implement with global and local exploration and exploitation capabilities. The implicit parallelism makes the algorithm less susceptible to getting trapped on a local optimum. The flexibility of the algorithm in controlling the balance between the global and local exploration of the search space assists the algorithm in overcoming the premature convergence issue. In general, for a minimization problem, the PSO algorithm consists of the following steps:

Step 1: Initialization

The PSO algorithm uses the following Eq. (2.41) and (2.42) to generate random positions (x) from the search space and random velocities (v) from the prespecified range for a pre-specified number of particles (N). Therefore, the sizes of the position and velocity matrices are $N \times D$, where D is the dimension of the optimization problem for example, the number of variables to be optimized. During the initialization stage, the PSO algorithm also sets the maximum number of iteration ($ItMax$), inertia weight (w), decrement constant (α), cognitive parameter (c_1), social (c_2) parameter values, and other required parameters.

$$x = U\left(low_j, up_j\right) \qquad \forall\, i\, \&\, j \qquad (2.41)$$

$$v = U\left(v_{\min,j}, v_{\max,j}\right) \qquad \forall\, i\, \&\, j \qquad (2.42)$$

Here, $i = 1, 2, 3, \ldots, N$ and $j = 1, 2, 3, \ldots, D$. The terms "up" and "low" refer to the upper and the lower boundaries of the parameters/variables to be optimized, respectively. Besides, $v_{\min}$ and $v_{\max}$ are the pre-specified minimum and maximum velocities of the particles, respectively. Finally, the term "U" stands for uniform distribution. The PSO algorithm uses the following equation to evaluate the values of the velocities where the value of M is also prespecified and vary from problem to problem:

$$v_{\max} = \frac{up_j - low_j}{M} \quad \text{and} \quad v_{\min} = -v_{\max} \qquad \forall\, j \qquad (2.43)$$

Step 2: Fitness evaluation and storing the individual and global best solutions

This step evaluates the fitness of initially generated particle positions based on the objective function of the optimization problem. Then, it stores each particle position and associated fitness as the individual best solution and individual best fitness. Besides, it evaluates the best solution from the individual best solutions and stores the particle position and associated fitness as the global best solution (x^{**}) and the global best fitness (J_{best}).

Step 3: Updating the inertia weights, velocities, and positions of the particles

This is the main step of the PSO algorithm where it updates the inertia weight, then the velocity of the particles using the updated inertia weight, particle positions, and individual (cognitive behavior) and global (social behavior) best solutions of the previous iteration. Finally, it adds the updated velocity to the particle positions of the previous iteration to get the new particle positions. Mathematical representations of the discussed operations are given below:

$$w(t) = \alpha * w(t-1) \qquad (2.44)$$

$$v_{j,k}(t) = w(t) \times v_{j,k}(t-1) + c_1 r_1 \times \left(x_{j,k}^{*}(t-1) - x_{j,k}(t-1)\right)$$
$$+ c_2 r_2 \times \left(x_{j,k}^{**}(t-1) - x_{j,k}(t-1)\right) \qquad (2.45)$$

$$x_{j,k}(t) = v_{j,k}(t) + x_{j,k}(t-1) \qquad (2.46)$$

Here, t and t-1 stand for the tth and $(t-1)^{th}$ iterations; j and k refer to the jth particle and the kth element of the jth particle; x^* and x^{**} stand for the individual and global best solutions of $(t-1)^{th}$ iteration. It is worth mentioning that the PSO algorithm sets a small value to α that is close but less than one; w is the inertia weight; and finally, c_1 and c_2 are assigned to some positive constant numbers.

Step 4: Checking the variables boundary

After updating the velocities and positions of the particles, their boundaries are checked as some of the elements of some of the particles might go beyond the search space. Among several boundary condition mechanisms, one is to set the elements to the respected violated bounds. The mentioned boundary checking mechanism can be expressed mathematically as follows [51]:

$$\text{if } x_{i,j} > up_j : x_{i,j} = up_j \text{ ;else if } x_{i,j} < low_j : x_{i,j} = low_j; \qquad \forall\, i\, \&\, j \qquad (2.47)$$

$$\text{if } v_{i,j} > v_{\max,j} : v_{i,j} = v_{\max,j} \text{ ;else if } v_{i,j} < v_{\min,j} : v_{i,j} = v_{\min,j}; \qquad \forall\, i\, \&\, j \qquad (2.48)$$

Step 5: Fitness evaluation and updating the individual and global best solutions

This step evaluates the fitness of updated particle positions. Then, it compares the current and previous fitness of each particle and updates the position and fitness of the particle if the current position shows better fitness over the previous one. Otherwise, it keeps the previous position and fitness. Finally, this step evaluates the best solution from the individual best solutions and stores the particle position and associated fitness as the x^{**} and J_{best}.

Step 6: Checking the termination criteria

The PSO algorithm checks the termination criteria after a prespecified number of iterations to avoid premature convergence. It terminates the optimization process if the values of the global best solution do not change for a pre-specified number of iterations or it reaches the maximum number of iterations. The technique then displays the global best position as the optimal solution to the optimization problem under investigation. Otherwise, it returns to Step 3. Fig. 2.7 shows a generic flowchart, and Alg. 2.F presents the pseudocode of the PSO technique.

Alg. 2.F: Pseudocode of the PSO algorithm

Input: PSO parameters and optimization problem information
1: Initialization of the initial positions (x) and velocities (v) of the particles
2: Fitness evaluation of the initial positions of the particles (x)
3: Storing each particle position and associated fitness as the individual best solution and fitness
4: Storing the best position and associated fitness as the global best solution and global best fitness
5: **for** iteration=2: *ItMax*, **do**
6:     Updating inertia weight (w) value, velocities (v), and positions (x) of the particles
7:     Checking and repairing the elements of the positions and velocities in case of any violation

8:	**for** particle=*1: N*, **do**
9:	Fitness evaluation of the updated position of the respective particle
10:	**if** current position fitness is better than the old position fitness of the respective particle, **then**
11:	Updating individual best position and associated fitness of the respective particle
12:	**end if**
13:	**end for**
14:	Evaluating the global best solution and the global best fitness from the updated individual solutions
15:	**if** the current global best fitness is better than the previous global best fitness, **then**
16:	Updating the global best position and global best fitness
17:	**end if**
18:	**if** termination criteria met, **then**
19:	Stopping the PSO algorithm
20:	**end if**
21:	**end for**
22:	Assigning the global best fitness and position as the optimal fitness and optimal solution
Output:	Optimal solution of the optimization problem under investigation.

2.3.3.2 Constriction factor particle swarm optimization

After the inception of the PSO algorithm, it has gone through many modifications and upgrades. Thus, many PSO variations are available in the literature, including bare-bones PSO, binary PSO, chaotic PSO, constriction factor PSO, fuzzy PSO, and quantum-behaved PSO [64–66]. The constriction factor PSO (CF-PSO) is one of the earlier modifications that addressed the monotonous decrease of the inertia weight of the original PSO algorithm. The monotonous decrease of the inertia weight, sometimes, may deteriorate the solution quality of the optimization problem. In response, the CF-PSO was initially proposed in Ref. [65] and was further polished in Ref. [67–69]. All CF-PSO steps are similar to the original PSO except Step 3 that updates the inertia weight, velocities, and positions of the particles. It employs the following equations to update the weight, velocities, and positions of the particles:

$$w(t) = w_{max} - \frac{w_{max} - w_{min}}{ItMax} \times t \tag{2.49}$$

$$v_{j,k}(t) = c_f \begin{pmatrix} w(t) \times v_{j,k}(t-1) + c_1 r_1 \times \left(x_{j,k}^*(t-1) - x_{j,k}(t-1)\right) \\ + c_2 r_2 \times \left(x_{j,k}^{**}(t-1) - x_{j,k}(t-1)\right) \end{pmatrix} \tag{2.50}$$

$$c_f = \frac{2}{\left|2 - \varphi - \sqrt{(\varphi^2 - 4\varphi)}\right|}; c_1 + c_2 = \varphi; |c_1| \le |c_2|; \varphi > 4 \tag{2.51}$$

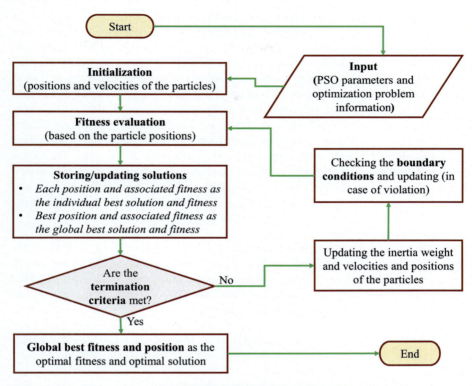

Fig. 2.7 Generic flowchart of the PSO algorithm. *PSO*, particle swarm optimization.

Here, w_{min} and w_{max} are the minimum and maximum values of the inertia weights; ϕ is the convergence characteristic control parameter. Though the values of the CF-PSO parameters vary from problem to problem, in general, their values are chosen as $w_{max} = 1.2$, $w_{min} = 0.1$, $4.05 \leq \phi \leq 4.15$, $2.00 \leq c_1 \leq 2.05$, and $c_2 = \phi - c_1$ [70–74].

2.3.3.3 Grey wolf optimization

The GWO technique is a recently proposed metaheuristic algorithm that mimics the grey wolf's strict hierarchy of social dominance and hunting mechanisms. In 2014, Mirjalili et al. [75] proposed an algorithm that consists of the following main hunting stages:

- Tracking, chasing, and getting closer to the prey/target.
- Encircling and harassing the prey/target until it stops moving.
- Attacking the prey/target.

The grey wolves (GW) prefer to stay mostly in a pack with an average size of 5–12. The pack leaders are termed as alphas (α) and consist of a male and a female GW. The alphas take decisions for the pack, including hunting, sleeping, and waking up. The second, third, and fourth levels in their hierarchy are termed as the betas (β), deltas (δ), and omegas (ω), respectively. The betas assist the alphas in taking and executing the decisions and other actions. They follow the commands of the alphas and dictate the other wolves to ensure the discipline of the pack. They can be either male or female, and they are considered as the best candidates to become the alphas when one alpha dies or retires. The deltas submit themselves to alphas and betas, and, at the same time, they dominate the omegas. The deltas consist of the hunters, sentinels, scouts, caretakers, and elders (retired alphas and betas). The omegas are the lowest-ranked GW, perform the scapegoat role, and always submit to all other prevailing GW in the pack. The GWO steps considering a minimization problem are presented below:

Step 1: Initialization

At the beginning of the GWO algorithm, it sets the number of the GW or search agents (*N*), maximum iteration number (*ItMax*), and the dimension of the search agents (*D*) associated with optimization problem dimension, and other required parameters. Based on the prespecified numbers, the GW or search agent positions are initialized randomly using the following equation:

$$x = U\left(low_j, up_j\right) \qquad \forall\, i\, \&\, j \qquad (2.52)$$

Here, $i = 1, 2, 3, \ldots, N$ and $j = 1, 2, 3, \ldots, D$. The terms "*up*" and "*low*" refer to the upper and the lower boundaries of the parameters/variables to be optimized, respectively.

Step 2: Fitness evaluation and assigning the hierarchy

The GWO algorithm evaluates the fitness of each solution after initialization or updating the positions of the search agents or the GW. Then, the position associated with the fittest solution (X_α) is assigned to the alpha GW. Likewise, the second-best solution (X_β) and the third-best solutions (X_δ) are assigned to the beta and delta GW, respectively. The remaining positions are assigned to the omega GW.

Step 3: Checking the termination criteria

After evaluating the fitness and assigning the best solutions to the alphas, betas, and deltas, the GWO algorithm checks the prespecified termination criterion (maximum number of iteration or no change in the best solution values for a certain number of iteration). It is worth noting that it starts checking the termination criteria after a pre-specified number of iterations to avoid premature convergence. Based on the mentioned criteria, the GWO algorithm either terminates the optimization process or goes to the next step (Step 4). In case of the process termination, the GWO assigns the alpha wolf position as the best solution to the optimization problem under investigation.

Metaheuristic optimization techniques

Step 4: Updating the GW positions to encircle and hunt the prey

This step updates GW positions using the following equations to surround and hunt the prey. Based on their satisfactory updated positions, the GW attacks the target.

$$\vec{x}(t+1) = \vec{x}_p(t) - \vec{A}.\vec{E} \qquad (2.53)$$

Here, $\vec{E} = |\vec{C}.\vec{x}_p(t) - \vec{x}(t)|$, $\vec{A} = 2.\vec{a}.\vec{r}_1 - \vec{a}$, and $\vec{C} = 2.\vec{r}_2$; 't' represents the iteration number; $\vec{x}_p$ and $\vec{x}$ refer to the location vectors of the targets and the GW; $\vec{A}$ and $\vec{C}$ are coefficient vectors; the elements of the $\vec{a}$ are linearly reduced from 2 to 0; and $\vec{r}_1$ and $\vec{r}_2$ are the vectors of randomly generated numbers in the range of [0, 1].

In the real situation, the GW have information about the location of the prey, and they update their positions to encircle and hunt the prey accordingly. However, in an abstract optimization problem, the users (GW) have no prior knowledge about the optimal solution (prey). Therefore, the GWO algorithm assumes that the alphas, betas, and deltas have better information as they resemble better fitness while mathematically mimicking the GW hunting behavior. Thus, the GWO algorithm updates the positions of all search agents or the GW based on the following equations instead of using Eq. (2.53):

$$\vec{x}(t+1) = \frac{\vec{x}_1 + \vec{x}_2 + \vec{x}_3}{3} \qquad (2.54)$$

Here, $\vec{x}_1 = \vec{x}_\alpha - \vec{A}_1.\vec{E}_\alpha$; $\vec{x}_2 = \vec{x}_\beta - \vec{A}_2.\vec{E}_\beta$; $\vec{x}_3 = \vec{x}_\delta - \vec{A}_3.\vec{E}_\delta$; $\vec{E}_\alpha = |\vec{C}_1.\vec{x}_\alpha - \vec{x}|$; $\vec{E}_\beta = |\vec{C}_2.\vec{x}_\beta - \vec{x}|$ and $\vec{E}_\delta = |\vec{C}_3.\vec{x}_\delta - \vec{x}|$

Based on the definition mentioned above, the $\vec{A}$ can take any arbitrary values. If such values are in the range of [−1, 1], the subsequent search agent position will be anywhere in between its present position and the position of the prey, and it will attack the target. However, the agents also diverge from each other to locate the prey if $\vec{A}$ takes a random value outside of the [−1, 1] domain. Thus, the algorithm puts stress on the exploration characteristics of search agents and makes global exploration possible by diversifying the positions of the search agents [75].

Step 5: Checking the boundary of the variables

The GWO algorithm checks the boundary of the variables in each iteration after updating the GW positions. In case of any violation, it repairs the positions to bring them back into the search space. To do so, it sets the parameters at the violated bound. The mentioned boundary checking mechanism can be expressed mathematically as follows [76]:

$$\text{if } x_{i,j} > up_j : x_{i,j} = up_j \text{;else if } x_{i,j} < low_j : x_{i,j} = low_j; \qquad \forall\, i\,\&\,j \qquad (2.55)$$

The updated positions are then sent to the fitness evaluation and hierarchy assigning step (Step 2). Fig. 2.8 shows a generic flowchart, and Alg. 2.G presents the pseudocode of the GWO algorithm.

Alg. 2.G: Pseudocode of the GWO algorithm

Input: GWO parameters and optimization problem information
1: Initialization of the search agents (GW) positions (X)
2: **for** *iteration=1: IterMax,* **do**
3:     Fitness evaluation of the search agents (GW)
4:     Selecting the alphas, betas, deltas, and omegas
5:     **if** termination criteria met, **then**
6:         Stopping the iterative process
7:     **else**
8:         Updating the search agents (GW) position using Eq. (2.54)
0:     Checking boundaries of the updated GW positions and repairing them in case of violation
10:     **end if**
11: **end for**
11: Assigning the position associated with the best fittest wolf (X_α) as the best solution.
Output: The best solution to the optimization problem under investigation.

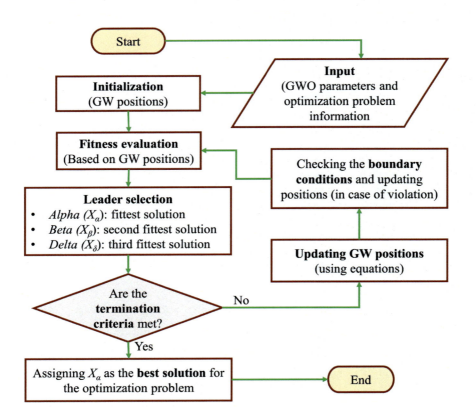

Fig. 2.8 Generic flowchart of the GWO algorithm. *GW*, grey wolves; *GWO*, grey wolf optimization.

2.3.4 Other metaheuristic techniques

Since the inception of the metaheuristic technique concept, many search optimization approaches have been explored and employed in many fields of science, engineering, business, and management [77–79]. In addition to their significant advantages, including simplicity, flexibility, the ability of local optima avoidance, and derivation-free mechanism over the classical optimization techniques; the no free lunch (NFL) theorem of Wolpert and Macready [80] inspired the researchers to search for new approaches and made the field very active. The NFL theorem logically proved that no metaheuristic technique is best suited for solving all optimization problems. A few of them may demonstrate superior performance for a specific set of problems, and at the same time, they may not be very effective for another set of problems [75]. A non-exhaustive list of the explored metaheuristic techniques is presented in Table 2.1, along with their year of development and developers' names. As can be seen, several new algorithms have been proposed almost every year in the 21st century.

Besides, it was also observed that the performance of many metaheuristic techniques had been improved by adding new operators or modifying the available operators to achieve better performance. For instance, the BSA of Ref. [59] was enhanced in Ref. [164], the GWO technique of Ref. [75] was updated in Ref. [165], the gravitational search algorithm of Ref. [103] was improved in Ref. [166], the firefly algorithm of Ref. [107] was updated in Ref. [167], and the cuckoo search algorithm of Ref. [106] was modified in Ref. [168]. Moreover, the researchers have also explored the potential of various combinations of metaheuristic techniques and proposed many hybrid metaheuristic techniques. The primary motivation of hybridizing different algorithms is to achieve better-performing systems by exploiting and combining the advantages of the individual algorithms that is, hybrids are believed to be benefited from the synergy [77,169,170].

Though the individual and hybrid metaheuristic algorithms have been employed successfully to solve many real-world optimization problems, most of them still do not have appropriate mathematical models to explain their working mechanisms [171]. Such metaheuristic techniques "tsunami" without proper explanation may lead the metaheuristics area away from scientific rigor [25,172]. Furthermore, several well-known scientific publishing outlets do not accept new algorithms that do not establish the basis of scientifically compelling argument, mathematical framework, and adequate validation [173,174].

2.4 Summary

The stochastic optimization techniques inspired by either natural, biological, or physical principles often offer near-optimal solutions in a computationally effective manner for any optimization problem, even the problems without proper mathematical models. Ease of implementation, parallelism, flexibility, and adaptability of the metaheuristic techniques, along with the NFL theorem, made them popular in solving real-world complex optimization problems. This chapter illustrated backgrounds,

Table 2.1 A nonexhaustive list of metaheuristic techniques.

Year	Ref.	Developer	Metaheuristic technique name
1989	[81]	D. B. Fogel and L. J. Fogel	Evolutionary programming
1999	[82]	M. Dorigo and G. D. Caro	Ant colony optimization
2000	[83]	L. N. D. Castro and F. J. V. Zuben	Clonal selection algorithm
2001	[84]	H. A. Abbass	Marriage in honey bees optimization
2001	[85]	Z. W. Geem, J. H. Kim, and G. V. Loganathan	Harmony search
2002	[86]	H. Beyer and H. Schwefel	Evolution strategies
2002	[87]	K.M. Passino	Bacterial foraging optimization
2003	[88]	B. Webster and P. J. Bernhard	Gravitational local search
2003	[89]	X. Li	Artificial fish-swarm algorithm
2004	[90]	H. F. Wedde, M. Farooq, and Y. Zhang	Beehive algorithm
2005	[91]	M. H. Roth	Termite algorithm
2006	[92]	O. K. Erol and I. Eksin	Big-bang big-crunch
2006	[93]	H. Du, X. Wu, and J. Zhuang	Small-world optimization algorithm
2006	[94]	M. Eusuff, K. Lansey, and F. Pasha	Shuffled frog-leaping algorithm
2006	[95]	S. Chu, P. Tsai, and J. Pan	Cat swarm optimization
2007	[96]	E. Atashpaz-Gargari, and C. Lucas	Imperialist competitive algorithm
2007	[97]	D. Karaboga and B. Basturk	Artificial bee colony optimization
2007	[98]	R. A. Formato	Central force optimization
2007	[99]	A. Mucherino and O. Seref	Monkey search
2007	[100]	P. C. Pinto, T. A. Runkler, and J. M. C. Sousa	Wasp swarm algorithm
2008	[101]	X. Lu and Y. Zhou	Bee collecting pollen algorithm
2008	[102]	D. Simon	Biogeography-based optimizer
2009	[103]	E. Rashedi, H. Nezamabadi-pour, and S. Saryazdi	Gravitational search algorithm
2009	[104]	Y. Shiqin, J. Jianjun, and Y. Guangxing	Dolphin partner optimization
2009	[105]	H. Shah-Hosseini	Water drops algorithm
2009	[106]	X. Yang and S. Deb	Cuckoo search
2010	[107]	X. Yang	Firefly algorithm
2010	[108]	X. Yang	Bat-inspired algorithm
2010	[109]	A. Kaveh and S. Talatahari	Charged system search algorithm
2011	[110]	H. Shah-Hosseini	Galaxy-based search algorithm
2011	[111]	B. Alatas	Artificial chemical reaction optimization
2012	[112]	H. Eskandar, A. Sadollah, A. Bahreininejad, M. Hamdi	Water cycle algorithm
2012	[113]	A. Kaveh and M. Khayatazad	Ray optimization
2012	[114]	X. Yang	Flower pollination algorithm

Table 2.1 (Cont'd)

Year	Ref.	Developer	Metaheuristic technique name
2012	[115]	F. F. Moghaddam, R. F. Moghaddam, and Mohamed Cheriet	Curved space optimization
2012	[116]	W. Pan	Fruit fly optimization
2012	[117]	A. H. Gandomi and A. H. Alavi	Krill herd algorithm
2012	[118]	A. Askarzadeh and A. Rezazadeh	Bird mating optimizer
2013	[119]	A. Hatamlou	Blackhole algorithm
2013	[120]	B. Xing and W. Gao	Invasive weed optimization
2013	[121]	X. Xie, J. Liu, and Z. Wang	Cooperative group optimization
2014	[122]	A. S. Eesa, A. M. A. Brifcani, and Z. Orman	Cuttlefish algorithm
2014	[123]	S. Mirjalili	Ant lion optimizer
2014	[124]	A. Kaveh and V. R. Mahdavi	Colliding bodies optimization
2015	[125]	S. Mirjalili	Dragonfly algorithm
2015	[126]	S. Mirjalili	Moth-flame optimization
2015	[127]	H. Shareef, A. A. Ibrahim, and A. H. Mutlag	Lightning search algorithm
2016	[128]	S. Mirjalili and A. Lewis	Whale optimization
2016	[129]	M. Onay	Fox hunting algorithm
2016	[130]	A. Askarzadeh	Crow search algorithm
2016	[131]	P. Savsani, V. Savsani	Passing vehicle search
2016	[132]	S. Mirjalili	Sine cosine algorithm
2016	[133]	T. R. Biyanto, H. Y. Fibrianto, G. Nugroho, A. M. Hatta, E. Listijorini, T. Budiati, and H. Huda	Duelist algorithm
2017	[134]	S. Saremi, S. Mirjalili, and A. Lewis	Grasshopper optimization
2017	[135]	A. Kaveh and A. Dadras	Thermal exchange optimization
2017	[136]	T. R. Biyanto, Matradji, S. Irawan, H. Y. Febrianto, N. Afdanny, A. H. Rahman, K. S. Gunawan, J. A. D. Pratama, and T. N. Bethiana	Killer whale algorithm
2017	[137]	A. Wedyan, J. Whalley, and A. Narayanan	Hydrological cycle algorithm
2017	[138]	G. Dhiman and V. Kumar	Spotted hyena optimizer
2018	[139]	A. Sadollah, H. Sayyaadi, A. Yadav	Neural network algorithm
2018	[140]	J. Zhang, M. Xiao, L. Gao, and Q. Pan	Queuing search algorithm
2019	[141]	F. A. Hashim, E. H. Houssein, M. S. Mabrouk, W. Al-Atabany, and S. Mirjalili	Henry gas solubility optimization
2019	[142]	S. Harifi, M. Khalilian, J. Mohammadzadeh, and S. Ebrahimnejad	Emperor penguins colony

(*continued*)

Table 2.1 (Cont'd)

Year	Ref.	Developer	Metaheuristic technique name
2019	[143]	R. Masadeh, B. A. Mahafzah, and A. Sharieh	Sea lion optimization
2019	[144]	A. A. Heidari, S. Mirjalili, H. Faris, I. Aljarah, M. Mafarja, and H. Chen	Harris hawks optimization
2019	[145]	S. Shadravan, H. R. Naji, and V. K. Bardsiri	Sailfish optimizer
2020	[146]	A. Faramarzia, M. Heidarinejad, B. Stephens, and S. Mirjalili	Equilibrium optimizer
2020	[147]	A. R. Moazzeni, E. Khamehchi	Rain optimization algorithm
2020	[148]	A. Faramarzi, M. Heidarinejad, S. Mirjalili, and A. H. Gandomi	Marine predators algorithm
2020	[149]	S. Kaur, L. K. Awasthi, A. L. Sangal, and G. Dhiman	Tunicate swarm algorithm
2020	[150]	S. Massan, A. I. Wagan, M. M. Shaikh	Dynastic optimization algorithm
2020	[151]	Q. Askari, M. Saeed, I. Younas	Heap-based optimizer
2020	[152]	A. Kaveh, A. Zaerreza	Shuffled shepherd optimization
2020	[153]	J. Chou, N. Nguyen	Forensic-based investigation
2020	[154]	Q. Askari, I. Younas, and M. Saeed	Political optimizer
2020	[155]	Y. Zhang and Z. Jin	Group teaching optimization
2020	[156]	A. Kaveh, A. D. Eslamlou	Water strider algorithm
2020	[157]	E. Bogar and S. Beyhanb	Adolescent identity search algorithm
2020	[158]	W. Al-Sorori and A. M. Mohsen	Caledonian crow learning algorithm
2020	[159]	H. Ghasemian, F. Ghasemian, and H. Vahdat-Nejada	Human urbanization algorithm
2020	[160]	O. K. Meng, O. Pauline, and S. C. Kiong	Carnivorous plant algorithm
2020	[161]	C. Taramasco, B. Crawford, R. Soto, E. M.Cortés-Toro, and R. Olivares	Vapor-liquid equilibrium algorithm
2021	[162]	D. Połap and M. Woźniak	Red fox optimization
2021	[163]	J. Chou and D. Truong	Artificial jellyfish search

working principles, flowcharts, and pseudocodes of several popular and widely adopted metaheuristic algorithms. Most of the illustrated algorithms will be applied to solve optimization problems for example, optimal phasor measurement placement problems and tuning the machine learning tools' critical parameters while developing intelligent fault diagnosis techniques for example, support vector machines in the subsequent chapters of this book. The readers will also find the interactive and easy to understand discussions on the illustrated algorithms useful to adopt any metaheuristic technique for their optimization problem. However, many metaheuristic techniques received criticism for not having proper mathematical frameworks to explain their working mechanisms despite having many advantages.

References

[1] S.E. De Leon-Aldaco, H. Calleja, J. Aguayo Alquicira, Metaheuristic optimization methods applied to power converters: a review, IEEE Trans. Power Electron. 30 (12) (Dec. 2015) 6791–6803, doi:10.1109/TPEL.2015.2397311

[2] J. Shi, Q. Zhang, A new cooperative framework for parallel trajectory-based metaheuristics, Appl. Soft Comput. J. 65 (2018) 374–386, doi:10.1016/j.asoc.2018.01.022.

[3] K.M. Sagayam, D.J. Hemanth, X.A. Vasanth, L.E. Henesy, C.C. Ho, Optimization of a HMM-based hand gesture recognition system using a hybrid cuckoo search algorithm. In: Hybrid Metaheuristics for Image Analysis, Springer International Publishing, 2018, pp. 87–114.

[4] J. Toutouh El Alamin, Natural computing for vehicular networks, Universidad de Málaga, Málaga, Spain, 2016.

[5] R. Amiri, J.M. Sardroud, B.G. De Soto, BIM-based Applications of metaheuristic algorithms to support the decision-making process: uses in the planning of construction site layout, Procedia Eng. 196 (2017) 558–564, doi:10.1016/j.proeng.2017.08.030.

[6] A.S. Azad, M.S. Md, J. Watada, P. Vasant, J.A.G. Vintaned, Optimization of the hydropower energy generation using meta-heuristic approaches: a review, Energy Rep. 6 (2020) 2230–2248, doi:10.1016/j.egyr.2020.08.009.

[7] M. Janga Reddy, D. Nagesh Kumar, Evolutionary algorithms, swarm intelligence methods, and their applications in water resources engineering: a state-of-the-art review, H2Open J 3 (1) (2020) 135–188, doi:10.2166/h2oj.2020.128.

[8] D.N. Kumar, Classical and advanced techniques for optimization, 2020. https://nptel.ac.in/content/storage2/courses/105108127/pdf/Module_1/M1L4slides.pdf. (Accessed 9 November 2020).

[9] M. Oliver, Practical guide to the simplex method of linear programming, 2020. http://math.jacobs-university.de/oliver/teaching/iub/spring2007/cps102/handouts/linear-programming.pdf. (Accessed 8 November 2020).

[10] Y.M.R. Aboelmagd, Linear programming applications in construction sites, Alexandria Eng. J. 57 (4) (2018) 4177–4187, doi:10.1016/j.aej.2018.11.006 .

[11] J.D. Dotson, Five areas of application for linear programming techniques, Sciencing (2018). https://sciencing.com/careers-use-linear-equations-6060294.html. (Accessed 8 November 2020).

[12] C. Lewis, Linear programming: theory and applications, 2008. https://www.whitman.edu/Documents/Academics/Mathematics/lewis.pdf. (Accessed 9 November 2020).

[13] J. Larrosa, A. Oliveras, and E. Rodríguez-Carbonell, Mixed integer linear programming: combinatorial problem solving (CPS), 2020. https://www.cs.upc.edu/~erodri/webpage/cps/theory/lp/milp/slides.pdf. (Accessed 9 November 2020).

[14] W.A. Kamal, D.W. Gu, I. Postlethwaite, MILP and its application in flight path planning, IFAC Proceedings Volumes (IFAC-PapersOnline) 38 (1) (2005) 55–60, doi:10.3182/20050703-6-cz-1902.02061.

[15] B.A. Yosuf, M. Musa, T. Elgorashi, and J. Elmirghani, Energy efficient distributed processing for IoT, 2020. http://arxiv.org/abs/2001.02974. (Accessed 9 November 2020).

[16] M. Shafiullah, M.I. Hossain, M.A. Abido, T. Abdel-Fattah, A.H. Mantawy, A modified optimal PMU placement problem formulation considering channel limits under various contingencies, Measurement 135 (2019) 875–885, doi:10.1016/J.MEASUREMENT.2018.12.039.

[17] X. Wang, J. Atkin, S. Bozhko, and C. Hill, Application of a MILP-based algorithm for power flow optimisation within more-electric aircraft electrical power systems, 2019, doi:10.23919/EPE.2019.8915388.
[18] B. Fu, C. Ouyang, C. Li, J. Wang, E. Gul, An improved mixed integer linear programming approach based on symmetry diminishing for unit commitment of hybrid power system, Energies 12 (5) (2019) 833, doi:10.3390/en12050833.
[19] R. Marinescu, R. Dechter, AND/OR Branch-and-Bound for Solving Mixed Integer Linear Programming Problems, Springer, Berlin, Heidelberg, 2005, p. 857.
[20] K. Genova, V. Guliashki, Linear integer programming methods and approaches-a survey, Cybern. Inf. Technol. 11 (1) (2011).
[21] J.W. Chinneck, Practical Optimization: A Gentle Introduction, Systems and Computer Engineering Carleton University, Ottawa, Canada, 2015.
[22] R.J. Vanderbei, Nonlinear programming and engineering applications, International Series in Operations Research and Management Science, 76, Springer, New York LLC, 2005, pp. 1–7.
[23] J. Bracken, G.P. Mccormick, Selected Applications of Nonlinear Programming, Research Analysis Corporation, McLean, Virginia, USA, 1968.
[24] G. Venter, Review of optimization techniques. In: Encyclopedia of Aerospace Engineering, John Wiley & Sons Ltd, New Jersey, USA, 2010.
[25] F. Glover, K. Sörensen, Metaheuristics, Scholarpedia 10 (4) (2015) 6532, doi:10.4249/scholarpedia.6532.
[26] S. Bandaru, K. Deb, Metaheuristic techniques, in: R.N. Sengupta, A. Gupta, J. Dutta (Eds.), Decision Sciences Theory and Practice, 1st ed., CRC Press, Florida, USA, 2016, pp. 709–766.
[27] N. Metropolis, A.W. Rosenbluth, M.N. Rosenbluth, A.H. Teller, E. Teller, Equation of state calculations by fast computing machines, J. Chem. Phys. 21 (6) (1953) 1087–1092, doi:10.1063/1.1699114.
[28] S. Kirkpatrick, C.D. Gelatt, M.P. Vecchi, Optimization by simulated annealing, Science 220 (4598) (1983) 671–680, doi:10.1126/science.220.4598.671.
[29] E. Aarts, J. Korst, Simulated Annealing and Boltzmann Machines: A Stochastic Approach to Combinatorial Optimization and Neural Computing, John Wiley & Sons, New Jersey, USA, 1991.
[30] Z. Varty, Simulated Annealing Overview, Lancaster, United Kingdom, 2017. https://www.lancaster.ac.uk/pg/varty/RTOne.pdf. (Accessed 16 November 2020).
[31] B. Suman, P. Kumar, A survey of simulated annealing as a tool for single and multiobjective optimization, J. Oper. Res. Soc. 57 (10) (2006) 1143–1160, doi:10.1057/palgrave.jors.2602068.
[32] F. Glover, Future paths for integer programming and links to artificial intelligence, Comput. Oper. Res. 13 (5) (1986) 533–549, doi:10.1016/0305-0548(86)90048-1.
[33] F. Glover, Tabu search—part i, ORSA J. Comput. 1 (3) (1989) 190–206, doi:10.1287/ijoc.1.3.190.
[34] F. Glover, Tabu search—part ii, ORSA J. Comput. 2 (1) (1990) 4–32, doi:10.1287/ijoc.2.1.4.
[35] A. Ebrahimzadeh, R. Khanduzi, A directed tabu search method for solving controlled Volterra integral equations, Math. Sci. 10 (3) (2016) 115–122, doi:10.1007/s40096-016-0185-x.
[36] E. Taillard, Robust taboo search for the quadratic assignment problem, Parallel Comput 17 (4–5) (1991) 443–455, doi:10.1016/S0167-8191(05)80147-4.
[37] R. Battiti, G. Tecchiolli, The reactive tabu search, ORSA J. Comput. 6 (2) (1994) 126–140, doi:10.1287/ijoc.6.2.126.

[38] S. Jayaswal, A comparative study of tabu search and simulated annealing for traveling salesman problem, 2000. http://www.eng.uwaterloo.ca/~sjayaswa/projects/MSCI703_project.pdf. (Accessed 16 November 2020).
[39] M.A. Abido, Optimal power flow using tabu search algorithm, Electr. Power Components Syst. 30 (5) (2002) 469–483, doi:10.1080/15325000252888425.
[40] A. Thesen, Design and evaluation of tabu search algorithms for multiprocessor scheduling, 1998.
[41] C.A. Anderson, K. Fraughnaugh, M. Parker, J. Ryan, Path assignment for call routing: an application of tabu search, Ann. Oper. Res. 41 (4) (1993) 299–312, doi:10.1007/BF02022997.
[42] M. Gendreau, An introduction to tabu search. In: Handbook of Metaheuristics, Kluwer Academic Publishers, London, UK, 2006, pp. 37–54.
[43] I. Dumitrescu, T. Stützle, Combinations of local search and exact algorithms, Lect. Notes Comput. Sci. 2611 (2003) 211–223, doi:10.1007/3-540-36605-9_20.
[44] C. Renman and H. Fristedt, A comparative analysis of a tabu search and a genetic algorithm for solving a university course timetabling problem, 2015. https://www.diva-portal.org/smash/get/diva2:810264/FULLTEXT01.pdf. (Accessed 17 November 2020).
[45] N.K.T. El-Omari, Sea lion optimization algorithm for solving the maximum flow problem, IJCSNS Int. J. Comput. Sci. Netw. Secur. 20 (8) (2020) 30, doi:10.22937/IJCSNS.2020.20.08.5.
[46] K. Jebari, M. Madiafi, Selection methods for genetic algorithms, Int. J. Emerg. Sci. 3 (4) (2013) 333–344.
[47] S. Rani, B. Suri, R. Goyal, On the effectiveness of using elitist genetic algorithm in mutation testing, Symmetry (Basel) 11 (9) (2019) 1145, doi:10.3390/sym11091145.
[48] F. Herrera, M. Lozano, J.L. Verdegay, Tackling real-coded genetic algorithms: operators and tools for behavioural analysis, Artif. Intell. Rev. 12 (4) (1998) 265–319, doi:10.1023/A:1006504901164.
[49] F. Herrera, M. Lozano, E. Pérez, A.M. Sánchez, P. Villar, Multiple crossover per couple with selection of the two best offspring: An experimental study with the BLX-α crossover operator for real-coded genetic algorithms, Lect. Notes Comput. Sci. 2527, 2002 392–401, doi:10.1007/3-540-36131-6_40.
[50] R. Peltokangas and A. Sorsa, Real-coded genetic algorithms and nonlinear parameter identification, 2008. http://jultika.oulu.fi/files/isbn9789514287862.pdf. (Accessed 13 November 2020).
[51] N. Padhye, K. Deb, P. Mittal, Boundary handling approaches in particle swarm optimization, Adv. Intel. Syst. Comput. 201, 2013 287–298, doi:10.1007/978-81-322-1038-2_25.
[52] R. Storn, K. Price, Differential Evolution-A Simple and Efficient Adaptive Scheme for Global Optimization Over Continuous Spaces, Berkeley, CA, 1995.
[53] R. Storn, K. Price, Differential evolution - a simple and efficient heuristic for global optimization over continuous spaces, J. Glob. Optim. 11 (4) (1997) 341–359, doi:10.1023/A:1008202821328.
[54] S. Das, P.N. Suganthan, Differential evolution: a survey of the state-of-the-art, IEEE Trans. Evol. Comput. 15 (1) (2011) 4–31, doi:10.1109/TEVC.2010.2059031.
[55] A.H. Al-Mohammed, M.A. Abido, M.M. Mansour, Optimal PMU placement for power system observability using differential evolution, 11th International Conference on Intelligent Systems Design and Applications, 2011, pp. 277–282, doi:10.1109/ISDA.2011.6121668.
[56] M. Ijaz, M. Shafiullah, M.A. Abido, Classification of power quality disturbances using Wavelet Transform and Optimized ANN, 2015 18th International Conference on

[57] A. Wagdy Mohamed and A. Wagady Mohamed, Differential evolution (DE): a short review, Robot. Autom. Eng. J.vol. 2, no. 1, 2018, doi:10.19080/RAEJ.2018.02.555579.
[58] V.P. Plagianakos, D.K. Tasoulis, M.N. Vrahatis, A review of major application areas of differential evolution, Studies in Computational Intelligence, 143, Springer, Berlin, Heidelberg, 2008, pp. 197–238, doi:10.1007/978-3-540-68830-3_8.
[59] P. Civicioglu, Backtracking search optimization algorithm for numerical optimization problems, Appl. Math. Comput. 219 (15) (2013) 8121–8144, doi:10.1016/j.amc.2013.02.017.
[60] X.S. Yang, S. Deb, S. Fong, X. He, Y.X. Zhao, From swarm intelligence to metaheuristics: nature-inspired optimization algorithms, Computer (Long. Beach. Calif.) 49 (9) (2016) 52–59, doi:10.1109/MC.2016.292.
[61] X.-S. Yang, Swarm-based metaheuristic algorithms and no-free-lunch theorems, in: R. Parpinelli (Ed.), Theory and New Applications of Swarm Intelligence, InTech Open, London, UK, 2012.
[62] K. Hussain, M.N.M. Salleh, S. Cheng, Y. Shi, On the exploration and exploitation in popular swarm-based metaheuristic algorithms, Neural Comput. Appl. 31 (11) (2019) 7665–7683, doi:10.1007/s00521-018-3592-0.
[63] R. Eberhart, J. Kennedy, A new optimizer using particle swarm theory, *MHS'95*. Proceedings of the Sixth International Symposium on Micro Machine and Human Science. IEEE, 1995, pp. 39–43, doi:10.1109/MHS.1995.494215.
[64] Y. Zhang, S. Wang, G. Ji, A comprehensive survey on particle swarm optimization algorithm and its applications, Math. Probl. Eng. 2015 (2015), doi:10.1155/2015/931256 Special.
[65] M. Clerc, The swarm and the queen: towards a deterministic and adaptive particle swarm optimization, Proceedings of the 1999 Congress on Evolutionary Computation, 3, 1999 http://ieeexplore.ieee.org/abstract/document/785513/. (Accessed 5 October 2018).
[66] M.A. Khanesar, M. Teshnehlab, and M. A. Shoorehdeli, A novel binary particle swarm optimization, 2007, doi:10.1109/MED.2007.4433821.
[67] R. Eberhart, Y. Shi, Comparing inertia weights and constriction factors in particle swarm optimization, Proceedings of the 2000 Congress on Evolutionary Computation, 1, 2000 84–88. http://ieeexplore.ieee.org/abstract/document/870279/. (Accessed 3 October, 2018).
[68] M. Clerc, J. Kennedy, The particle swarm-explosion, stability, and convergence in a multidimensional complex space, IEEE Trans. Evol. Comput. 6 (1) (2002) 58–73, doi:10.1109/4235.985692.
[69] J. Xue-ling, L. Ming, L. Wei, Constriction factor particle swarm optimization algorithm with overcoming local optimum, Comput. Eng. 1 (1) (2011) 1–7.
[70] A. Malekpour, A. Seifi, Application of constriction factor particle swarm optimization to optimum load shedding in power system, Mod. Appl. Sci. 4 (7) (2010) 188–196, doi:10.5539/mas.v4n7p188.
[71] R. Eberhart, Y. Shi, Particle swarm optimization: developments, applications and resources, Proceedings of the 2001 Congress on Evolutionary Computation, 1 (2001) 81–86. http://ieeexplore.ieee.org/abstract/document/934374/. (Accessed 13 September 2018).
[72] K. Parsopoulos, M. Vrahatis, Particle swarm optimization method for constrained optimization problems, Optimization 181 (6) (2011) 1153–1163, doi:10.1016/j.ins.2010.11.033.
[73] R.P. Patwardhan, S.L. Mhetre, Effect of constriction factor on minimization of transmission power loss using Particle Swarm Optimization, 2015 International Conference

on Energy Systems and Applications. IEEE, Oct. (2015) 152–157, doi:10.1109/ICESA.2015.7503330.
[74] M.N. Alam, K. Kumar, A. Mathur, Economic load dispatch considering valve-point effects using time varying constriction factor based particle swarm optimization, 2015 IEEE UP Section Conference on Electrical Computer and Electronics (UPCON). IEEE, 2015, pp. 1–6, doi:10.1109/UPCON.2015.7456740.
[75] S. Mirjalili, S. Mohammad, A. Lewis, Grey wolf optimizer, Adv. Eng. Softw. 69 (2014) 46–61, doi:10.1016/j.advengsoft.2013.12.007.
[76] S. Mirjalili, Grey wolf optimizer (GWO), MATLAB Central File Exchange, MATLAB, Massachusetts, USA, 2020.
[77] M. Abdel-Basset, L. Abdel-Fatah, A.K. Sangaiah, Metaheuristic algorithms: a comprehensive review. In: Computational Intelligence for Multimedia Big Data on the Cloud with Engineering Applications, Elsevier, Amsterdam, Netherlands, 2018, pp. 185–231.
[78] M. Shafiullah, M.J. Rana, M.S. Alam, M.A. Uddin, Optimal placement of Phasor Measurement Units for transmission grid observability, 2016 International Conference on Innovations in Science, Engineering and Technology (ICISET). IEEE, Chittagong, Bangladesh, 2016, pp. 1–4, doi:10.1109/ICISET.2016.7856492.
[79] K.G. Dhal, S. Ray, A. Das, S. Das, A survey on nature-inspired optimization algorithms and their application in image enhancement domain, Arch. Comput. Methods Eng. 26 (5) (2019) 1607–1638, doi:10.1007/s11831-018-9289-9.
[80] D.H. Wolpert, W.G. Macready, No free lunch theorems for optimization, IEEE Trans. Evol. Comput. 1 (1) (1997) 67–82, doi:10.1109/4235.585893.
[81] D.B. Fogel, L.J. Fogel, Evolutionary programming for voice feature analysis, Conference Record - Asilomar Conference on Circuits, Systems & Computers, 1, 1989 381–383, doi:10.1109/acssc.1989.1200817.
[82] M. Dorigo and G. Di Caro, "Ant colony optimization: A new meta-heuristic," Proceedings of the 1999 Congress on Evolutionary Computation, CEC 1999, 2 (1999), pp. 1470–1477, doi: 10.1109/CEC.1999.782657.
[83] L.N. De Castro, F.J. Von Zuben, The clonal selection algorithm with engineering applications. In: Workshop on Artificial Immune Systems and Their Applications, Workshop Proceedings, Las Vegas, USA, 2000, pp. 36–37.
[84] H.A. Abbass, MBO: marriage in honey bees optimization a haplometrosis polygynous swarming approach, Proceedings of the IEEE Conference on Evolutionary Computation, ICEC, 1, 2001 207–214, doi:10.1109/cec.2001.934391.
[85] Zong Woo Geem, Joong Hoon Kim, G.V. Loganathan, A new heuristic optimization algorithm: harmony search, Simulation 76 (2) (2001) 60–68, doi:10.1177/003754970107600201.
[86] H.-G. Beyer, H.-P. Schwefel, Evolution strategies – a comprehensive introduction, Nat. Comput. 1 (1) (2002) 3–52, doi:10.1023/A:1015059928466.
[87] K.M. Passino, Biomimicry of bacterial foraging for distributed optimization and control, IEEE Control Syst 22 (3) (2002) 52–67, doi:10.1109/MCS.2002.1004010.
[88] B. Webster and P. J. Bernhard, A local search optimization algorithm based on natural principles of gravitation, 2003. https://repository.lib.fit.edu/handle/11141/117. (Accessed 18 November 2020).
[89] X. Li, A New Intelligent Optimization-Artificial Fish Swarm, Zhejiang University, Zhejiang, China, 2003.
[90] H.F. Wedde, M. Farooq, Y. Zhang, BeeHive: an efficient fault-tolerant routing algorithm inspired by honey bee behavior. In: Lect. Notes Comput. Sci. 2004 83–94, doi:10.1007/978-3-540-28646-2_8 vol. 3172.

[91] M.H. Roth, Termite: a swarm intelligent routing algorithm for mobile wireless ad-hoc networks, Cornell University, Ithaca, New York, 2005.
[92] O.K. Erol, I. Eksin, A new optimization method: Big Bang-Big Crunch, Adv. Eng. Softw. 37 (2) (2006) 106–111, doi:10.1016/j.advengsoft.2005.04.005.
[93] H. Du, X. Wu, J. Zhuang, Small-world optimization algorithm for function optimization. In: Lect. Notes Comput. Sci. 4222 (2006) 264–273, doi:10.1007/11881223_33.
[94] H.M. Hasanien, Shuffled frog leaping algorithm for photovoltaic model identification, IEEE Trans. Sustain. Energy 6 (2) (2015) 509–515, doi:10.1109/TSTE.2015.2389858.
[95] S.-C. Chu, P. Tsai, J.-S. Pan, Cat Swarm Optimization, Springer, Berlin, Heidelberg, 2006, pp. 854–858.
[96] E. Atashpaz-Gargari, C. Lucas, Imperialist competitive algorithm: An algorithm for optimization inspired by imperialistic competition, 2007 IEEE Congress on Evolutionary Computation, CEC 2007, 2007, pp. 4661–4667, doi:10.1109/CEC.2007.4425083.
[97] D. Karaboga, B. Basturk, Artificial bee colony (ABC) optimization algorithm for solving constrained optimization problems. In: Lect. Notes Comput. Sci. 4529 (2007) 789–798, doi:10.1007/978-3-540-72950-1_77.
[98] R.A. Formato, Central force optimization: a new nature inspired computational framework for multidimensional search and optimization, Stud. Comput. Intell. 129 (2008) 221–238, doi:10.1007/978-3-540-78987-1_21.
[99] A. Mucherino, O. Seref, Monkey search: a novel metaheuristic search for global optimization, AIP Conference Proceedings, 953, (2007) 162–173, doi:10.1063/1.2817338.
[100] P.C. Pinto, T.A. Runkler, J.M.C. Sousa, Wasp swarm algorithm for dynamic MAX-SAT problems, Lect. Notes Comput. Sci. 4431 (2007) 350–357, doi:10.1007/978-3-540-71618-1_39.
[101] X. Lu, Y. Zhou, A novel global convergence algorithm: Bee collecting pollen algorithm. In: Lect. Notes Comput. Sci. 5227 (2008) 518–525, doi:10.1007/978-3-540-85984-0_62.
[102] D. Simon, Biogeography-based optimization, IEEE Trans. Evol. Comput. 12 (6) (2008) 702–713, doi:10.1109/TEVC.2008.919004.
[103] E. Rashedi, H. Nezamabadi-pour, S. Saryazdi, GSA: a gravitational search algorithm, Inf. Sci. (Ny). 179 (13) (2009) 2232–2248, doi:10.1016/j.ins.2009.03.004.
[104] S. Yang, J. Jiang, G. Yan, A dolphin partner optimization, Proceedings of the 2009 WRI Global Congress on Intelligent Systems, GCIS 2009, 1 (2009) 124–128, doi:10.1109/GCIS.2009.464.
[105] H. Shah-Hosseini, The intelligent water drops algorithm: A nature-inspired swarm-based optimization algorithm, Int. J. Bio-Inspired Comput. 1 (1–2) (2009) 71–79, doi:10.1504/IJBIC.2009.022775.
[106] X.S. Yang, S. Deb, Cuckoo search via Lévy flights, 2009 World Congress on Nature and Biologically Inspired Computing, NABIC 2009 - Proceedings, 2009, pp. 210–214, doi:10.1109/NABIC.2009.5393690.
[107] X.S. Yang, Firefly algorithm, stochastic test functions and design optimization, Int. J. Bio-Inspired Comput. 2 (2) (2010) 78–84, doi:10.1504/IJBIC.2010.032124.
[108] X.S. Yang, A new metaheuristic bat-inspired algorithm, Studies in Computational Intelligence, 284 (2010) 65–74, doi:10.1007/978-3-642-12538-6_6.
[109] A. Kaveh, S. Talatahari, A novel heuristic optimization method: charged system search, Acta Mech 213 (3–4) (2010) 267–289, doi:10.1007/s00707-009-0270-4.
[110] H.S. Hosseini, Principal components analysis by the galaxy-based search algorithm: a novel metaheuristic for continuous optimisation, Int. J. Comput. Sci. Eng. 6 (1/2) (2011) 132, doi:10.1504/ijcse.2011.041221.

[111] B. Alatas, ACROA: artificial chemical reaction optimization algorithm for global optimization, Expert Syst. Appl. 38 (10) (2011) 13170–13180, doi:10.1016/j.eswa.2011.04.126.
[112] H. Eskandar, A. Sadollah, A. Bahreininejad, M. Hamdi, Water cycle algorithm - A novel metaheuristic optimization method for solving constrained engineering optimization problems, Comput. Struct. 110–111 (Nov. 2012) 151–166, doi:10.1016/j.compstruc.2012.07.010.
[113] A. Kaveh, M. Khayatazad, A new meta-heuristic method: ray optimization, Comput. Struct. 112–113 (2012) 283–294, doi:10.1016/j.compstruc.2012.09.003.
[114] X.S. Yang, Flower pollination algorithm for global optimization, Lect. Notes Comput. Sci. 7445 (2012) 240–249, doi:10.1007/978-3-642-32894-7_27.
[115] F.F. Moghaddam, R.F. Moghaddam, M. Cheriet, "Curved space optimization: a random search based on general relativity theory," 2012, http://arxiv.org/abs/1208.2214. (Accessed 18 November 2020).
[116] W.T. Pan, A new fruit fly optimization algorithm: taking the financial distress model as an example, Knowledge-Based Syst 26 (2012) 69–74, doi:10.1016/j.knosys.2011.07.001.
[117] A.H. Gandomi, A.H. Alavi, Krill herd: a new bio-inspired optimization algorithm, Commun. Nonlinear Sci. Numer. Simul. 17 (12) (2012) 4831–4845, doi:10.1016/j.cnsns.2012.05.010.
[118] A. Askarzadeh, A. Rezazadeh, A new heuristic optimization algorithm for modeling of proton exchange membrane fuel cell: bird mating optimizer, Int. J. Energy Res. 37 (10) (2013) 1196–1204, doi:10.1002/er.2915.
[119] A. Hatamlou, Black hole: a new heuristic optimization approach for data clustering, Inf. Sci. (Ny). 222 (2013) 175–184, doi:10.1016/j.ins.2012.08.023.
[120] B. Xing, W.-J. Gao, Invasive weed optimization algorithm, Intell. Syst. Ref. Libr. 62 (2014) 177–181, doi:10.1007/978-3-319-03404-1_13.
[121] X.F. Xie, J. Liu, Z.J. Wang, A cooperative group optimization system, Soft Comput 18 (3) (2014) 469–495, doi:10.1007/s00500-013-1069-8.
[122] A.S. Eesa, A.M.A. Brifcani, Z. Orman, Cuttlefish algorithm – a novel bio-inspired optimization algorithm, Int. J. Sci. Eng. Res. 4 (9) (2014).
[123] S. Mirjalili, The ant lion optimizer, Adv. Eng. Softw. 83 (2015) 80–98, doi:10.1016/j.advengsoft.2015.01.010.
[124] A. Kaveh, V.R. Mahdavi, Colliding bodies optimization: A novel meta-heuristic method, Comput. Struct. 139 (Jul. 2014) 18–27, doi:10.1016/j.compstruc.2014.04.005.
[125] S. Mirjalili, Dragonfly algorithm: a new meta-heuristic optimization technique for solving single-objective, discrete, and multi-objective problems, Neural Comput. Appl. 27 (4) (May 2016) 1053–1073, doi:10.1007/s00521-015-1920-1.
[126] S. Mirjalili, Moth-flame optimization algorithm: A novel nature-inspired heuristic paradigm, Knowledge-Based Syst 89 (Nov. 2015) 228–249, doi:10.1016/j.knosys.2015.07.006.
[127] H. Shareef, A.A. Ibrahim, A.H. Mutlag, Lightning search algorithm, Appl. Soft Comput. J. 36 (Aug. 2015) 315–333, doi:10.1016/j.asoc.2015.07.028.
[128] S. Mirjalili, A. Lewis, The Whale Optimization Algorithm, Adv. Eng. Softw. 95 (May 2016) 51–67, doi:10.1016/j.advengsoft.2016.01.008.
[129] M. Onay, A New and Fast Optimization Algorithm: Fox Hunting Algorithm (FHA), 2016 International Conference on Applied Mathematics, Simulation and Modelling, (May 2016), pp. 153–156, doi:10.2991/amsm-16.2016.35.
[130] A. Askarzadeh, A novel metaheuristic method for solving constrained engineering optimization problems: Crow search algorithm, Comput. Struct. 169 (Jun. 2016) 1–12, doi:10.1016/j.compstruc.2016.03.001.
[131] P. Savsani, V. Savsani, Passing vehicle search (PVS): A novel metaheuristic algorithm, Appl. Math. Model. 40 (5–6) (Mar. 2016) 3951–3978, doi:10.1016/j.apm.2015.10.040.

[132] S. Mirjalili, SCA: A Sine Cosine Algorithm for solving optimization problems, Knowledge-Based Syst 96 (Mar. 2016) 120–133, doi:10.1016/j.knosys.2015.12.022.
[133] T.R. Biyanto, et al., Duelist algorithm: An algorithm inspired by how duelist improve their capabilities in a duel, Lecture Notes in Computer Science (including subseries Lecture Notes in Artificial Intelligence and Lecture Notes in Bioinformatics), vol. 9712 LNCS, Springer Verlag, 2016, pp. 39–47.
[134] S. Saremi, S. Mirjalili, A. Lewis, Grasshopper Optimisation Algorithm: Theory and application, Adv. Eng. Softw. 105 (Mar. 2017) 30–47, doi:10.1016/j.advengsoft.2017.01.004.
[135] A. Kaveh, A. Dadras, A novel meta-heuristic optimization algorithm: Thermal exchange optimization, Adv. Eng. Softw. 110 (Aug. 2017) 69–84, doi:10.1016/j.advengsoft.2017.03.014.
[136] T.R. Biyanto, et al., Killer whale algorithm: an algorithm inspired by the life of killer whale, Procedia Comput. Sci. 124 (2017) 151–157, doi:10.1016/j.procs.2017.12.141.
[137] A. Wedyan, J. Whalley, A. Narayanan, Hydrological cycle algorithm for continuous optimization problems, J. Optim. 2017 (2017) 1–25, doi:10.1155/2017/3828420.
[138] G. Dhiman, V. Kumar, Spotted hyena optimizer: a novel bio-inspired based metaheuristic technique for engineering applications, Adv. Eng. Softw. 114 (2017) 48–70, doi:10.1016/j.advengsoft.2017.05.014.
[139] A. Sadollah, H. Sayyaadi, A. Yadav, A dynamic metaheuristic optimization model inspired by biological nervous systems: neural network algorithm, Appl. Soft Comput. J. 71 (2018) 747–782, doi:10.1016/j.asoc.2018.07.039.
[140] J. Zhang, M. Xiao, L. Gao, Q. Pan, Queuing search algorithm: a novel metaheuristic algorithm for solving engineering optimization problems, Appl. Math. Model. 63 (2018) 464–490, doi:10.1016/j.apm.2018.06.036.
[141] F.A. Hashim, E.H. Houssein, M.S. Mabrouk, W. Al-Atabany, S. Mirjalili, Henry gas solubility optimization: a novel physics-based algorithm, Futur. Gener. Comput. Syst. 101 (2019) 646–667, doi:10.1016/j.future.2019.07.015.
[142] S. Harifi, M. Khalilian, J. Mohammadzadeh, S. Ebrahimnejad, Emperor penguins colony: a new metaheuristic algorithm for optimization, Evol. Intell. 12 (2) (2019) 211–226, doi:10.1007/s12065-019-00212-x.
[143] R. Masadeh, B.A. Mahafzah, A. Sharieh, Sea lion optimization algorithm, Int. J. Adv. Comput. Sci. Appl. 10 (5) (2019) 388–395, doi:10.14569/ijacsa.2019.0100548.
[144] A.A. Heidari, S. Mirjalili, H. Faris, I. Aljarah, M. Mafarja, H. Chen, Harris hawks optimization: Algorithm and applications, Futur. Gener. Comput. Syst. 97 (2019) 849–872, doi:10.1016/j.future.2019.02.028.
[145] S. Shadravan, H.R. Naji, V.K. Bardsiri, The sailfish optimizer: a novel nature-inspired metaheuristic algorithm for solving constrained engineering optimization problems, Eng. Appl. Artif. Intell. 80 (2019) 20–34, doi:10.1016/j.engappai.2019.01.001.
[146] A. Faramarzi, M. Heidarinejad, B. Stephens, S. Mirjalili, Equilibrium optimizer: a novel optimization algorithm, Knowledge-Based Syst 191 (2020) 105190, doi:10.1016/j.knosys.2019.105190.
[147] A.R. Moazzeni, E. Khamehchi, Rain optimization algorithm (ROA): a new metaheuristic method for drilling optimization solutions, J. Pet. Sci. Eng. 195 (2020) 107512, doi:10.1016/j.petrol.2020.107512.
[148] A. Faramarzi, M. Heidarinejad, S. Mirjalili, A.H. Gandomi, Marine predators algorithm: a nature-inspired metaheuristic, Expert Syst. Appl. 152 (2020) 113377, doi:10.1016/j.eswa.2020.113377.

[149] S. Kaur, L.K. Awasthi, A.L. Sangal, G. Dhiman, Tunicate swarm algorithm: a new bio-inspired based metaheuristic paradigm for global optimization, Eng. Appl. Artif. Intell. 90 (2020) 103541, doi:10.1016/j.engappai.2020.103541.

[150] S. ur R. Massan, A.I. Wagan, M.M. Shaikh, A new metaheuristic optimization algorithm inspired by human dynasties with an application to the wind turbine micrositing problem, Appl. Soft Comput. J. 90 (2020) 106176, doi:10.1016/j.asoc.2020.106176.

[151] Q. Askari, M. Saeed, I. Younas, Heap-based optimizer inspired by corporate rank hierarchy for global optimization, Expert Syst. Appl. 161 (2020) 113702, doi:10.1016/j.eswa.2020.113702.

[152] A. Kaveh, A. Zaerreza, Shuffled shepherd optimization method: a new Meta-heuristic algorithm, Eng. Comput. (Swansea, Wales) 37 (7) (2020) 2357–2389, doi:10.1108/EC-10-2019-0481.

[153] J.S. Chou, N.M. Nguyen, FBI inspired meta-optimization, Appl. Soft Comput. J. 93 (2020) 106339, doi:10.1016/j.asoc.2020.106339.

[154] Q. Askari, I. Younas, M. Saeed, Political optimizer: a novel socio-inspired meta-heuristic for global optimization, Knowledge-Based Syst 195 (2020) 105709, doi:10.1016/j.knosys.2020.105709.

[155] Y. Zhang, Z. Jin, Group teaching optimization algorithm: a novel metaheuristic method for solving global optimization problems, Expert Syst. Appl. 148 (2020) 113246, doi:10.1016/j.eswa.2020.113246.

[156] A. Kaveh, A.Dadras Eslamlou, Water strider algorithm: a new metaheuristic and applications, Structures 25 (2020) 520–541, doi:10.1016/j.istruc.2020.03.033.

[157] E. Bogar, S. Beyhan, Adolescent identity search algorithm (AISA): a novel metaheuristic approach for solving optimization problems, Appl. Soft Comput. J. 95 (2020) 106503, doi:10.1016/j.asoc.2020.106503.

[158] W. Al-Sorori, A.M. Mohsen, New caledonian crow learning algorithm: a new metaheuristic algorithm for solving continuous optimization problems, Appl. Soft Comput. J. 92 (2020) 106325, doi:10.1016/j.asoc.2020.106325.

[159] H. Ghasemian, F. Ghasemian, H. Vahdat-Nejad, Human urbanization algorithm: a novel metaheuristic approach, Math. Comput. Simul. 178 (2020) 1–15, doi:10.1016/j.matcom.2020.05.023.

[160] O.K. Meng, O. Pauline, S.C. Kiong, A carnivorous plant algorithm for solving global optimization problems, Appl. Soft Comput. (2020) 106833, doi:10.1016/j.asoc.2020.106833.

[161] C. Taramsco, B. Crawford, R. Soto, E.M. Cortés-Toro, R. Olivares, A new metaheuristic based on vapor-liquid equilibrium for solving a new patient bed assignment problem, Expert Syst. Appl. 158 (2020) 113506, doi:10.1016/j.eswa.2020.113506.

[162] D. Połap, M. Woźniak, Red fox optimization algorithm, Expert Syst. Appl. 166 (2021) 114107, doi:10.1016/j.eswa.2020.114107.

[163] J.S. Chou, D.N. Truong, A novel metaheuristic optimizer inspired by behavior of jellyfish in ocean, Appl. Math. Comput. 389 (2021) 125535, doi:10.1016/j.amc.2020.125535.

[164] H.C. Tsai, Improving backtracking search algorithm with variable search strategies for continuous optimization, Appl. Soft Comput. J. 80 (2019) 567–578, doi:10.1016/j.asoc.2019.04.032.

[165] M.H. Nadimi-Shahraki, S. Taghian, S. Mirjalili, An improved grey wolf optimizer for solving engineering problems, Expert Syst. Appl. 166 (2021) 113917, doi:10.1016/j.eswa.2020.113917.

[166] B. Yin, Z. Guo, Z. Liang, X. Yue, Improved gravitational search algorithm with crossover, Comput. Electr. Eng. 66 (2018) 505–516, doi:10.1016/j.compeleceng.2017.06.001.

[167] J. Wu, Y.G. Wang, K. Burrage, Y.C. Tian, B. Lawson, Z. Ding, An improved firefly algorithm for global continuous optimization problems, Expert Syst. Appl. 149 (2020) 113340, doi:10.1016/j.eswa.2020.113340.
[168] J. Li, Y. xiang Li, S. sha Tian, J. lin Xia, An improved cuckoo search algorithm with self-adaptive knowledge learning, Neural Comput. Appl. 32 (16) (2020) 11967–11997, doi:10.1007/s00521-019-04178-w.
[169] T.O. Ting, X.S. Yang, S. Cheng, K. Huang, Hybrid metaheuristic algorithms: Past, present, and future, Studies in Computational Intelligence, 585, Springer Verlag, Berlin, Germany, 2015, pp. 71–83.
[170] C. Blum, J. Puchinger, G.R. Raidl, A. Roli, Hybrid metaheuristics in combinatorial optimization: a survey, Appl. Soft Comput. J. 11 (6) (2011) 4135–4151, doi:10.1016/j.asoc.2011.02.032.
[171] X.S. Yang, S.F. Chien, T.O. Ting, Computational intelligence and metaheuristic algorithms with applications, Sci. World J. 2014 (2014), doi:10.1155/2014/425853.
[172] K. Sörensen, Metaheuristics-the metaphor exposed, 2012.
[173] Journal of Heuristics, Policies on heuristic search, 2020. https://www.springer.com/journal/10732/updates/17199246. (Accessed 22 November 2020).
[174] 4OR: A Quarterly Journal of Operations Research, 4OR Aims and scope, 2020. https://www.springer.com/journal/10288/aims-and-scope. (Accessed 22 November 2020).

Artificial intelligence techniques

3.1 Introduction

Unlike the natural intelligence of humans and animals, artificial intelligence attempts to understand the behaviors of the systems and build intelligent ones. The term "artificial Intelligence" (AI) is used to describe the machines or entities that mimic the cognitive functions of humans or animals, such as learning and problem solving, to achieve the prespecified goals effectively [1–3]. There are many definitions of AI, including the followings:

- "The automation of activities that we associate with human thinking, activities such as decision-making, problem-solving, and learning [4]."
- "The study of the computations that make it possible to perceive, reason, and act [5]."
- "A system's ability to interpret external data correctly, to learn from such data, and to use those learnings to achieve specific goals and tasks through flexible adaptation [6]."

3.1.1 Brief history of artificial intelligence

AI is a subfield of computer science that was formally introduced in 1956 through an academic conference in Hanover, New Hampshire, at Dartmouth College. Cognitive scientist John McCarthy and the other participants, including Marvin Minsky, Nathaniel Rochester, Claude Shannon, Herbert Simon, Julian Bigelow, Ray Solomonoff, and others, were quite optimistic about the future of AI. However, the journey to understand if the machines can think began much before that time. For instance, ancient Greek myths talked about intelligent robots, and Egyptian and Chinese engineers made automatons. The father of modern computing, Alan Mathison Turing, shortened World War II and brought British victory by leading the deciphering team to crack the Nazi code "unbreakable" enigma. In 1936, he proved that a universal calculator is possible that is currently known as the Turing machine. His central insight was that any problem that can be represented and solved is also solvable by a machine. W. McCulloch and W. Pitts publication of the year 1943 titled "*A Logical Calculus of the Ideas Immanent in Nervous Activity*" made the foundations for artificial neural networks (ANNs), the most widely used and popular AI technique. In 1950, Alan Mathison Turing introduced the Turing Test as a way of operationalizing a test of intelligent behavior by throwing the question, "Can machines think?" In the same year, Claude Shannon illustrated chess playing as search, and Isaac Asimov proposed three laws of robotics [7–12]. Arthur Lee Samuel popularized the term "machine learning" (ML) in 1959 while developing the first game-playing program to achieve sufficient skill to challenge a world champion. In 1962, Engelberger and Devol established the world's first robotics company, Unimation, in Danbury, Connecticut, USA [13]. A brief history of AI is illustrated in Fig. 3.1 [7–13].

In 1963, Thomas Evans demonstrated that the computer could solve IQ test analogy questions. Danny Bobrow explained that computers could understand natural language to solve algebra word problems in 1964. An artificial intelligence program (ELIZA, first chatbot) was written in 1965, and the first animated robot, Shakey, was developed in 1966. The first successful knowledge-based program for scientific reasoning was illustrated in 1967. Paul J. Werbos first described the training process of the ANN in 1974. Herbert Simon was awarded the Nobel Prize (Economic Sciences) in 1978 for his important AI work.

However, the AI journey was not smooth; it experienced a massive reduction in the public interest; hence, research funding between 1974 to 1980 is known as the first AI winter [7–12]. After the end of the first winter, IBM produced the first personal computer in 1981. During the mid-1980s, the ANNs were widely used. In 1985, Harold Cohen created the autonomous drawing program, Aaron. The collapse of the business community's fascination brought the second winter to the AI industry in 1987 that lasted till 1993. However, in the 1990s, significant AI advancements were achieved with substantial demonstrations in case-based reasoning, data mining, games, intelligent tutoring, machine learning, natural language understanding and translation, uncertain reasoning, virtual reality, and vision. Kasparov, a world-famous chess player, was defeated by the supercomputer, Deep Blue, in 1997. The first artificial intelligence player, Furby, was driven to the market in 1998. Honda produced a humanoid robot, ASIMO, in 2000.

In 2008, Google introduced "speech to search" for iPhone. Google artificial brain learned to detect cats. Apple introduced an intelligent personal assistant, Siri, in 2012. In 2014, Facebook created a deep learning (DL) facial recognition system, Deep Face, which shows the human level of performance. In the same year, Amazon released

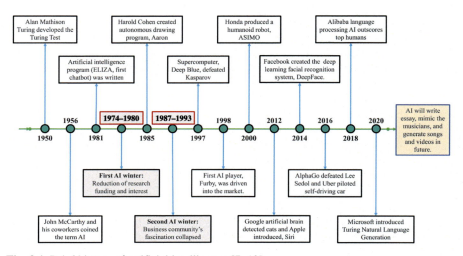

Fig. 3.1 Brief history of artificial intelligence [7–13].

its virtual assistant called Alexa. In 2016, the former Go champion, Lee Sedol, was defeated by Google's AlphaGo during Google DeepMind Challenge Match, and Uber piloted a self-driving car in Pittsburgh. Alibaba language processing AI outscored top humans in 2018. Microsoft introduced Turing Natural Language Generation (T-NLG) in 2020. With the tremendous progress of AI, it is expected that AI will write the essay, mimic the musician, and generate songs and videos soon [7–12].

3.1.2 Artificial intelligence is the new electricity

AI is transforming industries and businesses by creating huge economic impacts like the electricity that transformed almost everything 100 years ago. Therefore, professor Andrew Ng, the cofounder of Coursera, cofounder of Google Brain, and adjunct professor Stanford University, coined the term "Artificial Intelligence is the New Electricity" [14–16]. This disruptive technology has already revolutionized many sectors by automating the tasks and creating impacts from healthcare to agriculture, banking, entertainment, logistics, manufacturing, retail, and transportation. It shaped many other prominent fields, including autonomous planning and scheduling, game playing, machine translations, robotics, spam fighting, and speech recognition [1,17].

3.1.3 Artificial intelligence foundation and basic components

The foundation of modern AI has been built by combining many academic disciplines, including philosophy, psychology, linguistics, economics, mathematics, neurosciences, control theory, cybernetics, and computer engineering [1,18–23]. The AI users should have a basic understanding and knowledge of the mentioned fields and comprehensive domain expertise to become AI experts in their fields. However, the fundamental component of AI includes computer vision, cognitive computing, ML, DL, and natural language processing [1].

ML is the subset of AI that allows the machines to learn and improve their performance without being explicitly programmed; thus, they act intelligently. ML focuses on algorithm development that decides by analyzing the available data. In general, ML deals with structured and semistructured data, whereas AI deals with all kinds of data. ML techniques can be further classified as supervised learning, semi-supervised learning, unsupervised learning, and reinforcement learning. Supervised learning techniques deal with the labeled data and are usually employed for classification and regression problems. On the contrary, unsupervised learning techniques deal with the unlabeled data and make inferences based on the attributes of the training data. They are employed for data clustering and dimensionality reduction. The reinforcement learning techniques interact with data to maximize the rewards and minimize the penalties set by the programmer, which in turn provide the desired output [24]. DL, a subfield of ML techniques, learns through data processing employing ANNs. DL machines evaluate the outputs from the inputs by engaging multiple layers of ANN. They learn through positive and negative reinforcement of the tasks, and they process the data and reinforce the progress simultaneously. They

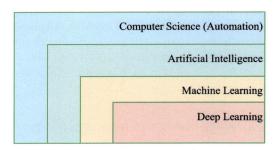

Fig. 3.2 Relationship of different subfields of AI with automation [25–27]. *AI*, artificial intelligence.

offer solutions to a large set of data, also known as big data, by structuring, connecting, and visualizing them to explore and expedite data insight to empower decision-making. They outperform the traditional shallow ML techniques while dealing with vast data [24]. Fig. 3.2 represents the relationship of different AI subfields under the automation process [25–27].

With time, the researchers are coming up with different sophisticated learning algorithms for AI techniques. The modifications and upgrades of the available learning algorithms are keys for the widespread adoption of AI technologies. Also, many supporting technologies are making momentum to their popularity and acceptability. Such supporting technologies include graphical processing units, the internet of things, advanced algorithms, application programming interface, communication infrastructure, and intelligent data processing [24]. It is expected that the AI market size will be 126 billion dollars in 2025 that was around 10 billion dollars only in 2018 [28]. The development of sophisticated learning algorithms and supporting technologies as per the need of our day-to-day life are the main enablers of such enormous market growth.

3.1.4 Limitations of artificial intelligence

Despite having so many advantages, AI has a few downsides [29–32]. For instance, it is relatively costly to implement, not as effective as humans at intuition-based decision making, and not creative like humans in some cases. Besides, it sometimes either overestimates or underestimates the appropriate actions. The application of AI with experts may lead to disastrous outcomes. It is also making humans lazy and at the same time creating the risk of unemployment. Also, as it learns from the data, data inaccuracies will be reflected in the results [17]. AI systems are trained to do clearly defined tasks and cannot be implemented for nontrained jobs. For instance, an AI system that detects health care fraud cannot accurately detect tax fraud or other kinds of scams.

The rest of the chapter is organized as follows: Section 3.2 discusses the most widely adopted AI and ML techniques, including ANN, extreme learning machines, support vector machines (SVMs), fuzzy logic models (FLMs), genetic programming (GP), and DL techniques. Section 3.3 demonstrates hybrid, ensemble, and metaheuristic algorithm-inspired AI techniques. Finally, Section 3.4 draws the concluding remarks of the chapter.

3.2 Artificial intelligence techniques

Since the inception of AI, many techniques have been explored, and a good number of them are being used behind the scenes that influence our everyday lives in various forms. This section briefly illustrates the popular, widely adopted, and efficient ML techniques, including ANN, extreme learning machine (ELM), FLM, GP, and SVM.

3.2.1 Artificial neural networks

In 1943, Warren McCulloch, a neurophysiologist, and Walter Pitts, a mathematician, modeled a simple neural network with electrical circuits by applying the biological processes in the brain to mathematical algorithms. Their combined work built the foundation of today's ANN [33]. ANN is the most powerful computational tool in the field of AI and ML. These brain-inspired computational systems intended to replicate human learning by analyzing the physical phenomena and decision-making processes of complex real-world problems. The structures, processing techniques, and learning abilities of the ANN are like biological neural networks. Due to their parallel computational capabilities and adaptiveness to external disturbances, they can efficiently deal with detection, classification, clustering, and regression problems [34–37].

Among different types of neural networks, the feedforward neural networks refer to the purest form of the ANN, where the multilayer perceptron (MLP) represents the simplest structure. The multilayer perceptron neural networks (MLP-NN) builds models through supervised learning algorithms by mapping the inputs onto the outputs of the problem under investigation. In general, the neural networks consist of three layers namely, input, hidden, and output layers, where each layer may have either one or several nodes. The nodes of each layer are linked to the nodes of the subsequent layer through weighting factors. A biasing factor is also added to the summation of the inputs of each node before being processed through different kinds of nonlinear activation functions. The processed output of each node is then transferred to the nodes of the subsequent layer. The process continues until the output layer nodes are reached [38–41].

Fig. 3.3 presents a simplified MLP-NN structure that consists of k number of inputs linked to n number of hidden nodes, and each hidden node is connected to m number of output nodes. Finally, each output node generates an output; therefore, this MLP-NN produces m outputs from k inputs. The MLP-NN is trained in two stages by adjusting its connecting weights (w_{ij}) and biases (b_{ij}). In the first stage, the inputs are propagated through the hidden nodes and the output nodes to predict the outputs

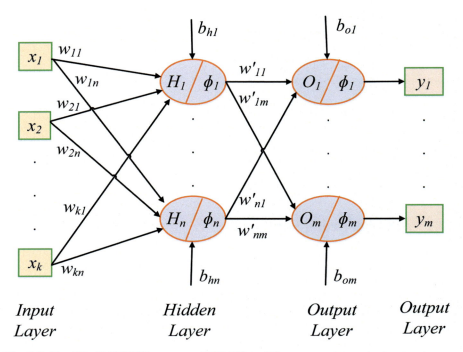

Fig. 3.3 Simplified MLP-NN structure. *MLP-NN*, multilayer perceptron neural networks.

by selecting the weights and biases randomly. Then, the estimated and the targeted outputs are compared, and their differences are estimated. The MLP-NN adjusts the connecting weights and biases in the second stage to minimize the evaluated differences. Finally, it terminates updating the weights and biases if the difference of actual and estimated outputs do not change for a certain number of epochs or if it reaches the prespecified number of epochs [42–44].

The neural network models employ different training algorithms to perform the learning processes to adjust the connecting weights and biases. Among many algorithms, gradient descent (GD) backpropagation, resilient backpropagation, Levenberg-Marquardt backpropagation, scaled conjugate gradient backpropagation, and one step secant (OSS) backpropagation are widely used. However, each algorithm has its pros and cons in terms of accuracy, convergence, speed, storage requirement, and other features, as can be found in [45]. Likewise, the neural network systems employ various types of squashing or activating functions, including linear, logistic sigmoid, softmax, tan sigmoid, ReLU, and many others, as graphically illustrated in Fig. 3.4 [45].

However, the traditional shallow feedforward neural networks suffer from several disadvantages, including vanishing gradient phenomenon, dissatisfactory generalization performance, overtraining, and iterative adjustment of the connecting

Artificial intelligence techniques

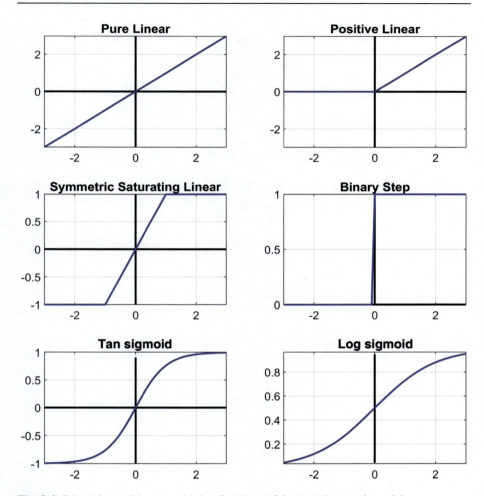

Fig. 3.4 Selected squashing or activation functions of the neural network models.

weights and biases [46]. In response, other types of shallow neural networks including ELMs, error correction (EC), improved second-order, K-nearest neighbor, logistic regression (LR), naïve Bayes (NB), SVMs, and other radial basis function (RBF) networks have been explored [46–48].

3.2.2 Support vector machines

In 1992, Boser et al. [49] introduced the SVMs that evolved into mighty and efficient supervised learning machines for analyzing data with time. They map data into the high-dimensional feature spaces *via* a nonlinear mapping by building optimal

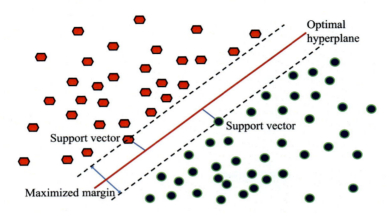

Fig. 3.5 SVM hyperplane, support vectors, and margin illustration [53–56]. *SVM*, support vector machine.

geometric hyperplanes for data separation. In SVM, the hyperplanes are the decision boundaries that decide the class of the observation points. The hyperplanes can just be a line, a plane, or something else based on the number of input features. The SVM employs different types of kernels to create separation surfaces by quantifying the similarities between various observations. Polynomial, linear splines, hyperbolic tangent, Gaussian RBF, Laplace RBF, and sigmoidal kernels are widely adopted SVM kernels [50]. Instead of all available data patterns, the SVM selects the data points closer to the separation surface (hyperplane) through several algorithms to optimize their performance, including linear and quadratic programming. Such data points are known as the support vectors, and they influence the position and orientation of the separation surface (hyperplane). The distance from the support vectors to the hyperplane is known as the margin, and the SVM algorithms maximize it to identify the optimal hyperplane [51–53]. Fig. 3.5 illustrates the idea of SVM hyperplane, support vectors, and margin for a two-class classification problem [53–56]. The SVM can easily avoid local optima and effectively deal with high dimensional spaces, even if the number of dimensions is larger than the number of samples. Additionally, the SVM requires less memory. The SVM is used for classification, outlier and novelty detection, regression, and feature reduction in various science and engineering applications due to their advantages and effectiveness in dealing with nonlinearly separable datasets [57–59].

Let us assume $x \in R^n$ is the input vector where R represents the real number and $f(x, w, b) \in R$ refers to estimated output in higher dimensional space using the nonlinear transformation $\Phi(x)$ from R^n. The following equation represents the mentioned phenomenon:

$$f(x,w,b) = w^T \cdot \Phi(x) + b \qquad (3.1)$$

Where, $w \in R^n$ refers to the weight vector, and $b \in R$ refers to the bias.

The SVM minimizes the following cost function while predicting the outputs from the available inputs [60]:

$$L\left(w, \xi, \xi_i^*\right) = \frac{\lambda}{2} w^T w + C \sum_{i=1}^{m} \xi_i + C \sum_{i=1}^{m} \xi_i^* \qquad (3.2)$$

Subject to:

$$f(x_i, w, b) - y_i \leq \varepsilon + \xi_i^*$$
$$y_i - f(x_i, w, b) \leq \varepsilon + \xi_i$$
$$C, \varepsilon, \xi_i, \xi_i^* \geq 0, i = 1, 2, 3, \ldots, p$$

Where, m refers to sample input numbers; ε is termination criterion tolerance; ξ_i and ξ_i^* are slack variables to handle nonseparable input data; λ and C control the penalty associated with the training errors, hence, known as the regularization coefficients (capacity constants). Finally, y_i refers to the actual or expected value.

The SVM uses the following equation to evaluate the optimal weights vector; hence, the effective solution for the problem under investigation by transforming the developed optimization formulation into the dual problem [61]:

$$w^* = \sum_{i=1}^{m} \left(\alpha_i - \alpha_i^*\right) \Phi(x_i) \qquad (3.3)$$

$$f(x, w, b) = \sum_{i,j=1}^{m} \left(\alpha_i - \alpha_i^*\right) \langle \Phi(x_i).\Phi(x_j)\rangle + b$$

$$or, f(x, w, b) = \sum_{i,j=1}^{m} \left(\alpha_i - \alpha_i^*\right) K(x_i, x_j) + b \qquad (3.4)$$

Subject to:

$$\left(\alpha - \alpha^*\right) 1 = 0$$
$$0 \leq \alpha_i, \alpha_i^* \leq C, i = 1, 2, 3, \ldots, m$$

Where, the constants α_i and α_i^* are the Lagrange multipliers (pushing and pulling forces) for estimation of $f(x, w, b)$ towards the target y_i; <.> denotes the dot product; $K(x_i, x_j)$ refers to the kernel function that is equivalent to the inner product of the vectors x_i and x_j in the feature space $\Phi(x_i)$ and $\Phi(x_j)$; and 1 is the unit vector [62]. Support vectors with multipliers $\alpha_i = 0$, $\alpha_i = C$, and $\alpha_i^* = 0$, $\alpha_i^* = C$ are called bound support vectors, whereas the rest are known as non-bound support vectors.

The following *Gaussian* function is widely adopted as the RBF kernel for the regression problem where an adjustable parameter (γ) of the particular kernel function are often employed [63]:

$$K(x_i, x_j) = e^{-\gamma \|x_i - x_j\|^2}; \gamma > 0 \qquad (3.5)$$

Despite having several advantages and better generalization performance, the SVM has a couple of weaknesses as well. They cannot effectively handle large, imbalanced, and noisy datasets. Besides, they need to transform multiclass classification problems into multiple binary classification problems as they were initially designed for binary classification problems. They also require larger training time. Finally, like other nonparametric techniques, the SVM does not demonstrate adequate transparency in results [57,64,65].

3.2.3 Extreme learning machines

ELM is another type of shallow feedforward neural network consisting of only one hidden layer and makes the data processing faster by removing several steps of the traditional feedforward neural networks. Besides, ELM employs Moore-Penrose generalize inverse to set the connecting weights and biases instead of the gradient-based backpropagation technique. It provides better generalization performance by effectively determining the global optima. Therefore, ELM is simple and easy to understand, imagine, and implement. Besides, ELM is employed to deal with the problems that require real-time retraining of the network. The mentioned properties make them popular in the research community since their inception in 2004 [66–69]. These single hidden layer feedforward neural networks were first proposed by Huang et al. [66] to solve optimization problems by developing direct links amongst several theories, including the ridge regression, matrix theory, linear system stability, and others. Unlike SVM and other traditional feedforward neural networks, ELM randomly chooses input weights and analytically evaluates output weights [70].

The ELM maps input data to a high dimensional feature space with a dimension size equal to the number of neurons in the hidden layer. For instance, if an ELM with $\tilde{S}$ number of neurons in the hidden layer deals with S number of distinct training samples and $S \geq \tilde{S}$. Then, it mathematically models the output $\{x_k, t_k\}_{k=1}^{S}$ using the following equation:

$$g_k = \sum_{i=1}^{\tilde{S}} \alpha_i f_i(z_i, b_i, x_k), k = 1, 2, 3, \dots, S \tag{3.6}$$

Where, $x_k = [x_{k1}, x_{k2}, \dots, x_{kq}]^T \in R^q$ refers to the input matrix with q number of inputs; $z_i = [z_{i1}, z_{i2}, \dots, z_{iq}]^T \in R^q$ refers to the input weight vector associated with weighting factors of the connected nodes of input and hidden layers; $\alpha_i = [\alpha_{i1}, \alpha_{i2}, \dots, \alpha_{ip}]^T$ refers to the output weight vector associated with weighting factors of the corresponding nodes of the hidden and output layers; b_i is the biasing factor for the ith neuron in the hidden layer; and $f_i(z_i, b_i, x_k) = z_i.x_k + b_i$ represents the activation function where $z_i.x_k$ denotes the inner product of z_i and x_k. Finally, $t_k = [tk_1, tk_2, \dots, tk_p]T \in R^p$ refers to the output matrix with p number of outputs.

Then, the ELM rewrites Eq. (3.6) as $G = F\alpha$ with the following definitions of F, G, and α [47]:

$$F(z_i, b_i, x_k) = \begin{bmatrix} f(z_1, x_1, b_1) & \cdots & f(z_{\tilde{S}}, x_1, b_{\tilde{S}}) \\ \vdots & \cdots & \vdots \\ f(z_1, x_S, b_1) & \cdots & f(z_{\tilde{S}}, x_S, b_{\tilde{S}}) \end{bmatrix}_{S \times \tilde{S}} \tag{3.7}$$

$$\alpha = \begin{bmatrix} \alpha_1^T \\ \vdots \\ \alpha_{\tilde{S}}^T \end{bmatrix}_{S \times \tilde{S}} \quad (3.8)$$

$$G = \begin{bmatrix} g_1^T \\ \vdots \\ g_S^T \end{bmatrix}_{S \times p} \quad (3.9)$$

Now, the F matrix is commonly known as the ELM hidden layer output matrix. It is an invertible squared matrix if several training samples in the dataset and the number of hidden neurons is equal for example, $S = \tilde{S}$. In such a case, the ELM can effectively approximate the training samples, meaning the z_i, α_i, and b_i values exist from where output T can be approximated for example, $F\alpha = T$. However, *the F matrix is usually non-squared*; hence, non-invertible in real-world problems as the number of samples is much larger than that of the number of ELM hidden neurons [71]. Therefore, appropriate values of the z_i, α_i, and b_i may not exist for which the output T can be approximated. In response, the ELM minimizes both training error and output weights norm by exploring appropriate values of the $\hat{z}_i, \hat{\alpha}_i$, and $\hat{b}_i$. To do so, the ELM considers the following equation as its objective function [47]:

$$\text{Minimize } F\alpha - T^2 \text{ and } \alpha \quad (3.10)$$

The ELM then evaluates the minimum norm of the output weights ($\tilde{\alpha}$) to confirm minimum training error and minimum norm of weights simultaneously by employing a minimal norm least-square solution. It achieves the best generalization performance and uniqueness of the solution through the mentioned actions [66]. For such purpose, the ELM uses the following equation:

$$\tilde{\alpha} = F^\dagger T \quad (3.11)$$

Where, $F^\dagger$ is the Moore-Penrose generalized inverse of the hidden layer output matrix of the network (F).

However, the ELM may lead towards overfitting risk due to their basis of the empirical risk minimization principle. Besides, the random selection of the input weights causes uncertainty in the learning and approximation processes of the ELM. Also, the wrong choice of the activation function may degrade their generalization performance, hence reducing overall accuracy. Considering the mentioned downsides of the ELM, a couple of improved versions of the ELM have been explored and reported in the literature [72–75].

3.2.4 Fuzzy logic model

In 1965, Lotfi Zadeh introduced the term fuzzy logic to deal with uncertainty other than randomness by resembling human reasoning [76]. For instance, an ambiguous and imprecise statement like "Sally is very tall" cannot be explained by the randomness of the probability theory but with the FLM. It is one of the powerful tools in

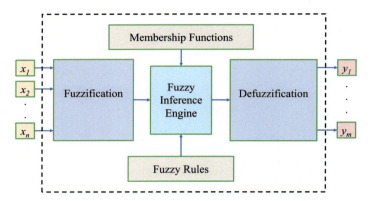

Fig. 3.6 Structure of the fuzzy logic model.

the age of cybernetic with graded membership. It deals with the vagueness, imprecision, and uncertainties of complex real-world problems using a set of "if-then" rules [77–79]. The FLM has been applied effectively in many real-life problems in various disciplines, including engineering, mathematics, computer science, medical and social sciences, business, public policy, and jurisdiction [80–82].

Fig. 3.6 shows a basic FLM structure that consists of three blocks fuzzification, fuzzy inference system (FIS), and defuzzification for converting the inputs (x_1, x_2, ..., and x_n) into the outputs (y_1 and y_2). Three conceptual components are involved in the FIS of the FLM: a rule base comprising of fuzzy rules, a database with definitions of the used membership functions, and a fuzzy inference engine to execute the inference procedure based on the developed rules to derive the outputs.

Fuzzification is the first step in the FLM that transforms the crisp quantities (numerical values) into linguistic variables (low, medium, high, etc.) represented by fuzzy sets. A fuzzy quantity is a generalized regular and real number that does not refer to a single value but a connected set of possible values. Each value can take the weight from the interval of 0 to 1. This weight is known as the fuzzy membership function and denoted by μ, which illustrates numeric numbers matching the fuzzy numbers [83,84]. There are several types of fuzzy numbers, including triangular, trapezoidal, differential piecewise quadratic, impulse, and gaussian bell shape numbers [85,86]. Fig. 3.7 illustrates the triangular and trapezoidal fuzzy numbers.

After fuzzification, the fuzzy inference engine derives the linguistic outputs based on the developed fuzzy rules and obtained fuzzy membership function. Mamdani-type, Sugeno-type, and Tsukamoto-type are widely adopted in the literature and industry among several FISs. Their main differences lie in the way of output determination [87]. Then, it follows a defuzzification process to infer the accurate crisp quantity (numerical values) from the linguistic variables (low, medium, high, etc.) according to the rules and the input data [88]. The FLM involves similar types of fuzzy numbers during the defuzzification process [89].

As stated earlier, the FLM resembles simple reasoning like human reasoning to deal with the uncertainty of inaccurate and incomplete data. Besides, it is customizable

Artificial intelligence techniques

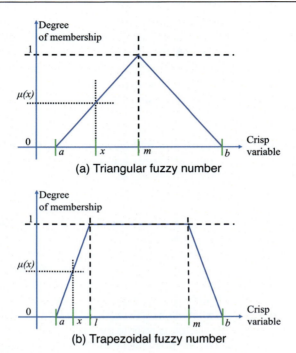

Fig. 3.7 Fuzzy numbers.

and straightforward; at the same time, it is robust and less expensive. However, it is tiresome to set exact rules and membership functions. It also makes the analysis difficult as the outputs can be interpreted in several ways and do not provide accurate results in some cases. Moreover, it creates confusion with the randomness of the probability theory [90]. To deal with the mentioned difficulties, the FLM is often combined with ANN to produce neuro-fuzzy systems where the FLM performs the inference mechanism under cognitive uncertainty. The ANN is responsible for adaptive learning, parallelism, and better generalization performance [91].

3.2.5 Genetic programming

In AI, GP is a biologically inspired ML technique. GP offers solutions to complex real-world problems by following the Darwinian principle of natural selection. Friedberg introduced the concept in the 1950s, and Cramer modified it by applying the genetic algorithm (GA) and tree structures in the 1980s. Finally, Koza solved the regression problems using GP in early 1990. Like the GA, the GP employs different operators such as the generation of the initial population, selection, crossover, and mutation to generate the optimal solution. However, though the genetic operators of the GP are similar to the operators of the GA, they encode the problems in different ways. The GA provides the solutions as the strings of numbers, whereas the GP provides the solutions, mathematical models, in terms of tree structures as mentioned earlier. Therefore, unlike the previously discussed ML techniques, including the ANN, SVM, ELM, and FLM, that provide the solution models of the problems under investigation as black boxes, the GP provides explicit mathematical models relating to the input variables [92–99]. Since its inception, it has been implemented for classification, feature extraction, and regression problems in engineering and science [100–102].

Two standard versions of GP are available in the literature, namely the single gene genetic programming (SGGP) and the multigene genetic programming (MGGP). The SGGP is the basic version of GP where each individual consists of a single tree, whereas in MGGP, each individual consists of more than one tree [96]. Different operators of SGGP and MGGP are illustrated below:

Step 1: Initialization

The GP starts with a prespecified number of randomly generated individuals (commonly known as the initial population) represented by tree structures based on three available techniques: ramped half and half, grow, and full [98]. Individuals of the initial population comprise of two types of genes, e.g., function set $\{+, \times, -, \div, \sin, \exp,$ etc.$\}$ that represents different arithmetic operators and terminal set $\{x_1, x_2, \ldots, x_n, c\}$ that represents the input variables, x_i, and numerical numbers, c. Fig. 3.8 shows two randomly generated tree structures or genes (G_1 and G_2) with their mathematical interpretation. The function set operators refer to the nodes, and the terminal set items refer to the tree leaves. In SGGP, the individual consists of only one gene (G_1 or G_2).

Conversely, in MGGP, the individual consists of more than one gene where the maximum allowable number of genes (G_{max}) in the individual is specified during initialization. If an individual of MGGP consists of two genes, e.g., G_1 or G_2, the individual will be interpreted as $w_0 + w_1G_1 + w_2G_2$, where w_0, w_1, and w_2 are weighting factors of the involved genes. During the initialization stage, the GP also sets other parameters like crossover and mutation probabilities, number of elites, and the maximum number of generations.

Step 2: Fitness evaluation and storing of the best solution

In this stage, the fitness of each generated individual is assessed, and the best individual is stored as the best solution to the problem under investigation.

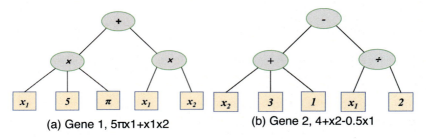

(a) Gene 1, 5πx1+x1x2   (b) Gene 2, 4+x2-0.5x1

Fig. 3.8 Randomly generated trees (genes) of GP. *GP*, genetic programming.

Step 3: Termination criteria checking

This step checks the termination criteria after a pre-specified number of generations to avoid premature convergence. It terminates the optimization process if the values of the best solution do not change for a pre-specified number of generations or it reaches the maximum number of generations. The technique then displays the best individual as the optimal solution to the optimization problem under investigation. Otherwise, it proceeds to the next step.

Step 3: Selection

Then, the GP selects well-doing individuals from the pool of initial/updated population. Several selection techniques are available, including proportionate fitness selection, greedy over selection, and tournament selection [98]. The fitness proportionate selection technique selects individuals based on their relative performance compared to the entire population. The greedy selection technique divides all individuals in the initial/updated population into two groups. Group I consists of the top 20% of individuals, and Group II consists of the rest. Then, it selects an equal number of individuals from each group using the fitness proportionate selection technique. This technique is biased to the better performers to reduce the number of generations [99]. Finally, the tournament selection technique selects the individuals comparing the best-fitted one from two stochastically selected individuals. It is comparatively more flexible and diverse, providing even lower-performing individuals chances in a few cases [95].

Step 4: Crossover

The GP generates a new population set from the selected individuals through crossover and mutation operations. The crossover is analogous to the sexual reproduction system of animals or humans, where two individuals (parents) produce another two individuals (offspring). As the offspring carry the characteristics of both parents, it is expected that their fitness will be better than the parents in most of cases. In general, a higher number of individuals (80 ~90%) are passed through the crossover operations, and the rest of the individuals are sent to the next step (mutation) directly. The SGGP randomly selects the non-leaf nodes from both parents as the crossover point and interchanges the subtrees, as illustrated in Fig. 3.9 and Fig. 3.10. This crossover is known as standard subtree or tree level, or low-level crossover.

On the other hand, the crossover operation is performed in two stages in MGGP. At the first stage, a high-level two-point crossover, also known as tree crossover, is

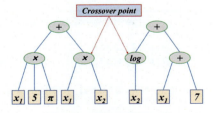

Fig. 3.9 Illustration of the SGGP crossover operation, parents, and crossover point. *SGGP*, single gene genetic programming.

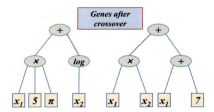

Fig. 3.10 Produced offspring after low-level crossover operation.

performed by exchanging genes of two selected parents to produce intermediate offspring. Table 3.1 illustrates the high-level two-point crossover operations of MGGP, considering $G_{max} = 4$. As can be observed, the second offspring contains more genes than G_{max}; therefore, the extra genes have been removed randomly. After this high-level crossover, the standard subtree crossover is also performed in MGGP.

Step 5: Mutation

The mutation operation adds new features in some cases and overcomes any loss of the characteristics in other cases. The SGGP randomly selects a nonleaf node from the tree (individual); then, it randomly replaces the arithmetic operator of that node from the function set, as illustrated in Fig. 3.11. The mutation operation of the MGGP is almost identical to SGGP, where the regular subtree mutation is performed after the selection of a single gene randomly from an individual. Then, the selected gene is replaced by the mutated gene of the individual.

The evolved individual may perform well compared to the original one; however, it will disappear from the race if it is not compatible. In general, a lower number of individuals (10 ~15%) are passed through the mutation operations, and most of the individuals are sent to the next generation directly. It is worth mentioning that, in both SGGP and MGGP, a prespecified number of elite individuals are sent to the next generation directly without passing through the crossover and mutation operations. After preparing the new set of individuals (population), the GP is returned to step 2.

Despite having several advantages, the GP has several limitations. For instance, it is almost impossible to find a unique solution for a single problem. It generates a massive number of possible solution trees that provide different mathematical models in different simulations. Sophisticated coding expertise is required to avoid errors

Table 3.1 High-level two-point (tree) crossover illustration.

Items	Parent to offspring transformation	
Selected parents	Parent 1: [G1 G2 G3 G4]	Parent 2: [G5 G6 G7 G8]
Starting of tree crossover	Parent 1: [G1 <G2 G3 G4>]	Parent 2: [G5 G6 <G7 G8>]
Generated intermediate offspring	Offspring 1: [G1 <G7 G8>]	Offspring 2: [<G2 G3 G4> G5 G6]
Polished intermediate offspring	Offspring 1: [G1 G7 G8]	Offspring 2: [G2 G4 G5 G6]

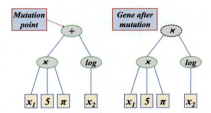

Fig. 3.11 SGGP mutation operation. *SGGP*, single gene genetic programming.

while replacing arithmetic operators during mutation. Besides, a small variation can sometimes bring a disastrous effect on the generated solutions [103].

3.2.6 Deep learning technique

DL techniques deal with a sufficiently large amount of data without pre-processing as they learn from various types of data representations (structured, unstructured, labeled, and unlabeled data). However, they are built on a foundation of traditional neural networks. DL architectures consist of input, hidden, and output layers, neurons in the hidden layers, activation functions, and connecting weights and biases like shallow neural networks. The fundamental difference lies in the number of hidden layers; the DL architectures have more than two hidden layers. The shallow neural networks have one or two hidden layers. Also, DL techniques adjust the connecting weights and biases during the training processes to make a better prediction by overcoming the bottlenecks and overfitting issues of the shallow neural network models. Besides, they execute feature engineering to enable faster learning without being explicitly programmed [104].

In DL, various architectures and algorithms have been reported in the literature since the inception of the concept in the 1990s. The most widely adopted DL architectures are listed below with a brief introduction [105–112]:

- **Recurrent neural network (RNN)**: In RNN, a uniform task is performed for every single sequence element with the output dependent on the previous computations. It might have connections amongst the nodes of the previous and same layers for transferring previous computations. The captured and stored information of the connected nodes calculates the outcome for subsequent nodes. Such feedback allows it to model the problems by maintaining memories in time. It is one of the foundational DL architectures from which other architectures and techniques have been developed [105–108].
- **Long short-term memory (LSTM)**: A special kind of RNN that includes a special memory cell to hold information for long or short periods as the functions of the inputs. It not only stores the last computed information but also holds important information for a longer period. A set of gates, for example, input, forget, and output gates, is used to determine when a piece of particular information enters the memory and when it is forgotten [107,108].
- **Convolutional neural network (CNN)**: It is the popular choice for different computer vision tasks, including object detection, image recognition, and instance segmentation. The CNN is made up of several layers where the early layers extract features from the images,

and later layers utilize them to produce the decisions. For instance, the CNN divides the images into receptive fields and feeds them into a convolutional layer for feature extraction. Then, it uses different types of pooling strategies in the next layer to reduce extracted features dimensionality through down-sampling without losing the essential information. The CNN repeats the mentioned process several times before feeding the final features into a fully connected MLP for classification purposes [105–107].

- **Deep belief network (DBN)**: It is a typical deep neural (multilayer) network consolidated with probability theory in statistics. It is a representation of a stack of restricted Boltzmann machine (RBM) where each pair of connected layers is an RBM. The nodes of the first and second layers of the pair are considered visible and hidden nodes, respectively. Therefore, in the first RBM, the input layer nodes are the visible nodes, and the first hidden layer nodes are the hidden nodes. Likewise, the first hidden layer nodes are the visible nodes, and the second hidden layer nodes are the hidden nodes in the second RBM. The process continues until the output layer is reached. Each DBN layer is pre-trained through this connection, and such pre-training is known as unsupervised learning. The supervised fine-tuning process is started using either GD learning or backpropagation to complete the training process [107,112].

- **Deep stacking network (DSN)**: It is also known as a deep convex network. However, it is different from traditional DL frameworks. It consists of a deep set of individual neural networks (modules) where each set (module) has a single input layer, a single hidden layer, and a single output layer. The DSN modules are stacked on top of each other, where the inputs of each module are the original input vector and output vector of the prior layer (except the first module). Eventually, it transforms the training problem into a set of problems to overcome the training complexity of the traditional DL networks [107,111].

As the DL neural networks overcome the bottlenecks of the traditional shallow neural networks, they have been employed in various engineering and science fields, including image processing, computer vision, natural language processing, feature extraction, pattern recognition, data augmentation, object tracking, regression, and others [106,113–115]. Despite having numerous advantages, DL neural networks require a substantial amount of data to develop better models. Like shallow neural networks, they are black boxes to the users for example, lack of transparency and flexibility [104].

3.3 Hybrid, ensemble, and other artificial intelligence techniques

This section briefly introduces several hybrids, ensembles, and other types of ML techniques, including adaptive neuro-fuzzy inference system (ANFIS), evolutionary algorithm-inspired AI techniques.

3.3.1 Adaptive neuro-fuzzy inference system

The ANFIS combines two basic ML techniques: the ANN and FLM. It uses the FLM to convert the provided inputs to targeted outputs with highly interconnected ANN processing elements. It was developed in the early 1990s as a universal estimator

Artificial intelligence techniques

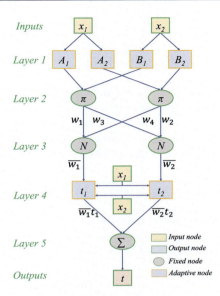

Fig. 3.12 ANFIS structure [119–123]. *ANFIS*, Adaptive neuro-fuzzy inference system.

[116]. It not only combines the advantages of the ANN and FIS but also avoids their shortcomings. Therefore, it provides better generalization performance, archives highly nonlinear mapping, and provides stable solutions [117–119]. Consequently, researchers employed it to solve classification, regression, and feature extraction problems of various fields of studies [119–123].

Fig. 3.12 presents a generalized structure of ANFIS that consists of five different layers, along with the input and output layers [119–123]. This figure demonstrates the prediction of a single output (t) from two distinct inputs (x_1 and x_2) employing two membership functions and two fuzzy rules. In the given ANFIS architecture, T_i^j represents the output of node i of the jth layer. Also, the nodes of layer one and layer four are adaptive, whereas the nodes of layer two, layer three, and layer five are fixed.

The first stage of the ANFIS is the fuzzification stage that obtains the fuzzy clusters from the provided inputs using the membership functions. The premise perimeters (p, q, and r in this case) assist in determining the nature and degree of the membership functions. In layer 1, the outputs of the nodes A_1 and A_2 for the input x_1 and the bell-shaped membership function can be calculated as [124–126]:

$$T_i^1(x_1) = \mu_{A_i}(x_1) = \frac{1}{1 + \left|\frac{x_1 - r_i}{p_i}\right|^{2q_i}}; \quad \text{for } i = 1, 2 \qquad (3.12)$$

Similarly, the outputs of nodes B_1 and B_2 for the input x_2 can be represented.

$$T_i^1(x_2) = \mu_{B_i}(x_2) = \frac{1}{1+\left|\frac{x_1-r_i}{p_i}\right|^{2q_i}}; \quad \text{for } i = 1, 2 \tag{3.13}$$

Other types of membership functions, including Gaussian, trapezoidal, and triangular membership functions, are also employed.

The second layer of ANFIS uses the membership values of the previous layer to evaluate the firing strengths (w_i) of the applied rules. Hence, this layer is also known as the rule layer. It uses the following equation to calculate the firing strengths (w_i):

$$T_i^2 = w_i = \mu_{A_i}(x_1) * \mu_{B_i}(x_2) \text{ where } i = 1, 2 \tag{3.14}$$

The next layer, also known as the normalization layer, evaluates the normalized firing strengths ($\overline{w_i}$) for all the applied rules using the following rules. The corresponding value of firing strength is divided by the summation of all firing strength.

$$T_i^3 = \overline{w_i} = \frac{w_i}{\sum_{k=1}^{4} w_k} \text{ where } i = 1, 2 \tag{3.15}$$

In the fourth layer, defuzzification is performed using the following equation where each node output is the product of the normalized firing strength and first-order polynomial:

$$T_i^4 = \overline{w_i} t_i = \overline{w_i}(m_i x_1 + n_i x_2 + o_i) \text{ where } i = 1, 2 \tag{3.16}$$

Where m_i, n_i, and o_i are the consequence parameters of the ANFIS model.

Finally, the overall output of the ANFIS is evaluated using the following equation in the fifth stage by summing up the node outputs of the previous step:

$$T_i^5 = \sum_i \overline{w_i} t_i \text{ where } i = 1, 2 \tag{3.17}$$

ANFIS model performance depends on successful training and parameter (both premise and consequence) selection. Therefore, both the derivative and metaheuristic-based approaches have been reported in the literature for tuning the ANFIS parameters [4]. Despite acceptance amongst the researchers from various fields, the ANFIS suffers from several limitations. They are computationally expensive while dealing with large inputs and comparatively complex problems. Besides, the unlimited number of epoch values can also become disastrous in some cases [123,127,128].

3.3.2 Ensemble techniques

Ensemble models aim to achieve better generalization performance by employing ML techniques parallel and combining their outputs. They try to manage the strengths and weaknesses of the used approaches while developing better-aggregated solutions for complex real-world problems. Besides, they can effectively deal with model nonlinearity and missing data. In the ensemble models, each ML technique is trained, tested, and validated individually, therefore, produces different or closer solutions. Then, it employs a wide range of processing techniques, including averaging, bagging (bootstrap aggregating), boosting, stacking, and voting to produce an ultimate solution from the generated solutions of the employed learning methods [129–133].

Among the mentioned processing techniques, averaging and voting are widely adopted for regression and classification purposes, respectively. These two techniques are simple and easy to implement and interpret. The voting technique for classifications can be further ramified into two categories namely, majority voting and weighted voting. In the majority voting technique, the ensemble models collect the predicted solutions of the employed methods and select a majority vote solution by providing equal rights to each technique. Conversely, the weighted voting technique gives more priority to some models and less emphasis to others as per their past performance and generalization accuracy. Similarly, an averaging technique for the regression problems can also be ramified into two categories: providing either equal weights or a different percentage of weights to the outputs of the employed methods [129–131].

Bagging improves model accuracy using decision trees by reducing the variance and eliminating the overfitting. It creates multiple samples from a given set of data with refitting using the bootstrap technique. Then, it builds multiple distinct prediction trees and finally generates an aggregated solution using the voting or averaging method. Boosting combines a set of models (decision trees) to obtain a higher accuracy than can convert the weak learners to healthy learners by reducing the biases. Finally, the stacking combines multiple prediction models by introducing the meta-learning concept, an asymptotically optimal learning system that reduces generalization errors. All of the mentioned ensemble techniques (bagging, boosting, and stacking) are applied in classification and regression problems [133].

Despite having multiple benefits, the ensemble models are computationally expensive as they employ several prediction techniques simultaneously. Besides, they involve complexities in determining the hyper-parameters of the used methods and interpretation of results [133].

3.3.3 Metaheuristic algorithm inspired artificial intelligence techniques

Metaheuristic algorithms are often used to tune the hyper-parameters of ML techniques to achieve better generalization performance while solving classification, clustering, feature extraction, anomaly detection, and regression problems. In general, the mentioned hyper-parameters are chosen randomly or on a trial-and-error basis, leading to the relatively lower performance of the ML tools. Among many metaheuristic algorithms, ant colony, artificial bee colony, backtracking search algorithm, bat algorithm, cuckoo search algorithm, differential evolution, firefly optimization, GA, gravitational search algorithm, grey wolf optimizer, particle swarm optimization, and sine cosine algorithm are widely adopted [131].

In ANN, metaheuristic algorithms are employed to tune the connecting weights and biases, select activation/squashing functions and training algorithms, choose the number of hidden layers and neurons, or adjust all the mentioned hyper parameters. Eventually, the better and appropriate neural network architectures are evolved [134–137]. On the other hand, their regularization coefficients, kernel option, and tolerance of termination criterion are tuned using metaheuristic algorithms for SVMs. Likewise, the kernel option and regularization coefficient of the extreme learning machines are adjusted [64,138,139]. Besides, the researchers suggested that the employment of metaheuristic algorithms might enhance the performances of the DL networks as their basic structures similar to that of the traditional neural networks [140–142].

In the FLM, the hyperparameters, including the number and shapes of fuzzy membership functions and the rule base, are critical in achieving the appropriate models for the problem under investigation. Therefore, like other ML tools, these parameters of the FLM are chosen using metaheuristic techniques [143–145]. Likewise, the premise and consequence parameters of the ANFIS models are also tuned using the nature-inspired metaheuristic algorithms [120,146]. It was evident that the researchers have also used metaheuristic algorithms along with the ensemble techniques to enhance the overall performance [147–149]. However, the excessive computational time for training is the only drawback of the metaheuristic algorithms-inspired ML tools [51].

3.3.4 Other artificial intelligence techniques

Apart from the discussed ML techniques, there are many more supervised, unsupervised, and reinforced ML tools have been reported in the literature, including linear regression [150], LR [151], NB [152], random forest [153], hidden Markov model [154], group method data handling [155], and functional network [156]. Other techniques, like case-based reasoning and knowledge-based models, have also been explored to solve many complex real-life problems [157–159].

3.4 Summary

The primary aim of AI is to develop systems that perform tasks intelligently by processing external inputs. Due to their adaptiveness, flexibility, and parallel computation capability, AI received enormous attention and has been employed to solve many real-world complex problems from various science and engineering fields. This chapter introduces the foundational concepts of several popular ML techniques (subset of AI), along with their advantages and disadvantages. Many of the discussed ML techniques are employed to diagnose, for example, detect, classify, and locate, faults in power system networks. The reader will also find the interactive and easy-to-understand discussions on the illustrated ML techniques useful for adopting the appropriate methods for complex real-world problems on anomaly detection, data augmentation, classification, regression, and many others.

Despite having enormous potential, the AI is still facing several challenges, including lack of transparency (black box), lack of generalized learning algorithms, obtaining a sufficiently large amount of data, improper model development based on available data without the proper understanding of the even insight, and others. Besides, they may pose safety and security risks if not designed with adequate attention, care, and knowledge of the problem. Furthermore, there is still a lack of regulations regarding the responsible party for damage caused by an AI-operated device or service.

References

[1] S.J. Russell, P. Norvig, Artificial Intelligence: A Modern Approach, Fourth, Prentice Hall, New Jersey, USA, 2020.
[2] JRC Technical Reports, AI watch defining artificial intelligence, 2020. doi:10.2760/382730.
[3] J. Krupansky, "Untangling the definitions of artificial intelligence, machine intelligence, and machine learning," 2017. https://jackkrupansky.medium.com/untangling-the-definitions-of-artificial-intelligence-machine-intelligence-and-machine-learning-7244882f04c7. (Accessed 9 December 2020).
[4] R.E. Bellman, An introduction to artificial intelligence: can computers think? 1978.
[5] P.H. Winston, Artificial Intelligence, Addison-Wesley, Massachusetts, USA, 1992.
[6] M. Haenlein, A. Kaplan, A brief history of artificial intelligence: on the past, present, and future of artificial intelligence, Calif. Manage. Rev. 61 (4) (2019), 1–10, doi:https://doi.org/10.1177/0008125619864925.
[7] W. Ertel, Introduction to Artificial Intelligence, Second, Springer International Publishing, Cham, 2017.
[8] M. Lim, History of AI winters, 2018. https://www.actuaries.digital/2018/09/05/history-of-ai-winters/. (Accessed 9 December 2020).
[9] M.M. Mijwil, History of artificial intelligence, 2015. https://www.researchgate.net/publication/322234922_History_of_Artificial_Intelligence. (Accessed 9 December 2020).
[10] J. Soma, Y. Shanker, History of artificial intelligence (AI), 2017. https://www.cag.edu.tr/d/l/21b425a6-4d68-49ca-bbf6-67e2919d6b4d. (Accessed 9 December 2020).
[11] G. Press, A very short history of artificial intelligence (AI), 1 (1) (2016), 1–1 https://www.forbes.com/sites/gilpress/2016/12/30/a-very-short-history-of-artificial-intelligence-ai/?sh=69cdb9286fba#41f142dd6fb. (Accessed 9 December 2020).
[12] V. Kaul, S. Enslin, S.A. Gross, History of artificial intelligence in medicine. In: *Gastrointestinal Endoscopy*, vol. 92, no. 4, Mosby Inc., Missouri, USA, 2020, pp. 807–812, doi:10.1016/j.gie.2020.06.040.
[13] O. Port, Invasion of the robots, 1997. https://www.bloomberg.com/news/articles/1997-03-02/invasion-of-the-robots. (Accessed 7 December 2020).
[14] S. Lynch, Andrew Ng: why AI is the new, 2017. https://www.gsb.stanford.edu/insights/andrew-ng-why-ai-new-electricity. (Accessed 9 December 2020).
[15] SyncedReview, Artificial intelligence is the new electricity — Andrew Ng, 2017. https://medium.com/syncedreview/artificial-intelligence-is-the-new-electricity-andrew-ng-cc132ea6264. (Accessed 9 December 2020).
[16] C. Jewell, Artificial intelligence: the new electricity, 2019. https://www.wipo.int/wipo_magazine/en/2019/03/article_0001.html. (Accessed 9 December, 2020).

[17] SAS, Artificial intelligence — what it is and why it matters, 2020. https://www.sas.com/en_sa/insights/analytics/what-is-artificial-intelligence.html. (Accessed 15 December 2020).
[18] Y.D. Mamuya, Y.-D. Lee, J.-W. Shen, M. Shafiullah, C.-C. Kuo, Application of machine learning for fault classification and location in a radial distribution grid, Appl. Sci. 10 (14) (2020) 4965, doi:10.3390/app10144965.
[19] M. Farhoumandi, Q. Zhou, M. Shahidehpour, A review of machine learning applications in IoT-integrated modern power systems, Electr. J. 34 (1) (2021) 106879, doi:10.1016/j.tej.2020.106879.
[20] H. Sun, H.V. Burton, H. Huang, Machine learning applications for building structural design and performance assessment: State-of-the-art review, J. Build. Eng. 33 (1) (2021) 101816, doi:10.1016/j.jobe.2020.101816.
[21] M. Shafiullah, M.A. Abido, A review on distribution grid fault location techniques, Electr. Power Components Syst. 45 (8) (2017) 807–824, doi:10.1080/15325008.2017.1310772.
[22] S.D. Ahmed, F.S.M. Al-Ismail, M. Shafiullah, F.A. Al-Sulaiman, I.M. El-Amin, Grid integration challenges of wind energy: a review, IEEE Access, 8, Institute of Electrical and Electronics Engineers Inc., New Jersey, USA, 2020, pp. 10857–10878, doi:10.1109/ACCESS.2020.2964896.
[23] M. Shafiullah, S.M. Rahman, M.G. Mortoja, B. Al-Ramadan, Role of spatial analysis technology in power system industry: an overview, Renew. Sustain. Energy Rev. 66 (Dec) (2016) 584–595, doi:10.1016/j.rser.2016.08.017.
[24] S. Otte, How does artificial intelligence work? 2020. https://www.innoplexus.com/blog/how-artificial-intelligence-works. (Accessed 15 December 2020).
[25] K.A. Carpenter, D.S. Cohen, J.T. Jarrell, X. Huang, Deep learning and virtual drug screening, Future Med. Chem. 10 (21) (2018) 2557–2567, doi:10.4155/fmc-2018-0314.
[26] M. Kumar, Machine learning, 2015. https://madhureshkumar.wordpress.com/2015/08/09/machine-learning/. (Accessed 15 December 2020).
[27] L. Danzig, Can you explain briefly about machine learning, deep learning, AI, neural network? 2019. https://www.quora.com/Can-you-explain-briefly-about-machine-learning-deep-learning-AI-neural-network. (Accessed 15 December 2020).
[28] S. Liu, AI market size 2018-2025, 2020. https://www.statista.com/statistics/607716/worldwide-artificial-intelligence-market-revenues/. (Accessed 13 January 2021).
[29] V. Advani, What is artificial intelligence? How does AI work, applications and future? 2020. https://www.mygreatlearning.com/blog/what-is-artificial-intelligence/. (Accessed 15 December 2020).
[30] D. Roe, A look at the downsides of artificial intelligence, 2020. https://www.reworked.co/information-management/a-look-at-the-downsides-of-artificial-intelligence/. (Accessed 14 December 2020).
[31] LiveTiles, 15 pros and 6 cons of artificial intelligence in the classroom, 2017. https://livetiles-global.com/pros-cons-artificial-intelligence-classroom/. (Accessed 14 December 2020).
[32] Edureka, What are advantages and disadvantages of artificial intelligence? 2020. https://www.edureka.co/blog/what-are-the-advantages-and-disadvantages-of-artificial-intelligence/#disadvantagesofai. (Accessed 14 December 2020).
[33] A. Krogh, What are artificial neural networks?, Nat. Biotechnol. 26 (2) Nature Publishing Group, Berlin, Germany, (2008). 195–197, doi:10.1038/nbt1386.
[34] A. Ali, et al., Review of online and soft computing maximum power point tracking techniques under non-uniform solar irradiation conditions, Energies 13 (12) (2020) 3256, doi:10.3390/en13123256.

[35] I. Świetlicka, A. Sujak, S. Muszyński, M. Świetlicki, The application of artificial neural networks to the problem of reservoir classification and land use determination on the basis of water sediment composition, Ecol. Indic. 72 (2017) 759–765, doi:10.1016/j.ecolind.2016.09.012.
[36] S. Masiur Rahman, A.N. Khondaker, M. Imtiaz Hossain, M. Shafiullah, M.A. Hasan, Neurogenetic modeling of energy demand in the United Arab Emirates, Saudi Arabia, and Qatar, Environ. Prog. Sustain. Energy 36 (4) (2017), doi:10.1002/ep.12558.
[37] M. Ismail Hossain, M. Shafiullah, M. Abido, Induction motor speed control employing LM-NN based adaptive PI controller, Renew. Energies Power Qual. J. 18 (2020) 1–6. http://www.icrepq.com/icrepq20/239-20-hossain.pdf. (Accessed 23 August 2020).
[38] S. Haykin, S. Haykin, S. Haykin, and S. Haykin, Neural networks and learning machines, 2009.
[39] A. Aljohani, A. Fataa, T. Sheikhoon, M. Shafiullah, M.A. Abido, Design and Implementation of an Intelligent SLG Fault Locator for Power Distribution Grid, 2019 International Conference on Control, Automation and Diagnosis (ICCAD'19), 2019, pp. 1–6.
[40] A. Aljohani, A. Aljurbua, M. Shafiullah, M.A. Abido, Smart fault detection and classification for distribution grid hybridizing ST and MLP-NN, 2018 15th International Multi-Conference on Systems, Signals & Devices (SSD), Hammamet, Tunisia, IEEE, 2018, pp. 1–5.
[41] M. Shafiullah, M.A. Abido, S-transform based FFNN approach for distribution grids fault detection and classification, IEEE Access 6 (1) (2018) 8080–8088, doi:10.1109/ACCESS.2018.2809045.
[42] D.Tien Bui, T.A. Tuan, H. Klempe, B. Pradhan, I. Revhaug, Spatial prediction models for shallow landslide hazards: a comparative assessment of the efficacy of support vector machines, artificial neural networks, kernel logistic regression, and logistic model tree, Landslides 13 (2) (2016) 361–378, doi:10.1007/s10346-015-0557-6.
[43] M. Ijaz, M. Shafiullah, M.A. Abido, Classification of power quality disturbances using Wavelet Transform and Optimized ANN, 2015 18th International Conference on Intelligent System Application to Power Systems (ISAP), Proceedings of the Conference on, 2015, pp. 1–6, doi:10.1109/ISAP.2015.7325522.
[44] M.J. Rana, M.S. Shahriar, M. Shafiullah, Levenberg–Marquardt neural network to estimate UPFC-coordinated PSS parameters to enhance power system stability, Neural Comput. Appl. 31 (4) (2019), doi:10.1007/s00521-017-3156-8.
[45] M. Shafiullah, M.A.M. Khan, S.D. Ahmed, PQ disturbance detection and classification combining advanced signal processing and machine learning tools, in: P. Sanjeevikumar, C. Sharmeela, J.B. Holm-Nielsen, P. Sivaraman (Eds.), Power Quality in Modern Power Systems, Academic Press, Massachusetts, USA, 2021, pp. 311–335.
[46] C. Cecati, J. Kolbusz, P. Rozycki, P. Siano, B.M. Wilamowski, A novel RBF training algorithm for short-term electric load forecasting and comparative studies, IEEE Trans. Ind. Electron. 62 (10) (2015) 6519–6529, doi:10.1109/TIE.2015.2424399.
[47] G.-B. Huang, H. Zhou, X. Ding, R. Zhang, Extreme learning machine for regression and multiclass classification, IEEE Trans. Syst. Man. Cybern. B. Cybern. 42 (2) (2012) 513–529, doi:10.1109/TSMCB.2011.2168604.
[48] H. Liu, B. Lang, Machine learning and deep learning methods for intrusion detection systems: a survey, Appl. Sci. 9 (20) (2019) 4396, doi:10.3390/app9204396.
[49] B.E. Boser, I.M. Guyon, V.N. Vapnik, A training algorithm for optimal margin classifiers, 5th annual workshop on Computational learning theory - COLT '92, Proceedings of the Conference on, 1992, pp. 144–152, doi:10.1145/130385.130401.

[50] Data Flair, "Kernel functions-introduction to SVM kernel & examples," 2021. https://data-flair.training/blogs/svm-kernel-functions/. (Accessed 3 January 2021).

[51] M. Shafiullah, M. Ijaz, M.A. Abido, Z. Al-Hamouz, Optimized support vector machine & wavelet transform for distribution grid fault location, 2017 11th IEEE International Conference on Compatibility, Power Electronics and Power Engineering (CPE-POWERENG), IEEE, 2017, pp. 77–82, doi:10.1109/CPE.2017.7915148.

[52] V. Kecman, I. Hadzic, Support vectors selection by linear programming, Proceedings of the International Joint Conference on Neural Networks, 5 (2000) 193–198, doi:10.1109/ijcnn.2000.861456.

[53] R. Pupale, "Support vector machines (SVM) — an overview," 2018. https://towardsdatascience.com/https-medium-com-pupalerushikesh-svm-f4b42800e989. (Accessed 5 January 2021).

[54] G. James, D. Witten, T. Hastie, R. Tibshirani, Support Vector Machines, Springer, New York, NY, 2013, pp. 337–372.

[55] R. Gholami, N. Fakhari, Support vector machine: principles, parameters, and applications. In: Handbook of Neural Computation, Elsevier Inc., Amsterdam, Netherlands, 2017, pp. 515–535.

[56] D.A. Pisner, D.M. Schnyer, Support vector machine. In: Machine Learning: Methods and Applications to Brain Disorders, Elsevier, New York, USA, 2019, pp. 101–121.

[57] J. Cervantes, F. Garcia-Lamont, L. Rodríguez-Mazahua, A. Lopez, A comprehensive survey on support vector machine classification: Applications, challenges and trends, Neurocomputing 408 (Sept) (2020) 189–215, doi:10.1016/j.neucom.2019.10.118.

[58] M. McGregor, "SVM machine learning tutorial — what is the support vector machine algorithm, explained with code examples," 2020. https://www.freecodecamp.org/news/svm-machine-learning-tutorial-what-is-the-support-vector-machine-algorithm-explained-with-code-examples/. (Accessed 5 January 2021).

[59] M.S. Shahriar, M. Shafiullah, M.J. Rana, Stability enhancement of PSS-UPFC installed power system by support vector regression, Electr. Eng. 100 (1) (2017) 1–12, doi:10.1007/s00202-017-0638-8.

[60] C.-C. Chang, C.-J. Lin, LIBSVM: a library for support vector machines, ACM Trans. Intell. Syst. Technol. 2 (3) (2011) 1–27, doi:10.1145/1961189.1961199.

[61] V. Vapnik, The Nature of Statistical Learning Theory, Springer Science & Business Media, Berlin, Germany, 2013.

[62] C.J.C. Burges, A tutorial on support vector machines for pattern recognition, Kluwer Academic Publishers, London, UK, 1998, pp. 1–43. https://www.di.ens.fr/~mallat/papiers/svmtutorial.pdf. (Accessed 22 May 2019).

[63] C.L. Chih-wei Hsu, Chih-chung Chang, A practical guide to support vector classification, Tech. Rep. 2020 (1) (2010) 1–16. http://citeseerx.ist.psu.edu/viewdoc/summary?doi=10.1.1.224.4115. (Accessed 20 February 2016).

[64] M. Shafiullah, M. Abido, T. Abdel-Fattah, Distribution grids fault location employing ST based optimized machine learning approach, Energies 11 (9) (2018) 2328, doi:10.3390/en11092328.

[65] L. Auria and R.A. Moro, "Support vector machines (SVM) as a technique for solvency analysis," 2008. https://core.ac.uk/download/pdf/188978526.pdf. (Accessed 5 January 2021).

[66] Guang-Bin Huang, Qin-Yu Zhu, Chee-Kheong Siew, Extreme learning machine: a new learning scheme of feedforward neural networks, 2004 IEEE International Joint Conference on Neural Networks, 2004, pp. 985–990, doi:10.1109/IJCNN.2004.1380068.

[67] K. Erdem, "Introduction to extreme learning machines," 2020. https://towardsdatascience.com/introduction-to-extreme-learning-machines-c020020ff82b. (Accessed 3 January 2021).

[68] M. Shafiullah, M.A. Abido, Z. Al-Hamouz, Wavelet-based extreme learning machine for distribution grid fault location, IET Gener. Transm. Distrib. 11 (17) (2017) 4256–4263, doi:10.1049/iet-gtd.2017.0656.

[69] M. Shafiullah, M.J. Rana, M.S. Shahriar, F.A. Al-Sulaiman, S.D. Ahmed, A. Ali, Extreme learning machine for real-time damping of LFO in power system networks, Electr. Eng. 103 (1) (2021) 279–292, doi:10.1007/s00202-020-01075-7.

[70] E. Cambria, et al., Extreme learning machines [trends & controversies], IEEE Intell. Syst. 28 (6) (2013) 30–59, doi:10.1109/MIS.2013.140.

[71] G. Huang, Q. Zhu, C. Siew, Extreme learning machine: theory and applications, Neurocomputing 70 (1–3) (2006) 489–501. Available http://www.sciencedirect.com/science/article/pii/S0925231206000385 (Accessed 29 February 2016).

[72] S. Lin, X. Liu, J. Fang, Z. Xu, Is extreme learning machine feasible? A theoretical assessment (part II), IEEE Trans. Neural Networks Learn. Syst. 26 (1) (2015) 21–34, doi:10.1109/TNNLS.2014.2336665.

[73] L. Mao, L. Zhang, X. Liu, C. Li, H. Yang, Improved extreme learning machine and its application in image quality assessment, Math. Probl. Eng. 2014 (3) (2014) 1–7, doi:10.1155/2014/426152.

[74] A.N. Jahromi, et al., An improved two-hidden-layer extreme learning machine for malware hunting, Comput. Secur. 89 (Feb) (2020) 101655, doi:10.1016/j.cose.2019.101655.

[75] F. Han, H.F. Yao, Q.H. Ling, An improved extreme learning machine based on particle swarm optimization, Lecture Notes in Computer Science (including subseries Lecture Notes in Artificial Intelligence and Lecture Notes in Bioinformatics) 6840, New York, USA, 2011, pp. 699–704, doi:10.1007/978-3-642-24553-4_92.

[76] L.A. Zadeh, Fuzzy sets, Inf. Control 8 (3) (1965) 338–353, doi:10.1016/S0019-9958(65)90241-X.

[77] F.O. Karray, C.W. De Silva, Soft Computing and Intelligent Systems Design : Theory, Tools, and Applications, Pearson/Addison Wesley, Massachusetts, USA, 2004.

[78] C. Von Altrock, Fuzzy Logic and NeuroFuzzy Applications Explained, Prentice Hall PTR, New Jersey, USA, 1995.

[79] H. Tanyildizi, Fuzzy logic model for prediction of mechanical properties of lightweight concrete exposed to high temperature, Mater. Des. 30 (6) (2009) 2205–2210, doi:10.1016/j.matdes.2008.08.030.

[80] H. Singh, et al., Real-life applications of fuzzy logic, Adv. Fuzzy Syst 2013 (Special Issue) (2013), 1–3, doi:10.1155/2013/581879.

[81] M.I. Hossain, S.A. Khan, M. Shafiullah, M.J. Hossain, Design and implementation of MPPT controlled grid connected photovoltaic system, 2011 IEEE Symposium on Computers & Informatics. IEEE, 2011, pp. 284–289, doi:10.1109/ISCI.2011.5958928.

[82] M.I. Hossain, M.S. Alam, M. Shafiullah, M. Al Emran, Asynchronous Induction Motor Speed Control Using Takagi-Sugeno Fuzzy Logic, 10th International Conference on Electrical and Computer Engineering (ICECE), Dhaka, IEEE, 2018, pp. 249–252, doi:10.1109/ICECE.2018.8636799.

[83] M. Hanss, Applied Fuzzy Arithmetic: An Introduction With Engineering Applications, Springer, Berlin Heidelberg, 2005.

[84] J.G. Dijkman, H. van Haeringen, S.J. de Lange, Fuzzy numbers, J. Math. Anal. Appl. 92 (2) (1983) 301–341, doi:10.1016/0022-247X(83)90253-6.
[85] M.J. Wierman, An Introduction to the Mathematics of Uncertainty, Creighton University, Nebraska, USA, 2010.
[86] S. Chakraverty, D.M. Sahoo, N.R. Mahato, S. Chakraverty, D.M. Sahoo, N.R. Mahato, Fuzzy numbers. In: Concepts of Soft Computing, Springer, Singapore, 2019, pp. 53–69.
[87] P. Kovac, D. Rodic, V. Pucovsky, B. Savkovic, M. Gostimirovic, Application of fuzzy logic and regression analysis for modeling surface roughness in face milliing, J. Intell. Manuf. 24 (4) (2013) 755–762, doi:10.1007/s10845-012-0623-z.
[88] C.-J. Chung, Y.-Y. Hsieh, H.-C. Lin, Fuzzy inference system for modeling the environmental risk map of air pollutants in Taiwan, J. Environ. Manage. 246 (Sept) (2019) 808–820, doi:10.1016/J.JENVMAN.2019.06.038.
[89] E. Kuram, B. Ozcelik, "Fuzzy logic and regression modelling of cutting parameters in drilling using vegetable based cutting fluids," 2013.
[90] S. Godil, M. Shamim, S. Enam, U. Qidwai, Fuzzy logic: a 'simple' solution for complexities in neurosciences, Surg. Neurol. Int. 2 (1) (2011), doi:10.4103/2152-7806.77177.
[91] Y. Yildirim, M. Bayramoglu, Adaptive neuro-fuzzy based modelling for prediction of air pollution daily levels in city of Zonguldak, Chemosphere 63 (9) (2006) 1575–1582, doi:10.1016/J.CHEMOSPHERE.2005.08.070.
[92] J.R. Koza, Genetic Programming: on the Programming of Computers by Means of Natural Selection, MIT Press, Massachusetts, USA, 1992.
[93] J.R. Koza, Genetic Programming II : Automatic Discovery of Reusable Programs, MIT Press, Massachusetts, USA, 1994.
[94] GPTIPS, An Open Source Genetic Programming Toolbox for Multigene Symbolic Regression, International MultiConference of Engineers and Computer Scientists (IMECS), Kowloon, Hong Kong, 2010, pp. 77–80.
[95] S.A. Razzak, M. Shafiullah, S.M. Rahman, M.M. Hossain, J. Zhu, A multigene genetic programming approach for modeling effect of particle size in a liquid–solid circulating fluidized bed reactor, Chem. Eng. Res. Des. 134 (2018) 370–381, doi:10.1016/J.CHERD.2018.04.021.
[96] M. Shafiullah, M. Juel Rana, M. Shafiul Alam, M.A. Abido, Online tuning of power system stabilizer employing genetic programming for stability enhancement, J. Electr. Syst. Inf. Technol 5 (3) (2018), 287–299, doi:10.1016/j.jesit.2018.03.007.
[97] A. Hossein Alavi, A. Hossein Gandomi, A robust data mining approach for formulation of geotechnical engineering systems, Eng. Comput. 28 (3) (2011) 242–274, doi:10.1108/02644401111118132.
[98] M. Walker, Introduction to genetic programming, Tech. Np Univ. Mont. 2001 (Oct) (2001) 1–9.
[99] J.R. Koza, D. Andre, F.H. Bennett, M.A. Keane, Genetic Programming III : Darwinian Invention and Problem Solving, Morgan Kaufmann, Massachusetts, USA, 1999.
[100] A.H. Gandomi, A.H. Alavi, C. Ryan (Eds.), Handbook of Genetic Programming Applications, Springer, New York, USA, 2015.
[101] K. Nag, N.R. Pal, Genetic programming for classification and feature selection, Studies in Computational Intelligence, 779, Springer Verlag, Berlin, Germany, 2019, pp. 119–141.
[102] M. Shafiullah, M.J. Rana, M.S. Shahriar, M.H. Zahir, Low-frequency oscillation damping in the electric network through the optimal design of UPFC coordinated PSS employing MGGP, Measurement 138 (May) (2019) 118–131, doi:10.1016/J.MEASUREMENT.2019.02.026.

[103] N. Choudhary, B. Singh, E. Gaurav Bagaria, Genetic programming: a study on computer language, Int. J. Innov. Eng. Technol. 3 (4) (2014) 203–207. http://ijiet.com/wp-content/uploads/2014/05/31.pdf. (Accessed 5 January 2021).

[104] V. Shchutskaya, Deep learning: strengths and challenges, 2018. https://indatalabs.com/blog/deep-learning-strengths-challenges?cli_action=1609949423.707. (Accessed 6 January 2021).

[105] A. Varangaonkar, Top 5 deep learning architectures, https://hub.packtpub.com/top-5-deep-learning-architectures/. (Accessed 8 January 2021).

[106] T.D. Akinosho, et al., Deep learning in the construction industry: a review of present status and future innovations, J. Build. Eng. 32 (2020) 101827, doi:10.1016/j.jobe.2020.101827.

[107] M. T. Jones, Deep learning architectures, 2017. https://developer.ibm.com/technologies/artificial-intelligence/articles/cc-machine-learning-deep-learning-architectures/. (Accessed 8 January 2021).

[108] J. Duan, H. ZuoCP, Y. Bai, J. Duan, M. Chang, B. Chen, Short-term wind speed forecasting using recurrent neural networks with error correction, Energy 217 (Feb) (2020) 119397, doi:10.1016/j.energy.2020.119397.

[109] A. Dhillon, G.K. Verma, Convolutional neural network: a review of models, methodologies and applications to object detection, Progress in Artificial Intelligence, 9, Springer, New York, USA, 2020, pp. 85–112, doi:10.1007/s13748-019-00203-0.

[110] A. Ajit, K. Acharya, and A. Samanta, "A review of convolutional neural networks," 2020, doi:10.1109/ic-ETITE47903.2020.049.

[111] B.L.S. da Silva, F.K. Inaba, E.O.T. Salles, P.M. Ciarelli, Fast deep stacked networks based on extreme learning machine applied to regression problems, Neural Netw. 131 (Nov) (2020) 14–28, doi:10.1016/j.neunet.2020.07.018.

[112] Y. Rizk, N. Hajj, N. Mitri, M. Awad, Deep belief networks and cortical algorithms: A comparative study for supervised classification, Applied Computing and Informatics, 15, Elsevier B.V., Amsterdam, Netherlands, 2019, pp. 81–93, doi:10.1016/j.aci.2018.01.004.

[113] D. Chen, P. Wawrzynski, Z. Lv, Cyber security in smart cities: A review of deep learning-based applications and case studies, Sustain. Cities Soc. 66 (Mar) (2021) 102655, doi:10.1016/j.scs.2020.102655.

[114] X. Yuan, J. Shi, L. Gu, A review of deep learning methods for semantic segmentation of remote sensing imagery, Expert Systems with Applications, 169, Elsevier Ltd, Amsterdam, Netherlands, 2021, p. 114417, doi:10.1016/j.eswa.2020.114417.

[115] E. Allibhai, "Building a deep learning model using keras," 2018.

[116] J.S.R. Jang, ANFIS: adaptive-network-based fuzzy inference system, IEEE Trans. Syst. Man Cybern. 23 (3) (1993) 665–685, doi:10.1109/21.256541.

[117] L. Zhang, J. Liu, J. Lai, Z. Xiong, Performance analysis of adaptive neuro fuzzy inference system control for mems navigation system, Math. Probl. Eng. 2014 (Special Issue) (2014) 1–7, doi:10.1155/2014/961067.

[118] R. I. Navarro, "Study of a neural network-based system for stability augmentation of an airplane annex 1 introduction to neural networks and adaptive neuro-fuzzy inference systems (ANFIS)," 2013.

[119] D.V. Lukichev, G.L. Demidova, A.Y. Kuzin, A.V. Saushev, Application of Adaptive Neuro Fuzzy Inference System (ANFIS) controller in servodrive with multi-mass object, 2018 25th International Workshop on Electric Drives: Optimization in Control of Electric Drives, IWED 2018 - Proceedings, 2018 (2018) 1–6, doi:10.1109/IWED.2018.8321388.

[120] M. Ilius Hasan Pathan, M. Juel Rana, M. Shoaib Shahriar, M. Shafiullah, M. Hasan Zahir, A. Ali, Real-time LFO damping enhancement in electric networks employing PSO optimized ANFIS, Invent 5 (4) (2020) 61, doi:10.3390/inventions5040061.

[121] S. Inyurt, M.R. Ghaffari Razin, Regional application of ANFIS in ionosphere time series prediction at severe solar activity period, Acta Astronaut 179 (1) (2021) 450–461, doi:10.1016/j.actaastro.2020.11.027.

[122] A. Al-Hmouz, J. Shen, R. Al-Hmouz, J. Yan, Modeling and simulation of an Adaptive Neuro-Fuzzy Inference System (ANFIS) for mobile learning, IEEE Trans. Learn. Technol. 5 (3) (2012) 226–237, doi:10.1109/TLT.2011.36.

[123] S. Gautam, P. Sihag, N.K. Tiwari, S. Ranjan, Neuro-fuzzy approach for predicting the infiltration of soil, Lect. Notes Civil Eng. 31 (1) (2019) 221–228.

[124] U. Çaydaş, A. Hasçalik, S. Ekici, An adaptive neuro-fuzzy inference system (ANFIS) model for wire-EDM, Expert Syst. Appl. 36 (3) (2009) 6135–6139, doi:10.1016/j.eswa.2008.07.019.

[125] M. Hossain, et al., Application of the hybrid ANFIS models for long term wind power density prediction with extrapolation capability, PLoS One 13 (4) (2018) e0193772, doi:10.1371/journal.pone.0193772.

[126] N. Talpur, M.N.M. Salleh, K. Hussain, An investigation of membership functions on performance of ANFIS for solving classification problems, IOP Conference Series: Materials Science and Engineering, 226 (1), 2017 012103, doi:10.1088/1757-899X/226/1/012103.

[127] R.K. Yadav, M. Balakrishnan, Comparative evaluation of ARIMA and ANFIS for modeling of wireless network traffic time series, Eurasip J. Wirel. Commun. Netw. 2014 (1) (2014) 15, doi:10.1186/1687-1499-2014-15.

[128] M.N.M. Salleh, N. Talpur, K. Hussain, Adaptive neuro-fuzzy inference system: overview, strengths, limitations, and solutions, Lecture Notes in Computer Science (including subseries Lecture Notes in Artificial Intelligence and Lecture Notes in Bioinformatics, 10387, LNCS, New York, USA, 2017, pp. 527–535, doi:10.1007/978-3-319-61845-6_52.

[129] T.B. Chandra, K. Verma, B.K. Singh, D. Jain, S.S. Netam, Coronavirus disease (COVID-19) detection in Chest X-Ray images using majority voting based classifier ensemble, Expert Syst. Appl., Maryland, USA 165 (2021) 113909, doi:10.1016/j.eswa.2020.113909.

[130] K.J. Assi, M. Shafiullah, K. Md Nahiduzzaman, U. Mansoor, Travel-to-school mode choice modelling employing artificial intelligence techniques: a comparative study, Sustainability 11 (16) (2019) 1–12, doi:10.3390/su11164484.

[131] M.M. Rahman, M. Shafiullah, S.M. Rahman, A.N. Khondaker, A. Amao, M.H. Zahir, Soft computing applications in air quality modeling: Past, present, and future, Sustainability (Switzerland), 12 (10), MDPI AG, Basel, Switzerland, 2020), p. 4045, doi:10.3390/SU12104045.

[132] H. Werbin-Ofir, L. Dery, E. Shmueli, Beyond majority: label ranking ensembles based on voting rules, Expert Syst. Appl. 136 (12) (2019) 50–61, doi:10.1016/J.ESWA.2019.06.022.

[133] M.H.D.M. Ribeiro, L. dos Santos Coelho, Ensemble approach based on bagging, boosting and stacking for short-term prediction in agribusiness time series, Appl. Soft Comput. J. 86 (1) (2020) 105837, doi:10.1016/j.asoc.2019.105837.

[134] Y. Shin, Z. Kim, J. Yu, G. Kim, S. Hwang, Development of NOx reduction system utilizing artificial neural network (ANN) and genetic algorithm (GA), J. Clean. Prod. 232 (Sept) (2019) 1418–1429, doi:10.1016/j.jclepro.2019.05.276.

[135] M.S. Shahriar, M. Shafiullah, M.J. Rana, A. Ali, A. Ahmed, S.M. Rahman, Neurogenetic approach for real-time damping of low-frequency oscillations in electric networks, Comput. Electr. Eng. 83 (5) (2020) 1–14, doi:10.1016/j.compeleceng.2020.106600.

[136] D. Devikanniga, K. Vetrivel, N. Badrinath, Review of meta-heuristic optimization based artificial neural networks and its applications, J. Phys. Conf. Ser. 1362 (1) (2019) 12074, doi:10.1088/1742-6596/1362/1/012074.
[137] A. Ali, et al., Investigation of MPPT techniques under uniform and non-uniform solar irradiation condition–a retrospection, IEEE Access 8 (1) (2020) 127368–127392, doi:10.1109/access.2020.3007710.
[138] L. Wu, G. Huang, J. Fan, X. Ma, H. Zhou, W. Zeng, Hybrid extreme learning machine with meta-heuristic algorithms for monthly pan evaporation prediction, Comput. Electron. Agric. 168 (12) (2019) 105115, doi:10.1016/j.compag.2019.105115.
[139] J. Zhou, et al., Optimization of support vector machine through the use of metaheuristic algorithms in forecasting TBM advance rate, Eng. Appl. Artif. Intell. 97 (Oct) (2021) 104015, doi:10.1016/j.engappai.2020.104015.
[140] L.M. Rasdi Rere, M.I. Fanany, A.M. Arymurthy, Metaheuristic algorithms for convolution neural network, Comput. Intell. Neurosci. 2016 (Special Issue) (2016) 1–13, doi:10.1155/2016/1537325.
[141] Z. Tian, S. Fong, Survey of meta-heuristic algorithms for deep learning training. In: Optimization Algorithms - Methods and Applications, InTech, London, UK, 2016.
[142] H. Chiroma, et al., Nature Inspired Meta-heuristic Algorithms for deep learning: recent progress and novel perspective, Adv. Intell. Syst. Comput. 943 (2020) 59–70, doi:10.1007/978-3-030-17795-9_5.
[143] M. Nikolić, M. Šelmić, D. Macura, J. Ćalić, Bee colony optimization metaheuristic for fuzzy membership functions tuning, Expert Syst. Appl. 158 (Nov) (2020) 113601, doi:10.1016/j.eswa.2020.113601.
[144] S. Farajdadian, S.M.H. Hosseini, Optimization of fuzzy-based MPPT controller via metaheuristic techniques for stand-alone PV systems, Int. J. Hydrogen Energy 44 (47) (2019) 25457–25472, doi:10.1016/j.ijhydene.2019.08.037.
[145] G. García-Gutiérrez, et al., Fuzzy logic controller parameter optimization using metaheuristic cuckoo search algorithm for a magnetic levitation system, Appl. Sci. 9 (12) (2019) 2458, doi:10.3390/app9122458.
[146] I.M. El-Hasnony, S.I. Barakat, R.R. Mostafa, Optimized ANFIS model using hybrid metaheuristic algorithms for parkinson's disease prediction in IoT environment, IEEE Access 8 (1) (2020) 119252–119270, doi:10.1109/ACCESS.2020.3005614.
[147] R. Malhotra, M. Khanna, Particle swarm optimization-based ensemble learning for software change prediction, Inf. Softw. Technol. 102 (Oct) (2018) 65–84, doi:10.1016/j.infsof.2018.05.007.
[148] D. Li, L. Luo, W. Zhang, F. Liu, F. Luo, A genetic algorithm-based weighted ensemble method for predicting transposon-derived piRNAs, BMC Bioinformatics 17 (1) (2016) 329, doi:10.1186/s12859-016-1206-3.
[149] R.J. Kuo, C.H. Mei, F.E. Zulvia, C.Y. Tsai, An application of a metaheuristic algorithm-based clustering ensemble method to APP customer segmentation, Neurocomputing 205 (Sept) (2016) 116–129, doi:10.1016/j.neucom.2016.04.017.
[150] D. Maulud, A.M. Abdulazeez, A review on linear regression comprehensive in machine learning, J. Appl. Sci. Technol. Trends 1 (4) (2020) 140–147, doi:10.38094/jastt1457.
[151] A.L. Lynam, et al., Logistic regression has similar performance to optimised machine learning algorithms in a clinical setting: application to the discrimination between type 1 and type 2 diabetes in young adults, Diagnostic Progn. Res. 4 (1) (2020) 6, doi:10.1186/s41512-020-00075-2.
[152] M. Singh, M. Wasim Bhatt, H.S. Bedi, U. Mishra, Performance of bernoulli's naive bayes classifier in the detection of fake news, Mater. Today Proc. 2020 (1) (2020), 1–4, doi:10.1016/j.matpr.2020.10.896.

[153] A. Sarica, A. Cerasa, A. Quattrone, Random forest algorithm for the classification of neuroimaging data in alzheimer's disease: a systematic review," front, Aging Neurosci 9 (Oct) (2017) 329, doi:10.3389/fnagi.2017.00329.
[154] B. Mor, S. Garhwal, A. Kumar, A systematic review of hidden markov models and their applications, Arch. Comput. Methods Eng. 1 (2) (2020) 3, doi:10.1007/s11831-020-09422-4.
[155] A. Barron, "Predicted squared error: a criterion for automatic model selection," 1984, pp. 87–103. https://www.bibsonomy.org/bibtex/2ee006afab96760d835578376bdc5f655/idsia. (Accessed 23 May 2019).
[156] E. Castillo, Functional networks, Neural Process. Lett. 7 (3) (1998) 151–159, doi:10.1023/A:1009656525752.
[157] H.Y.A. Abutair, A. Belghith, Using case-based reasoning for phishing detection, Procedia Comput. Sci. 109 (1) (2017) 281–288, doi:10.1016/j.procs.2017.05.352.
[158] B. Raza, Y.J. Kumar, A.K. Malik, A. Anjum, M. Faheem, Performance prediction and adaptation for database management system workload using case-based reasoning approach, Inf. Syst. 76 (1) (2018) 46–58, doi:10.1016/j.is.2018.04.005.
[159] G. Blondet, J. Le Duigou, N. Boudaoud, A knowledge-based system for numerical design of experiments processes in mechanical engineering, Expert Syst. Appl. 122 (1) (2019) 289–302, doi:10.1016/j.eswa.2019.01.013.

Advanced signal processing techniques for feature extraction

4.1 Introduction

In the early 19th century, the signal processing techniques (SPTs) was not popular in analyzing power system transients as their underlying mathematics was partly understood. However, the advent of microprocessors and their communication systems opened new opportunities for the digital SPT in many areas, including information processing, medical diagnostics, seismology, and power system engineering [1–5]. In power system engineering, the advanced SPT provides the best characterization and analysis of the recorded signals from the power system networks. Such characterization facilitates a better understanding of the system behavior needed to monitor, control, and protect the electric grids [4].

In power system research, detection, and classification of power quality (PQ) disturbances are prominent issues where different SPT have successfully deployed. For instance, PQ disturbances were classified using a wide range of SPTs, including the Fourier transform (FT) [6], Gabor–Wigner transform [7], Hilbert-Huang transform [8], Kalman filtering [9], short-time FT [10], ST [11], Teager energy operator [12], Tsallis wavelet entropy [13], wavelet transform (WT) [14], and Winger-Ville time-frequency distribution [15]. Likewise, fault diagnosis, another critical topic for the power system researchers, where different signal processing schemes like Gabor transform [16], Hilbert–Huang transform [17], multivariate empirical mode decomposition [18], ST [19], short-time FT [20], and WT [21] were successfully utilized. Other critical issues in power system research including fault prediction in gas-turbine power units [22], induction motors [23], nuclear reactor power control [24], photovoltaic system [25], power electronics converters [26], power insulators [27], rotating machinery [28], and wind turbine blade pitch system [29] were also investigated using SPTs. Based on the presented discussion, it is evident that SPT is a prevalent and widely adopted tool in analyzing the complex behavior of the power system transients. Such tools will be essential and integral part of future smart grids as the complexities of the grids are increasing day by day with the rise of uncertainty in electricity generation, consumption, and price. Besides, the deployment of advanced measurement infrastructures in electric networks like phasor measurement units will facilitate the usage and exploration of the advanced SPTs [4,30,31].

This chapter briefly introduces the evolution of advanced SPTs along with their advantages and disadvantages. It then illustrates two advanced techniques: the discrete wavelet transform (DWT) and the ST. It also illustrates step-by-step feature extraction processes from the recorded signals after an illustration of the targeted SPTs. Besides, it demonstrates a systematic approach for selecting the essential features from the extracted features to be fetched for the artificial intelligence tools to develop the intelligent fault diagnosis scheme in power system networks.

The remaining parts of this chapter are organized as follows: Section 4.2 briefly introduces the evolution of advanced SPTs along with their advantages and disadvantages. Section 4.3 presents step-by-step feature extraction processes from the recorded signals using two advanced SPTs: DWT and ST. Finally, an overall summary of this chapter is presented in Section 4.4.

4.2 Signal processing techniques

Among many signal-processing techniques, the FT is the most widely used one for transforming the time-domain signals into the frequency domain signals. It retrieves the global frequency content of the stationary signals. The continuous-time FT deals with continuous-time signals. In contrast, the discrete Fourier transform (DFT) derives the frequency-domain (spectral) representations of the finite discrete signals in the time domain. In the case of many samples, the DFT becomes computationally expensive. Also, DFT suffers from aliasing and leakage problems. However, the fast Fourier transform, an efficient and easy to implement algorithm, is employed to reduce the DFT computation burden. However, it cannot effectively deal with the nonstationary signals and loses temporal information. Besides, it is sensitive to noise that requires additional filtering [1,32]. Dennis Gabor proposed the short-time Fourier transform (STFT) to adapt the FT to analyze the nonstationary signals. The STFT uses a small sampling window of the regular interval and decomposes the signals into the frequency domain. However, it creates a resolution problem between frequency and time. A good frequency resolution may result in poor time resolution and vice versa [1,14,32,33]. To overcome the resolution issue, the researchers introduce advanced SPT as WT, ST, and filter banks [4].

4.3 Wavelet transform

The introduction of the WT overcomes the drawbacks of DFT and STFT. The WT offers a window that varies in time to obtain better frequency or time resolution according to requirements. It uses a small window at a high frequency and a larger window at a lower frequency. Consequently, it becomes a multiresolution analysis that decomposes a signal into a series of wavelet components [34]. The decomposed components of the signals could be sharp or smooth, regular or irregular, and symmetrical or asymmetrical that contains characteristic signatures of the signals. Figs. 4.1 and 4.2 illustrate the concept and differences of STFT and WT.

The multiresolution analysis feature of the WT makes it a very effective and powerful tool to analyze the transients of the electric power systems. It is also capable of extracting useful features in both time and frequency domains from the recorded signals. Therefore, the WT has replaced earlier transforms for many engineering applications [35].

Advanced signal processing techniques for feature extraction

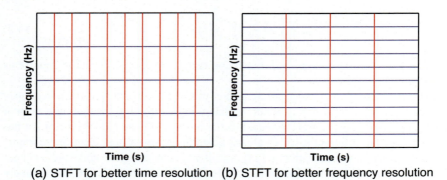

Fig. 4.1 Time and frequency resolution of STFT. *STFT*, short-time Fourier transform.

4.3.1 Continuous wavelet transform

The continuous wavelet transform (CWT) divides continuous-time functions into wavelets by convoluting the input data sequence with a set of functions generated from the mother wavelets, usually known as the family of wavelets. The family of wavelets is the dilated and translated version of mother wavelets. If $\psi(t)$ is the mother wavelet, then the family of wavelets could be generated using the equation (4.1), and CWT of a signal *x(t)* can be presented using the Eq. (4.2) as [36–39]:

$$\psi_{\tau,a}(t) := \frac{1}{\sqrt{a}} \psi\left(\frac{t-\tau}{a}\right), a \in R^{+*}, \tau \in R \,\&\, a \neq 0 \qquad (4.1)$$

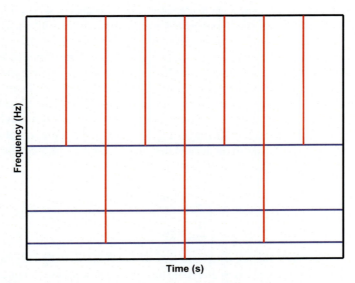

Fig. 4.2 Time and frequency resolution of WT. *WT*, wavelet transform.

$$X_{CWT}(a,\tau) = \frac{1}{\sqrt{a}} \int_{-\infty}^{\infty} x(t)\overline{\psi}\left(\frac{t-\tau}{a}\right)dt \qquad (4.2)$$

Where, a and τ are the scaling and translation factors, respectively, and $\overline{\psi}$ represents the complex conjugate.

Application of CWT includes the signal analysis, self-similarity, and time-frequency relationship. Additionally, CWT is very efficient in determining the damping ratio of the oscillating signals and very resistant to the noise in the signals. With time, it has been employed in different research areas, including power system electromyography [37], examining oil spatial variation [38], oscillation detection [39], and fluid mechanics [40]. Despite having popularity, the CWT adds redundancy and is computationally burdensome; thus, it cannot be employed for signal analysis in real-time (online). Besides, the original signals cannot be reconstructed from their CWT coefficients. Finally, it does not provide phase information of the signals under investigation [41].

4.3.1.1 Discrete wavelet transform

DWT is obtained by discretizing CWT. The researchers use DWT in many power system applications because of its simplicity and lower computational time [42,43]. DWT of a signal $x(k)$ can be represented using the following equation [43,44]:

$$X_{DWT}(m,n) = \frac{1}{\sqrt{a^m}} \sum_{k=-\infty}^{\infty} x(k)\overline{g}\left(\frac{k-n\tau}{a^m}\right). \qquad (4.3)$$

Where, $\overline{g}(*)$ represents the complex conjugate of the family of wavelets obtained from the mother wavelet that has been scaled by a and translated by τ.

Input signals are passed through a series of low pass (LP) and high pass (HP) filters to get the nonuniform frequency bands decomposition coefficients. The LP and HP filters are linked to each other and recognized as quadrature mirror filters. The output of HP gives the detailed coefficients, whereas the output of LP gives the approximate coefficients. According to Nyquist's rule, half of the samples are discarded after each stage as half of the input signal frequencies have been removed. Sample reduction at each stage effectively decreases the time resolution by half as the full signal is represented by half of the samples. The signal can be reliably identified at higher stages compared to the lower one in time. The output of LP is again passed through HP and LP filters to get second-level coefficients. Fig. 4.3 illustrates the concept of DWT where $c(n)$ and $d(n)$ are the approximate and detailed coefficients of quadrature mirror filters and can be represented by the following equations [45,46]:

$$c(n) = \sum_{k=-\infty}^{\infty} x(k)h(k-2n) \qquad (4.4)$$

$$d(n) = \sum_{k=-\infty}^{\infty} x(k)g(k-2n) \qquad (4.5)$$

Application of DWT includes the analyses of nonstationary signals, PQ disturbances, harmonic analysis, faults of electrical machinery, power system fault analysis, synchronization of power electronic converters with the grids, electrocardiogram, communication signals, and image denoising [45–48].

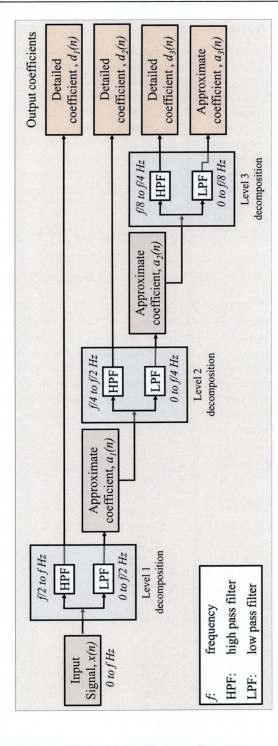

Fig. 4.3 Three-level decomposition tree of the discrete wavelet transform.

4.3.2 Stockwell transform

As mentioned earlier, the STFT is widely used for time-frequency analysis, but a good time resolution may result in low-frequency resolution and vice versa as it uses fixed sampling windows. On the contrary, the WT uses relatively bigger windows at lower frequencies and smaller windows at higher frequencies, which overcomes the resolution problem of STFT [14]. In addition, WT is quite efficient in extracting both frequency and time domain information. However, WT is sensitive to noise and does not contain phase information of the non-stationary signals [49–51]. In response, Stockwell et al. [50] combined the advantages of both STFT and WT. ST is a new transform that provides full-time-frequency decomposition and perfectly upholds the phase information and absolutely referenced frequency. Besides, ST is perfectly invertible and immune to system noise [1]. The ST of a given function $x(t)$ can be defined as [49–51]:

$$S(\tau, f) = S\{x(t)\} = \int_{-\infty}^{\infty} x(t)\, w(\tau - t, f)\, e^{-j2\pi ft}\, dt \qquad (4.6)$$

Where, $w(t, f) = \dfrac{|f|}{\sqrt{2\pi}} e^{-\frac{t^2 f^2}{2}}$ is the Gaussian window function; "t" and "τ" are both time variables, and "f" is the frequency. The signal $x(t)$ is decomposed into both temporal (τ) and frequency (f) components. The decomposed signal's coefficients cover the temporal axis and create full resolutions for each designated frequency by selecting possible values of τ that refers to the center of the window function. The ST applies different window functions, including the Gaussian window, bi-Gaussian window, and hyperbolic window, depending on the applications [49]. The window sizes are adjusted by different "f" values to realize multiresolution over different frequencies over the temporal axis.

Eq. (4.6) can be rewritten as [49–51]:

$$S(\tau, f) = \int_{-\infty}^{\infty} X(\alpha + f)\, e^{-\frac{2\pi^2 \alpha^2}{f^2}}\, e^{j2\pi \alpha \tau}\, d\alpha;\ f \ne 0 \qquad (4.7)$$

Where, $X(\alpha + f) = \int_{-\infty}^{\infty} x(t)\, e^{-j2\pi(\alpha + f)t}\, dt$.

Let the discrete-time series of $x(t)$ is denoted by $x[kT]$, $k = 0, 1, 2, \ldots, N-1$ with a time sampling interval of T. The DFT can be written as:

$$X\left[\frac{n}{NT}\right] = \frac{1}{N} \sum_{k=0}^{N-1} x[kT]\, e^{-\frac{j2\pi nk}{N}} \qquad (4.8)$$

Using the above equations and making $\tau = pT$ and $f = \dfrac{n}{NT}$, the ST of discrete $x[kT]$ can be written as:

$$S\left[pT, \frac{n}{NT}\right] = \sum_{m=0}^{N-1} X\left[\frac{m+n}{NT}\right] e^{-\frac{2\pi^2 m^2}{n^2}}\, e^{\frac{j2\pi mp}{N}};\ n \ne 0 \qquad (4.9)$$

Where, $m, n, p = 0, 1, 2, \ldots, N-1$ and N is the total number of samples.

The discrete inverse of the ST can be obtained as:

$$x[kT] = \sum_{n=0}^{N-1} \left\{ \frac{1}{N} \sum_{p=0}^{N-1} S\left[pT, \frac{n}{NT}\right] \right\} e^{\frac{j2\pi nk}{N}} \quad (4.10)$$

The ST output is an $M \times N$ matrix usually known as S-matrix, whose columns pertain to time and rows pertain to the fquency. Each element of S-matrix represents a complex number. The following equation calculates the energy matrix of a signal from the S-matrix.

$$E_{M \times N} = |S_{M \times N}|^2 \quad (4.11)$$

Hence, the ST-based feature extraction method produces new matrices, namely S_{cmax}, S_{rmax}, and E_{cmax}, from the S-matrices and the E-matrices. The S_{cmax}-matrices contain the maximum absolute values of the S-matrices columns. In contrast, the S_{rmax}-matrices contain the maximum absolute values of the rows of the S-matrices, and the E_{cmax}-matrices contain the absolute maximum values of the columns of the energy matrices. The ST-based feature extraction also produces another matrix called $S_{c\text{-phase-max}}$ that contains the phase angles of the S-matrix associated with the elements of S_{cmax}. The ST-based feature extraction method eventually applies standard statistical techniques on the produced matrices to extract essential features from the decomposed three-phase current signals. Among many features of the produced matrices, the amplitude, gradient of amplitude, mean value, standard deviation, entropy, skewness, kurtosis, time of occurrence, and energy of different harmonics are widely used features for the analysis of PQ transients [52–55]. A brief introduction of the selected statistical measures is presented in the appendix of this book. The ST has drawn significant attention from the research of various fields of study for analyzing PQ events [11], power system faults [52], structural damage identification [56], and electrocardiogram [57].

4.4 Feature extraction illustration

Advanced SPTs such as DWT and ST are proven methods for efficiently analyzing electric power system transients. The feature extraction process of the SPTs figures out the distinctive parameters of any electric waveform by retaining the significant information and the fundamental characteristics of the signal. The following parts of this section illustrate the feature extraction processes of DWT and ST in detail for the recorded signals from the four-node test distribution feeder presented in Chapter 10 (Fig. 10.1).

4.4.1 DWT-based feature extraction

In DWT, the temporal analysis is performed with a contracted high-frequency version of the prototype wavelet, while frequency analysis is performed with a dilated low-frequency version of the same wavelet. The prototype wavelet is known as the mother wavelet that is an effectively limited duration waveform with a zero-average value. Scholars proposed different variations of mother wavelets namely, Haar wavelet, Coifman wavelet, Daubechies wavelet, Mallat wavelet, Meyer wavelet, and Morlet

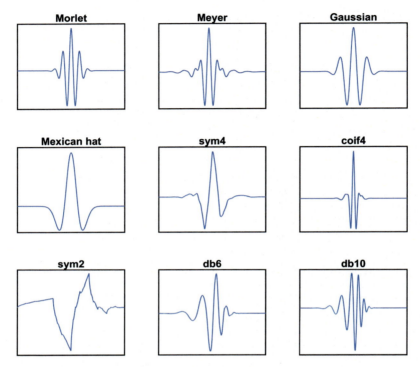

Fig. 4.4 Different mother wavelets.

wavelet (Fig. 4.4). This book uses the Daubechies mother wavelet formulated by the Belgian mathematician Ingrid Daubechies in 1988.

For the illustration of the DWT-based feature extraction process, this chapter first applied an AG (phase A to ground) fault 7 km away from the sending end or the data measurement bus with a fault resistance of 15 Ω on a four-node test distribution feeder presented in chapter 10 (Fig. 10.1). Detailed information about the test feeder can also be found in Ref. [19,43,48,52]. The three-phase faulty current signals from the sending end are recorded. For the accurate recording of the faulty current signals, advanced measurement devices like phasor measurement units can be employed. After recording the faulty current signals, this chapter decomposed them into seven levels employing the *Daubechies*-four (*db4*) mother wavelet that ended up with one approximate and seven detailed coefficients. Then, it evaluates six different statistical features: entropy, energy, skewness, kurtosis, mean, and standard deviation of all decomposed detail and approximate coefficients for further use as inputs of the machine learning tools. The following steps provide detailed procedures for the feature extraction process:

Step 1: Recording of the faulty current signals

In the first step of DWT based feature extraction process, the three-phase faulty current signals (noisy and noise-free) were recorded from the sending end or the data measurement bus. Fig. 4.5 presents the recorded noise-free three-phase faulty current

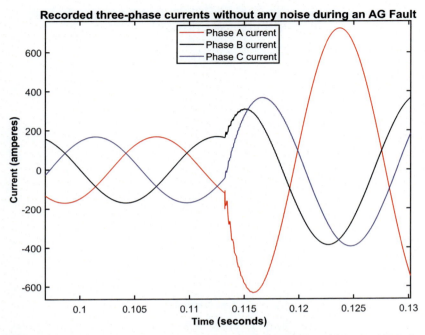

Fig. 4.5 Recorded three-phase faulty current signal from the sending end for an AG fault without any noise.

signals for two cycles where one cycle was recorded before the occurrence of the fault, and the other cycle was recorded after the fault. Figs. 4.6, 4.7 and 4.8 present the three-phase faulty current signals with 40 dB, 30 dB, and 20 dB signal-to-noise ratio (SNR), respectively. All mentioned signals (noisy and noise-free) were recorded with a sampling frequency of 100 kHz (~1667 samples/cycle).

Step 2: Decomposition of the recorded faulty current signals

After recording, the faulty current signals were decomposed into seven levels employing *Daubechies-four* (*db4*) mother wavelet that provides one approximate and seven detailed coefficients. Figs. 4.9, 4.10, and 4.11 present the shape of the detailed and the approximate coefficients related to the current signals of phases *A*, phase *B*, and phase *C*, respectively, for the noise-free measurement. These figures present twenty-one detailed (seven from each phase current) and three approximate (one from each phase current) coefficients that lead towards twenty-four coefficients. It can be observed that the DWT decomposed coefficients are series of numbers. They contain the characteristics signatures of the applied fault for example, presence or absence of the fault and its type and location.

Step 3: Evaluation of the statistical measures from the decomposed coefficients

In this step, six statistical measures (entropy, energy, skewness, kurtosis, mean, and standard deviation) for all twenty-four detailed and approximate coefficients were evaluated. Therefore, a total of 144 (~6*8*3) statistical measures, also known as features, were assessed from the recorded three-phase faulty current signals through

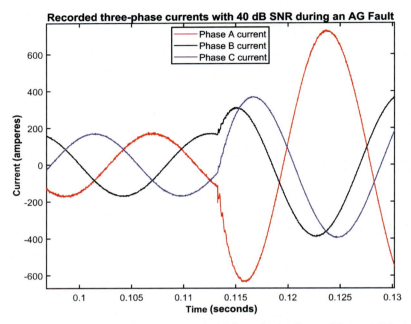

Fig. 4.6 Recorded three-phase faulty current signal from the sending end for an AG fault with 40 dB SNR. *SNR*, signal-to-noise ratio.

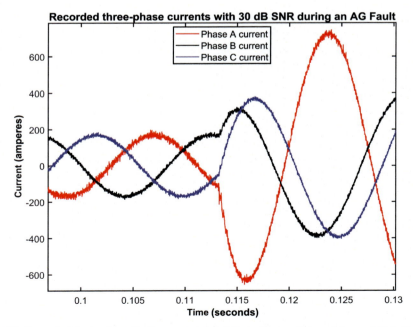

Fig. 4.7 Recorded three-phase faulty current signal from the sending end for an AG fault with 30 dB SNR. *SNR*, signal-to-noise ratio.

Advanced signal processing techniques for feature extraction 111

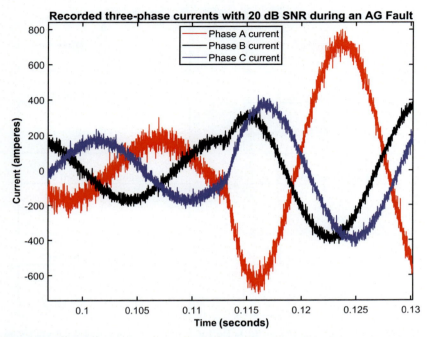

Fig. 4.8 Recorded three-phase faulty current signal from the sending end for an AG fault with 20 dB SNR. *SNR*, signal-to-noise ratio.

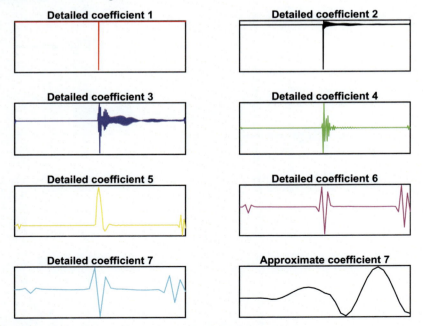

Fig. 4.9 DWT extracted coefficients (detailed and approximate) of phase A current for an AG fault (noise-free measurement). *DWT*, discrete wavelet transform.

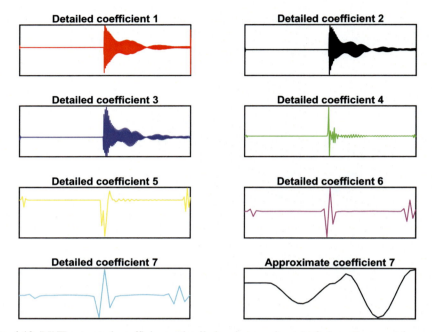

Fig. 4.10 DWT extracted coefficients (detailed and approximate) of phase B current for an AG fault (noise-free measurement). *DWT*, discrete wavelet transform.

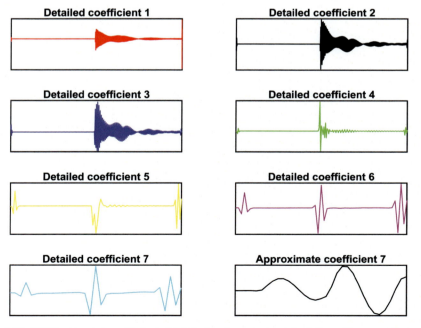

Fig. 4.11 DWT extracted coefficients (detailed and approximate) of phase C current for an AG fault (noise-free measurement). *DWT*, discrete wavelet transform.

7-level DWT decomposition. A brief introduction of the selected statistical measures is presented in the Appendix of this Book. All extracted features of the selected case (AG fault: 15Ω fault resistance and 7 km away from the sending end or the data measurement bus) of the four-node test distribution feeder of Chapter 10 are presented in Table 4.1.

Step 4: Generation of a wide range of fault scenarios

Likewise, the intelligent fault diagnosis scheme generates and records a wide range of fault scenarios by changing the fault information for example, fault resistance, fault inception time, and fault type. Besides, it mimics the dynamic behavior of loading conditions of the power system network by generating the system load randomly from the specified range. Then, the DWT evaluates the mentioned 144 features for each faulty case. The intelligent fault diagnosis scheme also keeps track of the fault distances and types for future use to train and test the machine learning tools. It uses the extracted features as the inputs and the tracked distances and types as the outputs of the machine learning tools. Similarly, the intelligent fault diagnosis scheme also creates a similar input matrix for non-faulty data by varying loading conditions to distinguish the faulty signals from their nonfaulty counterparts. Consequently, this book deals with three different types of problems: power system fault detection, classification, and location employing the DWT extracted features.

4.4.2 Stockwell transform-based feature extraction

The ST based feature extraction scheme is illustrated in the following steps:

Step 1: Generation of S-matrices and E-matrices

Like the DWT approach, the ST decomposes the previously mentioned recorded three-phase faulty current signals and creates three S-matrices and three E-matrices. The elements of the produced S-matrices are complex numbers whose columns pertain to time and rows pertain to frequency information. Thus, the ST scheme creates four vectors: S_{cmax}, S_{rmax}, E_{cmax}, and $S_{c\text{-phase-max}}$ for each phase as discussed in 4.2.2. Eventually, it creates 12 vectors from the recorded three-phase current signal.

Step 2: Extraction of statistical measures

After generation of S_{cmax}, S_{rmax}, E_{max}, and $S_{c\text{-phase-max}}$ matrices, the standard deviations, maximum values, minimum values, mean values, entropy, skewness, and kurtosis of each matrix are evaluated. Eventually, the ST based feature extraction scheme evaluates the following 28 features for each phase current:

- Feature F_1 to F_4: standard deviations of the matrices
- Feature F_5 to F_8: maximum values of the matrices
- Feature F_9 to F_{12}: minimum values of the matrices.
- Feature F_{13} to F_{16}: mean values of the matrices.
- Feature F_{17} to F_{20}: entropy of the matrices.
- Feature F_{21} to F_{24}: skewness of the matrices.
- Feature F_{25} to F_{28}: kurtosis of the matrices.

Table 4.2 presents the ST scheme extracted features for the previously mentioned fault case (AG fault: 15Ω fault resistance and 7 km away from the sending end or the data measurement bus) of the four-node test distribution feeder of Chapter 10. The ST-based signal processing scheme extracted a total of 84 (=28*3) features from the recorded three-phase faulty current.

Table 4.1 Discrete wavelet transform extracted features of an AG fault from the measured three-phase current signals.

Item		Entropy	Skewness	Kurtosis	Standard deviation	Mean	Energy
Detailed coefficient (d_1)	Phase A	1.4976	−32.5014	1108.8768	1.7114	−0.057521	0.001031
	Phase B	1.5598	−0.1920	16.6789	0.0276	0.000010	0.000001
	Phase C	1.5695	−1.7787	55.9907	0.0309	0.000010	0.000001
Detailed coefficient (d_2)	Phase A	2.2053	−23.9796	650.1211	1.2793	−0.048190	0.000289
	Phase B	2.1669	0.3929	16.8156	0.4166	0.000011	0.000082
	Phase C	2.1515	0.3921	16.5456	0.4184	−0.000904	0.000100
Detailed coefficient (d_3)	Phase A	1.9924	−2.2966	43.8068	2.3129	0.011036	0.000475
	Phase B	1.9969	0.9832	16.4626	2.0756	−0.003021	0.001029
	Phase C	1.9824	0.9642	16.1609	2.0851	0.007243	0.001247
Detailed coefficient (d_4)	Phase A	2.2924	−0.6113	47.6174	8.6073	−0.076154	0.003337
	Phase B	2.4007	3.9518	81.1994	3.0198	0.093838	0.001106
	Phase C	2.2389	3.7964	76.9475	3.0615	0.085530	0.001364
Detailed coefficient (d_5)	Phase A	2.1403	4.7454	27.6425	9.7060	1.522948	0.002235
	Phase B	2.6547	−4.3255	27.2543	2.1667	−0.312051	0.000299
	Phase C	2.5139	−1.6484	15.5580	2.9013	−0.303301	0.000636
Detailed coefficient (d_6)	Phase A	2.0558	−0.5245	10.7602	16.5177	0.060938	0.003330
	Phase B	3.9208	−0.5971	16.0280	7.9287	−0.373520	0.002069
	Phase C	3.0369	−0.2594	9.0901	10.9839	−0.222270	0.004756
Detailed coefficient (d_7)	Phase A	1.1409	−0.5796	8.2435	93.2936	3.133496	0.058685
	Phase B	1.8014	0.6240	10.4637	44.3270	−2.190346	0.035692
	Phase C	1.4663	0.3512	6.3131	53.3033	−1.285318	0.061814
Approximate coefficient (d_7)	Phase A	0.9834	0.4129	2.9931	3831.2968	−393.06208	99.930618
	Phase B	0.9673	−0.4088	2.3075	2310.9384	413.454735	99.959722
	Phase C	0.9940	−0.1405	2.6126	2143.7249	−20.572864	99.930082

Table 4.2 Stockwell transform extracted features of an AG fault from the measured three-phase current signals.

Item		Standard deviation	Maximum Values	Minimum Values	Mean Values	Entropy	Skewness	Kurtosis	
S_{cmax}	Phase A	0.8606	210.6619	208.2280	209.4449	0.00	2.02E−06	1.5000
	Phase B	0.3335	126.3545	125.4114	125.8827	0.00	0.0013	1.5000
	Phase C	0.3731	130.1035	129.0483	129.5757	0.00	0.0007	1.5000
S_{rmax}	Phase A	6.4216	210.6619	15.8145	31.6656	0.00	14.7143	381.2957
	Phase B	3.5990	126.3545	7.8737	15.7903	0.00	18.7791	549.2361
	Phase C	3.6756	130.1035	7.9656	15.8995	0.00	19.3174	573.1629
$S_{c\_phase\_max}$	Phase A	0.0004	−2.4327	−2.4338	−2.4332	0.00	−0.0045	1.5000
	Phase B	0.0022	0.0925	0.0863	0.0894	1.45	−0.0039	1.5000
	Phase C	0.0017	1.3101	1.3053	1.3077	0.00	0.0043	1.5000
E_{cmax}	Phase A	360.5107	44378.4247	43358.9006	43867.9216	0.00	0.0031	1.5000
	Phase B	83.9554	15965.4575	15728.0316	15846.5607	0.00	0.0033	1.5000
	Phase C	96.6955	16926.9096	16653.4549	16789.9966	0.00	0.0029	1.5000

Step 3: Selection of most useful features for fault diagnosis

Likewise, 500 similar AG faults on the random locations of the distribution line are applied, varying the loading condition, fault resistance, and inception angle, which lead towards a matrix of 500*84. Then, the procedure employs a systematic trial and error process to select the most useful features to diagnose, e.g., detect, classify, and locate, different types of faults in electric power system networks. To do so, all 28 ST extracted features (F_1 to F_{28}) associated with phase A (faulty phase) of all the applied faulty scenarios should be considered. Thus, it selects a matrix of 500*28 to remove the insignificant along with the highly correlated features. It was observed from the selected matrix that features F_{17} to F_{20} were zero for most of the cases. At the same time, the correlation factors of the features F_5 & F_6, F_9 & F_{13}, F_{21} & F_{24}, F_{25} & F_{28}, and F_8, F_{12} & F_{16} were exactly 1.00. Similarly, the correlation factors of the features F_{21} & F_{25}, F_5 & F_9, F_{11} & F_{15}, F_5 & F_8, F_{22} & F_{26}, and F_7 & F_{11} were 0.9999, 0.9997, 0.9996, 0.9965, 0.9958, and 0.9957, respectively. The feature selection scheme kept one feature and removed the other one in case of their unity and close to unity correlation factors as they were representing the redundant information. It also removed the features with most of the zero entities as they did not carry any significant information. Therefore, the selection scheme removed the features F_6, F_7, F_8, F_9, F_{12}, F_{13}, F_{15}, F_{16}, F_{17}, F_{18}, F_{19}, F_{20}, F_{24}, F_{25}, F_{26} and F_{28} and kept the remaining features (F_1, F_2, F_3, F_4, F_5, F_{10}, F_{11}, F_{14}, F_{21}, F_{22}, F_{23}, and F_{27}). Finally, it ended up with 12 features for each phase current that lead towards a total of 36 features for the recorded three-phase faulty current signal. Table 4.3 summarizes the full process of the removal and selection of features.

Step 4: Generation of a wide range of fault scenarios and storing of data

Like DWT, the ST-based fault diagnosis scheme keeps track of the fault distances and types for future use to train and test the intelligent fault diagnosis techniques.

Table 4.3 Summary of Stockwell transform extracted feature removal and selection process.

Selected Features	Comments	Action
F_5 & F_6	Correlation factor = 1.00	Remove feature F_6
F_8, F_{12} & F_{16}	Correlation factor = 1.00	Remove features F_{12} & F_{16}
F_9 & F_{13}	Correlation factor = 1.00	Remove feature F_{13}
F_{21} & F_{24}	Correlation factor = 1.00	Remove feature F_{24}
F_{25} & F_{28}	Correlation factor = 1.00	Remove feature F_{28}
F_{21} & F_{25}	Correlation factor = 0.9999	Remove feature F_{25}
F_5 & F_9	Correlation factor = 0.9997	Remove feature F_9
F_{11} & F_{15}	Correlation factor = 0.9996	Remove feature F_{15}
F_5 & F_8	Correlation factor = 0.9965	Remove feature F_8
F_{22} & F_{26}	Correlation factor = 0.9958	Remove feature F_{26}
F_7 & F_{11}	Correlation factor = 0.9957	Remove feature F_7
F_{17}, F_{19} & F_{20}	Most of the cases, they are zero.	Remove all of them
F_{27}	It is almost constant (~1.5) throughout the simulation.	Remove F_{27}
The selected features: F_1, F_2, F_3, F_4, F_5, F_{10}, F_{11}, F_{14}, F_{18}, F_{21}, F_{22}, and F_{23}		

The intelligent fault diagnosis scheme uses the evaluated and selected features as the inputs and the tracked distances and types as the outputs of the machine learning tools. It is worth mentioning that the ST-based approach also creates a similar input matrix for non-faulty data by varying the loading conditions to distinguish the faulty signals from their non-faulty counterparts. Consequently, this book deals with three different types of problems: power system fault detection, classification, and location employing the ST extracted features. Detailed and comparative results of DWT and ST-based machine learning techniques in diagnosing power system faults are presented in Chapters 10 and 11.

4.5 Summary

This chapter briefly introduces the importance of advanced SPTs in analyzing power system transients. It provides the necessary background of the evolution of the advanced signal processing methods and their special characteristics, advantages, and disadvantages. Finally, it illustrates step-by-step feature extraction and selection procedures employing two of the most advanced SPTs: the DWT and the ST. It also sheds light on how the extracted features will be fetched into artificial intelligence techniques to develop signal processing-based intelligent fault diagnosis schemes for electric power system networks.

References

[1] P. Khetarpal, M.M. Tripathi, A critical and comprehensive review on power quality disturbance detection and classification, Sustain. Comput. Informatics Syst. 28 (2020) 100417, doi:10.1016/j.suscom.2020.100417.
[2] W.A. Mousa, Advanced Digital Signal Processing of Seismic Data, Cambridge University Press, Cambridge, United Kingdom, 2019.
[3] J. Ruiz-Alzola, C. Alberola-López, C.F. Westin, Advanced signal processing methods for biomedical imaging, Int. J. Biomed. Imaging 2013 (2013), doi:10.1155/2013/696878.
[4] Z. Uddin, A. Ahmad, A. Qamar, M. Altaf, Recent advances of the signal processing techniques in future smart grids, Human-Centric Computing and Information Sciences, 8, Springer, Berlin Heidelberg, Dec. 01 2018, p. 2, doi:10.1186/s13673-018-0126-9.
[5] T.Y. Vega, V.F. Roig, H.B. San Segundo, Evolution of signal processing techniques in power quality, 2007 9th Int. Conf. Electr. Power Qual. Util, EPQU, 2007, doi:10.1109/EPQU.2007.4424200.
[6] S. Vanga and S. N. V. Ganesh, Comparison of Fourier transform and wavelet packet transform for quantification of power quality, 2012, doi:10.1109/APCET.2012.6302048.
[7] S.H. Cho, G. Jang, S.H. Kwon, Time-frequency analysis of power-quality disturbances via the Gabor-Wigner transform, IEEE Trans. Power Deliv. 25 (1) (2010) 494–499, doi:10.1109/TPWRD.2009.2034832 Jan.
[8] M. A. Rodriguez, J. F. Sotomonte, J. Cifuentes, and M. Bueno-Lopez, Classification of power quality disturbances using hilbert huang transform and a multilayer perceptron neural network model, 2019, doi:10.1109/SEST.2019.8849114.

[9] S. Swain, B. Subudhi, Grid synchronization of a PV system with power quality disturbances using unscented Kalman filtering, IEEE Trans. Sustain. Energy 10 (3) (2019) 1240–1247, doi:10.1109/TSTE.2018.2864822.

[10] D.O. Anggriawan, E. Wahjono, I. Sudiharto, A.A. Firdaus, D. Novita Nurmala Putri, A. Budikarso, Identification of Short Duration Voltage Variations Based on Short Time Fourier Transform and Artificial Neural Network, IES 2020 - International Electronics Symposium: The Role of Autonomous and Intelligent Systems for Human Life and Comfort, 2020, pp. 43–47, doi:10.1109/IES50839.2020.9231815.

[11] M. Shafiullah, M.A.M. Khan, S.D. Ahmed, PQ disturbance detection and classification combining advanced signal processing and machine learning tools, in: P. Sanjeevikumar, C. Sharmeela, J.B. Holm-Nielsen, P. Sivaraman (Eds.), Power Quality in Modern Power Systems, First. Eds. Academic Press, London, United Kingdom, 2021, pp. 311–335.

[12] V.S. Mahalle, G.N. Bonde, S.S. Jadhao, S.R. Paraskar, Teager energy operator: a signal processing approach for detection and classification of power quality events, Proceedings of the 2nd International Conference on Trends in Electronics and Informatics, ICOEI 2018, (Nov. 2018), pp. 1109–1114, doi:10.1109/ICOEI.2018.8553703.

[13] X. Wu, et al., Power quality disturbance detection of DC distribution network based on tsallis wavelet entropy, 2019 IEEE PES Innovative Smart Grid Technologies Asia, ISGT 2019, (May 2019), pp. 2395–2399, doi:10.1109/ISGT-Asia.2019.8881279.

[14] M. Ijaz, M. Shafiullah, M.A. Abido, Classification of power quality disturbances using Wavelet Transform and Optimized ANN, 2015 18th International Conference on Intelligent System Application to Power Systems (ISAP), Proceedings of the Conference on, (Sep. 2015), pp. 1–6, doi:10.1109/ISAP.2015.7325522.

[15] J. Deng, W. Song, Power quality analysis of short-time disturbance based on SPWVD. 2019, doi:10.1109/PHM-Qingdao46334.2019.8942882.

[16] Z. Moravej, M. Movahhedneya, M. Pazoki, Gabor transform-based fault location method for multi-terminal transmission lines, Meas. J. Int. Meas. Confed. 125 (2018) 667–679, doi:10.1016/j.measurement.2018.05.027.

[17] D. Wang, M. Hou, Travelling wave fault location algorithm for LCC-MMC-MTDC hybrid transmission system based on Hilbert-Huang transform, Int. J. Electr. Power Energy Syst. 121 (2020) 106125, doi:10.1016/j.ijepes.2020.106125.

[18] S. Liu, Y. Sun, L. Zhang, P. Su, Fault diagnosis of shipboard medium-voltage DC power system based on machine learning, Int. J. Electr. Power Energy Syst. 124 (2021) 106399, doi:10.1016/j.ijepes.2020.106399.

[19] M. Shafiullah, M. Abido, T. Abdel-Fattah, Distribution grids fault location employing ST based optimized machine learning approach, Energies 11 (9) (2018) 2328, doi:10.3390/en11092328.

[20] Y.M. Yeap, A. Ukil, Fault detection in HVDC system using short time fourier transform, IEEE Power and Energy Society General Meeting, 2016 (2016), doi:10.1109/PESGM.2016.7741323.

[21] M. Parsi, P. Crossley, P.L. Dragotti, D. Cole, Wavelet based fault location on power transmission lines using real-world travelling wave data, Electr. Power Syst. Res. 186 (2020) 106261, doi:10.1016/j.epsr.2020.106261.

[22] G. Rigatos, N. Zervos, D. Serpanos, V. Siadimas, P. Siano, M. Abbaszadeh, Fault diagnosis of gas-turbine power units with the derivative-free nonlinear Kalman Filter, Electr. Power Syst. Res. 174 (2019) 105810, doi:10.1016/j.epsr.2019.03.017.

[23] P. Gangsar, R. Tiwari, Signal based condition monitoring techniques for fault detection and diagnosis of induction motors: A state-of-the-art review, Mech. Syst. Signal Process. 144 (2020) 106908, doi:10.1016/j.ymssp.2020.106908.

[24] E. Hatami, H. Salarieh, N. Vosoughi, Design of a fault tolerated intelligent control system for a nuclear reactor power control: Using extended Kalman filter, J. Process Control 24 (7) (2014) 1076–1084, doi:10.1016/j.jprocont.2014.04.012.
[25] I.S. Kim, On-line fault detection algorithm of a photovoltaic system using wavelet transform, Sol. Energy 126 (2016) 137–145, doi:10.1016/j.solener.2016.01.005.
[26] L. Kou, C. Liu, G. wei Cai, Z. Zhang, Fault Diagnosis for power electronics converters based on deep feedforward network and wavelet compression, Electr. Power Syst. Res. 185 (2020) 106370, doi:10.1016/j.epsr.2020.106370.
[27] S.F. Stefenon, et al., Wavelet group method of data handling for fault prediction in electrical power insulators, Int. J. Electr. Power Energy Syst. 123 (2020) 106269, doi:10.1016/j.ijepes.2020.106269.
[28] M. Jalayer, C. Orsenigo, C. Vercellis, Fault detection and diagnosis for rotating machinery: A model based on convolutional LSTM, Fast Fourier and continuous wavelet transforms, Comput. Ind. 125 (2021) 103378, doi:10.1016/j.compind.2020.103378.
[29] S. Cho, M. Choi, Z. Gao, T. Moan, Fault detection and diagnosis of a blade pitch system in a floating wind turbine based on Kalman filters and artificial neural networks, Renew. Energy 169 (2021) 1–13, doi:10.1016/j.renene.2020.12.116.
[30] A.V. Christopher, S. Rex, A. Kumar.J, Smart grid challenges and signal processing based solutions - A Literature survey, Int. J. Eng. Trends Technol. 37 (4) (2016) 189–196, doi:10.14445/22315381/ijett-v37p232.
[31] S. Zhong, R. Broadwater, S. Steffel, Wavelet based load models from AMI data, arXiv (2015). http://arxiv.org/abs/1512.02183. (Accessed 10 February 2021).
[32] A. Swetha, A. Radhakrishna, Signal processing techniques for power system fault analysis-a review, Int. J. Innov. Res. Sci. Eng. Technol. 4 (8) (2015) 7606–7610, doi:10.15680/IJIRSET.2015.0408089.
[33] V.S. Yendole, K.A. Dongre, Power system fault analysis using signal processing technique-a review, Int. J. Innov. Res. Electr. Electron. Instrum. Control Eng. ISO 6 (5) (2018) 64–67, doi:10.17148/IJIREEICE.2018.6514.
[34] U.N. Khan, Signal processing techniques used in power quality monitoring, 2009 8th International Conference on Environment and Electrical Engineering (EEEIC), 2009, pp. 1–4. http://eeeic.eu/proc/papers/23.pdf. (Accessed 27 November 2015).
[35] A.G. Phadke, J.S. Thorp, Computer Relaying for Power Systems, John Wiley & Sons, New Jersey, United States, 2009.
[36] C. Gu, VFTO spectrum analysis method based on continuous wavelet transform, 10th International Conference on Communications, Circuits and Systems, ICCCAS 2018, 2018, pp. 369–372, doi:10.1109/ICCCAS.2018.8768965.
[37] A. Belkhou, A. Achmmad, and A. Jbari, Classification and diagnosis of myopathy EMG signals using the continuous wavelet transform. 2019, doi:10.1109/EBBT.2019.8742051.
[38] A. Biswas, B.C. Si, Application of continuous wavelet transform in examining soil spatial variation: A review, Mathematical Geosciences, 43, Springer, New York, United States, 2011, pp. 379–396, doi:10.1007/s11004-011-9318-9.
[39] M. Ismail Hossain, M. Abido, M. Shafiul Alam, M. Shafiullah, M. Al Emran, F.S. Hossain, Low-frequency inter-area mode detection in power system using continuous wavelet transform, 2018 International Conference on Innovations in Science, Engineering and Technology (ICISET), 2018, pp. 299–304, doi:10.1109/ICISET.2018.8745658.
[40] A. Rinoshika, H. Rinoshika, Application of multi-dimensional wavelet transform to fluid mechanics, Theoretical and Applied Mechanics Letters, 10, Elsevier Ltd, Amsterdam, Netherlands, 2020, pp. 98–115, doi:10.1016/j.taml.2020.01.017.

[41] National Instruments, Continuous Wavelet Transform (Advanced Signal Processing Toolkit), LabVIEW 2010 Advanced Signal Processing Toolkit Help, 2010. https://zone.ni.com/reference/en-XX/help/371419D-01/lvasptconcepts/wa_cwt/. (Accessed 14 February 2021).

[42] A. Borghetti, S. Corsi, C.A. Nucci, M. Paolone, L. Peretto, R. Tinarelli, On the use of continuous-wavelet transform for fault location in distribution power systems, Int. J. Electr. Power Energy Syst. 28 (9) (2006) 608–617, doi:10.1016/j.ijepes.2006.03.001.

[43] M. Shafiullah, M.A. Abido, Z. Al-Hamouz, Wavelet-based extreme learning machine for distribution grid fault location, IET Gener. Transm. Distrib. 11 (17) (2017) 4256–4263, doi:10.1049/iet-gtd.2017.0656.

[44] Q. Chen, G. Nicholson, J. Ye, C. Roberts, Fault diagnosis using discrete wavelet transform (DWT) and artificial neural network (ANN) for a railway switch, Proceedings - 2020 Prognostics and Health Management Conference, PHM-Besancon 2020, 2020, pp. 67–71, doi:10.1109/PHM-Besancon49106.2020.00018.

[45] S. D. Hole and C. A. Naik, Power quality events' classification employing discrete wavelet transform and machine learning, 2020, doi:10.1109/ICMICA48462.2020.9242894.

[46] J. Barros, R.I. Diego, Analysis of harmonics in power systems using the wavelet-packet transform, IEEE Trans. Instrum. Meas. 57 (1) (2008) 63–69, doi:10.1109/TIM.2007.910101.

[47] J. Olkkonen, Discrete Wavelet Transforms - Theory and Applications, InTech, London, United Kingdom, 2011.

[48] Y.D. Mamuya, Y.-D. Lee, J.-W. Shen, M. Shafiullah, C.-C. Kuo, Application of machine learning for fault classification and location in a radial distribution grid, Appl. Sci. 10 (14) (2020) 4965, doi:10.3390/app10144965.

[49] Y. Wang, The tutorial: S transform, Graduate Institute of Communication Engineering National Taiwan University, Taipei (2010), 2010 1–23. http://citeseerx.ist.psu.edu/viewdoc/download?doi=10.1.1.694.8727&rep=rep1&type=pdf. (Accessed 18 June 2016).

[50] R.G. Stockwell, L. Mansinha, R.P. Lowe, Localization of the complex spectrum: the S transform, IEEE Trans. Signal Process. 44 (4) (1996) 998–1001, doi:10.1109/78.492555.

[51] L. Mansinha, R.G. Stockwell, R.P. Lowe, Pattern analysis with two-dimensional spectral localisation: Applications of two-dimensional S transforms, Phys. A Stat. Mech. its Appl. 239 (1–3) (1997) 286–295, doi:10.1016/S0378-4371(96)00487-6.

[52] M. Shafiullah, M.A. Abido, S-transform based FFNN approach for distribution grids fault detection and classification, IEEE Access 6 (2018) 8080–8088, doi:10.1109/ACCESS.2018.2809045.

[53] N. Roy, K. Bhattacharya, Detection, classification, and estimation of fault location on an overhead transmission line using S-transform and neural network, Electr. Power Components Syst. 43 (4) (2015) 461–472, doi:10.1080/15325008.2014.986776.

[54] S. Mishra, C.N. Bhende, B.K. Panigrahi, Detection and classification of power quality disturbances using S-transform and probabilistic neural network, IEEE Trans. Power Deliv. 23 (1) (2008) 280–287, doi:10.1109/TPWRD.2007.911125.

[55] K.R. Krishnanand, P.K. Dash, M.H. Naeem, Detection, classification, and location of faults in power transmission lines, Int. J. Electr. Power Energy Syst. 67 (2015) 76–86, doi:10.1016/J.IJEPES.2014.11.012 .

[56] A. Katunin, Identification of structural damage using S-transform from 1D and 2D mode shapes, Meas. J. Int. Meas. Confed. 173 (2020) 108656, doi:10.1016/j.measurement.2020.108656.

[57] A. Bajaj, S. Kumar, A robust approach to denoise ECG signals based on fractional Stockwell transform, Biomed. Signal Process. Control 62 (2020) 102090, doi:10.1016/j.bspc.2020.102090.

Improved optimal phasor measurement unit placement formulation for power system observability

5.1 Introduction

The ever-increasing demand for electricity along with the emergence of deregulated electricity markets and the grid integration of renewable energy resources, drives the power system networks to be operated closer to their stability and thermal limits [1–3]. Mentioned items generate uncertainties for the market players, including load aggregators and independent system operators. As a result, the online estimation and monitoring of power system states are becoming essential for their safe, secure, and reliable operation [4]. In response, PMU placement all over the networks could offer a promising solution as the PMUs provide precise and time-synchronized phasor measurements (current and voltage phasors) compared to the traditional supervisory control and data acquisition (SCADA) system [5–7]. It is worth noting that the recent integration of electrical measurement devices with the global positioning system (GPS) technology facilitates the commercial production of PMUs [8–10]. Deployed PMU in the power system networks are time-stamped by the GPS satellite, and PMU collected data are sent to phasor data concentrator, local and central monitoring, and control centers through the inter-control center communications protocol (ICCP), as depicted in Fig. 5.1 [11–15].

The entire power system network can be observed and controlled by placing PMUs on each electric grid bus. However, such placement on each bus is costly due to high capital and operational cost. Besides, such placement cannot be achieved due to the absence of communication infrastructure at a few buses as the power system networks are spread over the vast terrain. At the same time, PMU placement on each bus is not even required as ideally one PMU can measure all adjacent branches current phasors, and PMU installed bus voltage phasor, which paves the way for calculating the voltage phasors of the adjacent buses utilizing Kirchhoff's laws and branch parameters [16–18]. Thus, the development of a useful and practical methodology is necessary to realize full network observability through a minimal number of PMU deployments.

Like other optimization problems, the researchers formulated OPP problems and solved them for power system networks (both transmission and distribution) utilizing both traditional mathematical methods [16–22] and metaheuristic [23–25] techniques. Traditional mathematical methods are commonly employed to solve linear or linearized mathematical equations. These methods formulate the OPP problems as integer linear programming problems. Nevertheless, the constraints of the ILP problems must be appropriately defined to obtain the optimum solutions. In addition, these

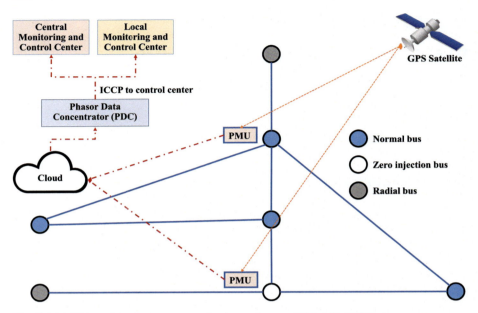

Fig. 5.1 PMU-based power system monitoring and control [11–15]. *PMU*, phasor measurement unit.

methods usually deliver a single solution while many optimum solutions may exist. On the other hand, the metaheuristic techniques incorporate the unobservable buses of the power system networks in the objective functions and obtain several solutions before choosing the best compromise one [16–19].

The presence of passive measurements in the electric networks and the possible power system contingencies, for example, line outage and PMU loss cases, have been overlooked in most of the existing OPP formulations. Only a few pieces of literature considered such contingencies while developing effective and efficient OPP formulations [19–21]. Besides, the standard OPP formulations did not consider PMU channel limitations, which indicates that a single PMU can measure PMU installed bus voltage phasor and all adjacent branches current phasors. Generally, the PMUs are manufactured with a certain number of channels in reality; hence, their manufacturing and production costs vary accordingly. The PMU channel limitations in the OPP formulations have been considered in recent years [26–29]. Furthermore, most of the available literature did not consider the attainment of the maximum number of measurements with the placement of a minimal number of PMUs.

Considering the mentioned notes, this chapter presents an improved OPP formulation for the electric grids (both transmission and distribution networks), intending to minimize the essential number of PMUs and maximize the number of measurements by guaranteeing full network observability. The formulation also integrates the presence of passive measurements or zero injection bus (ZIB) in the power system networks and channel limitations of the installed PMUs. Additionally, the formulation is extended to incorporate several power system contingencies. Then, this chapter investigates the efficacy of the presented formulation on several IEEE benchmarked transmission and distribution networks employing the grey wolf optimization (GWO) algorithm, a highly efficient metaheuristic technique. Besides, it uses the mixed-integer linear programming (MILP) approach to solve the OPP formulation for comparison purposes. Finally, a comparative analysis of the obtained results with referenced works confirms the effectiveness of the considered formulation and the employed solution methodology in achieving more measurements by deploying an equal or a smaller number of PMUs.

The remaining parts of this chapter are organized as follows: Section 5.2 demonstrates the OPP formulation for power system observability considering PMU channel limit and power system contingency. Section 5.3 illustrates the network observability and measurement redundancy with example. Sections 5.4 and 5.5 present the obtained results for transmission and distribution networks observability, respectively. Finally, an overall summary of this chapter is presented in Section 5.6.

5.2 Optimal phasor measurement unit placement formulation for power system observability

Since the restructuring process of electricity grids is pushing them to operate at their highest capacity, the real-time monitoring of the grids has become one of the essential tasks for the system operators to ensure a safe and reliable power supply to their customers. Such real-time monitoring of the grids can be achieved by placing an adequate number of advanced measurement devices throughout the networks. In response, power system researchers developed various measurement devices for example, PMU and SCADA placement methods for power system observability.

5.2.1 General optimal phasor measurement unit placement formulation

Among several approaches, the numerical and topological methods are commonly employed for power system observability. Inherent complexities of the numerical methods and the comparative advantages of the topological methods lead researchers to choose the later one [21]. The topological approach presents the OPP formulation of a power system network with N_B number of buses as [19,23,24]:

Minimize $F(x)$

$$F(x) = \sum_{i=1}^{N_B} C_{PMU} y_i \qquad (5.1)$$

Subjected to:

$$f_o = AY \geq b \qquad (5.2)$$

Where,

$$A = \left[a_{ij} \right]_{N_B \times N_B}$$

$$Y = \left[y_1, y_2, \ldots\ldots, y_{N_B} \right]^T_{N_B \times 1}$$

$$b = \left[1, 1, \ldots\ldots, 1 \right]^T_{N_B \times 1}$$

$$f_o = \left[f_1, f_2, \ldots\ldots, f_{N_B} \right]^T_{N_B \times 1}$$

C_{PMU}: A single PMU manufacturing and installation cost

y_i: A binary number to indicate PMU installation at bus "i" where the value "0" and "1" refer to the absence and presence of PMU, respectively.

A: A binary network connectivity matrix and its elements "a_{ij}" can be defined as:

$$a_{ij} = \begin{cases} 1 \text{ if } i = j \\ 1 \text{ if buses i and j are connected} \\ 0 \text{ otherwise} \end{cases}$$

f_o: Observability vector where the nonzero and zero entries denote the observability and unobservability of the corresponding buses, respectively.

This chapter follows the well-established topological observability rules while developing the improved OPP formulation for electric networks. The rules are summarized as follows considering adequate channels of the installed PMU at the beginning [19]:

Rule 1: Every single PMU can measure the voltage phasor of the installed bus and the current phasors of all connected branches. Then, the voltage phasors of the connected buses can be computed employing Kirchhoff's law and branch parameters. Hence, every single PMU can observe the PMU installed bus and connected buses.

Rule 2: The following inequality illustrates the observability constraint of a ZIB that forms a set of "n" numbers of buses along with its adjacent buses:

$$\sum_{q=1}^{n} f_q \geq n-1 \qquad (5.3)$$

Rule 3: If "m" number of ZIBs are linked directly or via a single non-ZIB between two buses and all the ZIBs along with their adjacent buses form a set of "n" numbers of buses, the observability constraint of the entire set is represented by the following inequality:

$$\sum_{q=1}^{n} f_q \geq n-m \qquad (5.4)$$

Rule 4: If a particular bus is contained in at least one set of ZIB, then the observability constraint f_q associated with that bus is retained unaltered for one set and replaced by zero for others.

Rule 5: If a particular bus is not part of any ZIB set, the observability constraint f_q associated with that bus should be retained unaltered.

These rules illustrated above modify the f_o vector of Eq. (5.2) by updating the bus connectivity matrix. A detailed explanatory example can be found in [18] that illustrates the transformation of the bus connectivity matrix after incorporating the presence of the network passive measurements or ZIB.

5.2.2 Phasor measurement units channel limit incorporation

The OPP formulation illustrated in Section 5.2.1 considers the PMU with an adequate number of channels that can measure the current phasors of all adjacent branches; hence, it can observe all adjacent buses. The PMUs, however, are manufactured with a certain number of channels that control their costs. Such channel limitation plays a crucial role in determining the required number of PMUs for complete network observability. This section incorporates PMU channel limitation in the OPP formulation. If a particular PMU with L number of channels is installed at bus k of an electric network that is connected to N_k number of buses and $L \geq N_k$; then, the installed PMU will observe the PMU installed bus and all connected buses as stated in Rule 1. Therefore, the row of the connectivity matrix associated with bus k should be kept unaltered. On the contrary, if $L < N_k$, the respective row will be replaced by L combinations of N_k number of rows (R_k) for the kth bus that can be achieved from the following equation [18]:

$$R_k = \begin{cases} \dfrac{N_k!}{(N_k - L)! \, L!} & N_k > L \\ 1 & N_k \leq L \end{cases} \qquad (5.5)$$

Now, the formulated OPP problem is well defined with the incorporation of the passive measurements or ZIB and PMU channel limitations that can be solved using the MILP technique. The built-in MILP function in different software packages for example, MATLAB, delivers only one solution amongst all possible solutions. However, the metaheuristic technique can facilitate the achievement of all possible solutions through a small adjustment in the cost function. The following equation represents the mentioned adjustment on the developed OPP formulation:

Minimize $F(x)$

$$F(x) = \sum_{i=1}^{N_B} C_{\text{PMU}} y_i + \sum_{i=1}^{N_B} P_i u_i \qquad (5.6)$$

Where,

P_i: Imposed penalty factor if the ith bus is not observable. It is set to a very high value, for instance, $P_i = C_{\text{PMU}} N_B$ as the main objective of the OPP formulation is to guarantee complete grid observability.

u_i: A binary number that indicates the observability or un-observability of the ith bus where the values "1" and "0" represent bus unobservability and observability, respectively.

Another minor adjustment in the cost function can assist in determining the solution that provides the solution with the maximum number of measurements. It is noteworthy to mention that such modification does not disturb complete network observability; instead deploys an equal number of PMUs in the network under investigation. It just rearranges the locations of PMUs to ensure the maximum number of measurements. However, more measurements enhance the monitoring reliability of the power system network due to the availability of redundant measurements from a few buses. The following equation represents the mentioned adjustment for achieving more measurements and can be solved using metaheuristic optimization algorithms:

Minimize $F(x)$

$$F(x) = \sum_{i=1}^{N_B} C_{\text{PMU}} y_i + \sum_{i=1}^{N_B} P_i u_i - \sum_{i=1}^{N_B} Q_i v_i \tag{5.7}$$

Where,

v_i: Difference between expected and obtained values of f_i.

Q_i: Incentive factor for the ith bus if observed more than required. It is set to a low value, for instance, $Q_i = 1/(C_{\text{PMU}} N_B)$ as the incentive should not be as high as the PMU cost.

5.2.3 Power system contingencies incorporation

Finally, the developed OPP formulation incorporates power system contingencies, including the single line outage and the single PMU loss cases. During a single line outage case, each bus of the electric grid must be observed by at least two PMUs; therefore, the associated element of the vector b of Eq. (5.2) should be "2." However, radial line termination points of the network are known as radial buses (RB). Their observability from a single PMU is considered sufficient as they are connected to the grids through a single line only. Besides, the outage of the radial lines does not affect the observability of the remaining portion of the power system network. Thus, the RB associated b vector elements should be "*1*" even during the single line outage case [18–20]. In the same way, the complete observability of an electric network will be ensured during a single PMU loss case if all buses of the network are observed by at least two PMUs simultaneously. Thus, all the elements of vector b should be "2" for such contingency case [18–20].

5.3 Network observability and measurement redundancy illustration

The MILP and the GWO techniques are employed in the MATLAB environment to solve the developed OPP formulation. Detailed discussions on these techniques are presented in Chapter 2. This section demonstrates the complete network observability of the considered OPP formulation employing both solution methodologies. It also confirms the superiority of the metaheuristic technique over the MILP technique in achieving better measurement redundancy. Fig. 5.2 presents a seven-bus test network topology as adopted

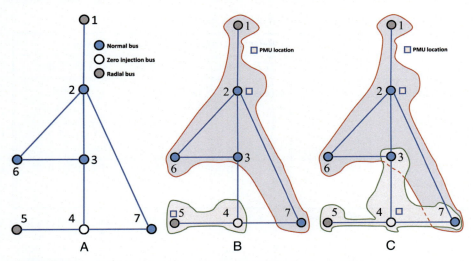

Fig. 5.2 The seven-bus test system topology, PMU placement, and achieved measurement illustration. *GWO*, grey wolf optimization; *MILP*, mixed-integer linear programming; *PMU*, phasor measurement unit.

from Ref. [19]. The adopted topology has four normal buses (bus # 2, 3, 5, and 7), two RBs (bus # 1 and 5), and one ZIB (bus # 4), as can be observed from Fig. 5.2A.

(A) Test network→(B) PMU placement with MILP→(C) PMU placement with GWO

The formulated OPP was applied on this test network using the MILP technique that placed PMUs on Bus # 2 and Bus # 5 to ensure full network observability, as illustrated in Fig. 5.2B. It can be seen that five buses (Bus # 1, 2, 3, 6, and 7) can be observed from the PMU placed at Bus # 2. Likewise, two buses (Bus # 4 and 5) can be observed from the PMU placed at Bus # 5. Therefore, the MILP technique ensured full network observability by placing two PMUs while none of the buses have been observed more than once. Obtained results using the GWO algorithm have been presented in Fig. 5.2C, which shows that with an equal number of PMU placement, two buses (Bus # 3 and 4) can be observed twice. In contrast, remaining buses were observed at least once. In summary, the GWO algorithm achieved nine measurements with two redundant measurements, while the MILP technique achieved seven measurements without any redundant measurements. Therefore, it can be concluded that the GWO metaheuristic technique can provide an equal or greater number of measurements than the MILP technique with the same number of installed PMUs. This observation will be more evident in the subsequent sections.

5.4 Transmission network observability

The efficacy of the modified OPP formulation has been investigated on four IEEE benchmarked transmission networks for example, 14-bus, 30-bus, 39-bus, and 57-bus networks. The required information regarding the test networks is summarized in

Table 5.1 IEEE benchmark transmission networks specifications [17–19].

	Selected transmission networks			
	IEEE 14-bus	IEEE 30-bus	IEEE 39-bus	IEEE 57-bus
# Number of ZIB (location: bus number)	1 (bus 7)	6 (buses 6, 9, 22, 25, 27, 28)	12 (buses 1, 2, 5, 6, 9, 10, 11, 13, 14, 17, 19, 22)	15 (buses 4, 7, 11, 21, 22, 24, 26, 34, 36, 37, 39, 40, 45, 46, 48)
# Number of RB (location: bus number)	1 (bus 8)	3 (buses 11, 13, 26)	9 (buses 30, 31, 32, 33, 34, 35, 36, 37, 38)	1 (bus 33)

RB, radial buses; ZIB, zero injection bus.

Table 5.1, and their detailed information can be explored from Ref. [30]. It is noteworthy that the buses related to any type of generation or load are regarded as the ZIB, and the remaining buses are considered as the non-ZIB.

Various simulation cases have been provided to illustrate the efficacy of the considered OPP formulation using the GWO and the MILP techniques. The objective is to attain the maximum number of measurements with the deployment of a minimal number of PMUs subjected to full grid observability. The investigated cases are given as follows:

1. Complete network observability at normal operating mode.
2. Complete network observability under contingency cases.
3. Complete network observability considering PMU channel limit.

5.4.1 Complete network observability at normal operating mode

This section provides the results achieved from the OPP formulation developed for the selected benchmark transmission networks at normal operating mode with and without considering the presence of ZIB. Table 5.2 shows the required number of PMUs, their locations, installation costs, and the needed and achieved number of measurements without considering passive measurements or ZIB in the networks. It is worth mentioning that the summation of the final "b" vector elements yields the required number of measurements. On the contrary, the summation of the elements of the production of the final "A" and "Y" matrices yields the obtained number of measurements. It can be observed that the GWO approach achieved more measurements than that of the MILP approach for most of the investigated cases with the same number of PMU placement by rearranging their locations. For example, GWO and MILP achieved 19 and 14 measurements, respectively, by installing 4 PMUs at different network buses of the IEEE 14-bus test network while it was required to achieve 14 measurements for complete network observability. Thus, the GWO achieved five redundant measurements while MILP did not accomplish any. This example

Table 5.2 Required number of PMUs, locations, and installation cost for the IEEE benchmark transmission network under normal operating mode without considering ZIB.

Test network	Method	Required PMU	Installation cost (unit)	Measurements Achieved	Measurements Required	PMU locations (# Bus)
IEEE 14-bus	GWO	4	3.643	19	14	2, 6, 7, 9
	MILP		4.000	14		2, 8, 10, 13
IEEE 30-bus	GWO	10	9.333	50	30	2, 3, 6, 9, 10, 12, 15, 19, 25, 27
	MILP		9.833	35		1, 5, 8, 10, 11, 12, 19, 23, 26, 29
IEEE 39-bus	GWO	13	12.667	52	39	2, 6, 9, 10, 11, 14, 17, 19, 20, 22, 23, 25, 29
	MILP		12.872	44		2, 6, 9, 12, 14, 17, 22, 23, 29, 32, 33, 34, 37
IEEE 57-bus	GWO	17	16.737	72	57	1, 4, 6, 9, 15, 20, 24, 28, 31, 32, 36, 38, 39, 41, 47, 50, 53
	MILP		16.877	64		2, 6, 12, 19, 22, 25, 27, 32, 36, 38, 41, 45, 46, 50, 52, 54, 57

GWO, grey wolf optimization; MILP, mixed-integer linear programming; PMUs, phasor measurement units; ZIB, zero injection bus.

demonstrates better efficacy of the metaheuristic method over the MILP method in obtaining more measurements; hence, increasing measurement redundancy. Besides, the cost of installing the PMUs has been reduced for the GWO method over the MILP method as the measurement redundancy problem has been integrated with the objective function of Eq. (5.7).

However, the GWO approach is computationally more burdensome than the MILP as the former one is iterative and needs to update grey wolves' positions in each iteration. Such computational burden is acceptable as the OPP problems are considered as power system planning problems and solved offline. Table 5.3 presents similar observations (required number of PMUs, their locations, cost of installation, and achieved/required number of measurements) considering the presence of passive measurements or ZIB in the tested networks. It can be observed that the incorporation of the ZIB reduces the required number of PMUs for full network observability significantly for both solution strategies. For instance, the IEEE 57-bus test network

Table 5.3 Required number of PMUs, locations, and installation cost for the IEEE benchmark transmission network under normal operating mode considering ZIB.

Test network	Method	Required PMU	Installation cost (unit)	Measurements Achieved	Measurements Required	PMU locations (# bus)
IEEE 14-Bus	GWO	3	2.857	15	13	2, 6, 9
	MILP		2.857	15		2, 6, 9
IEEE 30-Bus	GWO	6	5.767	31	24	2, 4, 10, 12, 15, 18
	MILP		5.933	26		2, 4, 12, 17, 19, 24
IEEE 39-Bus	GWO	7	6.897	31	27	2, 6, 16, 20, 23, 25, 29
	MILP		6.923	30		2, 3, 16, 20, 23, 25, 29
IEEE 57-Bus	GWO	11	10.895	48	42	1, 4, 9, 20, 25, 29, 32, 38, 51, 53, 56
	MILP		10.930	46		1, 6, 13, 20, 22, 25, 29, 32, 51, 54, 56

GWO, grey wolf optimization; MILP, mixed-integer linear programming; PMUs, phasor measurement units; ZIB, zero injection bus.

required only 11 PMUs after incorporating the ZIB for entire grid observability, while it needed 17 PMUs without the ZIB incorporation. Similar observations can be seen for other test networks as well. Therefore, the integration of the ZIB to the considered OPP formulation has been justified. However, like the previous case (Table 5.2), the GWO achieved more measurements, which increased the measurement redundancy over the MILP method by repositioning the same number of PMUs in the networks under investigation.

5.4.2 Complete network observability under contingency

This section provides the results achieved from the OPP formulation developed under contingency cases with and without considering the presence of ZIB or passive measurements in the selected test networks. Table 5.4 presents the required number of PMUs, their locations, and installation costs, and the needed/achieved number of measurements without accounting for the presence of ZIB. It can be seen that the required number of PMUs has been increased compared to the observations of Table 5.2 as power system contingencies have been considered. For example, it requires 8 and 9 PMUs for full grid observability of the IEEE 14-bus network under single PMU loss and single line outage cases, respectively. However, for the same network, only 4 PMUs were required at normal operating mode. Table 5.5 presents results obtained

Table 5.4 Required number of PMUs and their installation cost for the IEEE benchmark transmission networks under contingencies without considering ZIB.

Test network	Contingency	Method	Required PMU	Installation cost (unit)	Measurements Achieved	Measurements Required
IEEE 14-bus	Single line outage	GWO	8	7.286	37	27
		MILP		7.571	33	
	Single PMU loss	GWO	9	8.214	39	28
		MILP		8.357	37	
IEEE 30-bus	Single line outage	GWO	18	17.267	79	57
		MILP		17.767	64	
	Single PMU loss	GWO	21	20.167	85	60
		MILP		20.567	73	
IEEE 39-bus	Single line outage	GWO	19	18.769	78	69
		MILP		18.897	73	
	Single PMU loss	GWO	28	27.539	96	78
		MILP		27.539	96	
IEEE 57-bus	Single line outage	GWO	32	31.737	128	113
		MILP		31.807	124	
	Single PMU loss	GWO	33	32.719	130	114
		MILP		32.772	127	

GWO, grey wolf optimization; MILP, mixed-integer linear programming; PMUs, phasor measurement units; ZIB, zero injection bus.

Table 5.5 Required number of PMUs and their installation cost for the IEEE benchmark transmission networks under contingencies considering ZIB.

Test network	Contingency	Method	Required PMU	Installation cost (unit)	Measurements Achieved	Measurements Required
IEEE 14-bus	Single line outage	GWO	7	6.500	33	26
		MILP		6.500	33	
	Single PMU loss	GWO	7	6.500	33	26
		MILP		6.500	33	
IEEE 30-bus	Single line outage	GWO	12	11.733	55	47
		MILP		11.900	50	
	Single PMU loss	GWO	13	12.700	57	48
		MILP		12.867	52	
IEEE 39-bus	Single line outage	GWO	12	11.949	52	50
		MILP		11.949	52	
	Single PMU loss	GWO	16	15.846	60	54
		MILP		15.897	58	
IEEE 57-bus	Single line outage	GWO	21	20.9123	88	83
		MILP		20.947	86	
	Single PMU loss	GWO	22	21.877	91	84
		MILP		21.930	88	

GWO, grey wolf optimization; MILP, mixed-integer linear programming; PMUs, phasor measurement units; ZIB, zero injection bus.

employing GWO and MILP approaches considering the presence of ZIB for the selected networks under contingency cases. It also required more PMUs for complete grid observability than the required PMUs given in Table 5.3 as the contingencies are considered. Due to the incorporation of ZIB, Table 5.5 requires a smaller number of PMUs than that of Table 5.4 for the same network under the same contingency case. Therefore, the incorporation of the passive measurements into the OPP formulation has been justified again even under contingency cases.

The GWO metaheuristic technique outperformed the MILP approach in achieving more measurements with the placement of an equal number of PMUs for all selected networks with contingency cases. The discussions of Tables 5.4 and 5.5 would be similar to that of Tables 5.2 and 5.3 as similar observations and deductions can be made. Hence, these tables (Tables 5.4 and 5.5) are not discussed in detail. Also, for the sake of brevity, the PMU locations are not presented. However, the presented results have shown the ability of the GWO method in attaining more or at least an equal number of measurements by guaranteeing the full grid observability for the chosen test transmission networks without and with the incorporation of the passive measurements. As a result, it can be deduced that stochastic optimization methods are more beneficial for obtaining better measurement redundancy to the power system decision-makers over the linear programming-based methods.

5.4.3 Complete network observability considering phasor measurement units channel limit

This section presents the OPP formulation results considering the PMU channel limitations. For brevity, this section refrained from providing repetitive tables and discussions; instead, all the cases have been summarized in Table 5.6 for the normal operating mode and in Table 5.7 for contingency cases. The PMU installation costs were not shown as they can be calculated using Eq. (5.7). However, these tables (Tables 5.6 and 5.7) present the corresponding and available results of referenced works for comparison purposes. It can be observed from most of the investigated cases that the GWO obtained the same or more number of measurements than those of the referenced works in most of the cases. Therefore, the presented results justified the efficacy of the presented OPP formulation and the employed solution methodology.

The single PMU loss considering the availability of only one PMU channel case requires the maximum number of PMUs for full grid observability with all the investigated networks as it represents the most severe contingency case with the minimum number of PMU channels (Table 5.7). Besides, it is also observed from Tables 5.6 and 5.7 that with the increase of the number of PMU channels, the required number of PMUs decreases for all investigated networks. Nonetheless, the presented results reveal that the installation of PMUs with more than four channels does not reduce the required PMUs for the selected transmission networks; instead, it increases the overall cost as with the number of PMU channels increases. Therefore, it is neither economical nor necessary to produce or install PMUs with more than four channels for the networks investigated in this chapter.

Table 5.6 Comparisons among different techniques considering channel limits of PMU under normal operating mode.

Channel limit	Method	\multicolumn{8}{c}{IEEE test transmission network}
		\multicolumn{2}{c}{14-bus}	\multicolumn{2}{c}{30-bus}	\multicolumn{2}{c}{39-bus}	\multicolumn{2}{c}{57-bus}
		RP[a]	AM[b]	RP	AM	RP	AM	RP	AM
1	GWO	7	14	12	24	14	28	21	42
	ILP [20]	7	-	12	-	14	-	21	-
	BIP [26]	7	-	14	-	-	-	21	-
	BIP [22]	7	-	14	-	-	-	23	-
2	GWO	5	15	8	24	9	27	14	42
	ILP [20]	5	-	8	-	9	-	14	-
	BIP [26]	5	-	9	-	-	-	14	-
	BIP [22]	5	-	9	-	-	-	16	-
3	GWO	4	16	7	27	8	31	12	45
	CF-PSO [29]	4	16	7	27	8	31	12	45
	ILP [20]	4	-	7	-	8	-	12	-
	BIP [26]	4	-	7	-	-	-	12	-
	BIP [22]	4	-	8	-	-	-	13	-
4	GWO	3	15	6	28	7	31	11	46
	CF-PSO [29]	3	15	6	30	7	31	11	47
	ILP [20]	3	-	6	-	7	-	11	-
	BIP [26]	3	-	7	-	-	-	11	-
	BIP [22]	3	-	7	-	-	-	12	-
5	GWO	3	15	6	30	7	31	11	47
	CF-PSO [29]	3	15	6	30	7	31	11	47
	ILP [20]	3	-	6	-	7	-	11	-
	BIP [26]	3	-	7	-	-	-	11	-
	BIP [22]	3	-	7	-	-	-	12	-

AM, achieved measurement; BIP, binary integer programming; CF-PSO, constriction factor particle swarm optimization; GWO, grey wolf optimization; ILP, integer linear programming; PMU, phasor measurement unit; RP, required number of PMU.

5.5 Distribution network observability

The efficacy of the developed OPP formulation is also tested on three IEEE standard test distribution networks/feeders, namely IEEE 13-node, 34-node, and 37-node feeders employing the GWO and MILP approaches. Table 5.8 summarizes the required specifications of the test feeders. The detailed data of these networks can be found in Ref. [31]. In distribution feeders, there are two types of loads: the spot and distributed loads. The spot loads are connected to the buses while the distributed loads are

Table 5.7 Comparisons among different techniques considering channel limits of PMU under contingency cases.

| Channel limit | Operating mode | Method | \multicolumn{8}{c}{IEEE test transmission network} |
			14-bus RP[a]	14-bus AM[b]	30-bus RP	30-bus AM	39-bus RP	39-bus AM	57-bus RP	57-bus AM
1	Single line outage	GWO	13	26	24	48	25	50	42	84
		ILP [20]	13	-	23	-	23	-	42	-
	Single PMU loss	GWO	13	26	24	48	27	54	42	84
		CF-PSO [29]	13	26	24	48	27	54	42	84
		ILP [20]	13	-	24	-	27	-	42	-
2	Single line outage	GWO	9	27	16	48	17	51	28	84
		ILP [20]	9	-	15	-	15	-	28	-
	Single PMU loss	GWO	9	27	16	48	19	56	28	84
		ILP [20]	9	-	17	-	20	-	29	-
3	Single line outage	GWO	7	28	13	50	13	51	23	84
		CF-PSO [29]	7	28	13	50	13	51	23	84
		ILP [20]	7	-	13	-	12	-	23	-
	Single PMU loss	GWO	7	28	13	50	16	55	24	86
		CF-PSO [29]	7	28	13	50	16	55	24	86
		ILP [20]	7	-	14	-	17	-	24	-
		CGA [28]	8	-	17	-	-	-	30	-
4	Single line outage	GWO	7	32	12	51	12	52	21	85
		CF-PSO [29]	7	32	12	51	12	52	21	85
		ILP [20]	7	-	12	-	12	-	21	-
	Single PMU loss	GWO	7	32	12	51	16	62	22	87
		CF-PSO [29]	7	32	12	51	16	62	22	87
		ILP [20]	7	-	13	-	16	-	22	-
		CGA [28]	8	-	17	-	-	-	30	-
5	Single line outage	GWO	7	33	12	52	12	52	21	89
		CF-PSO [29]	7	33	12	52	12	52	21	89
		ILP [20]	7	-	11	-	12	-	21	-
	Single PMU loss	GWO	7	33	13	54	16	60	22	91
		CF-PSO [29]	7	33	13	54	16	60	22	91
		ILP [20]	7	-	13	-	16	-	22	-
		CGA [28]	8	-	17	-	-	-	30	-

AM, achieved measurement; CF-PSO, constriction factor particle swarm optimization; CGA, cellular genetic algorithm; GWO, grey wolf optimization; ILP, integer linear programming; PMU, phasor measurement unit; RP, required number of PMU.

Table 5.8 IEEE benchmark distribution networks specifications [5,31].

	Test distribution networks		
	IEEE 13-node	**IEEE 34-node**	**IEEE 37-node**
# Number of ZIB (Location: bus number)	3 (buses 633, 680, 684)	5 (buses 812, 814, 850, 852, 888)	11 (buses 702, 703, 704, 705, 706, 707, 708, 709, 710, 711, 775)
# Number of RB (Location: bus number)	6 (buses 611, 634, 646, 652, 675, 680)	9 (buses 810, 822, 826, 838, 840, 848, 856, 864, 890)	15 (buses 712, 718, 722, 724, 725, 728, 729, 731, 732, 735, 736, 740, 741, 742, 775)

RB, radial buses; ZIB, zero injection bus.

scattered between two buses. Like transmission networks, the buses associated with any load or generation are considered as the non-ZIB and others as the ZIB.

To minimize the required number of PMUs and maximize the number of measurements while guaranteeing entire grid observability, similar cases have been considered for the chosen test distribution feeders as done earlier for the transmission test systems.

5.5.1 Complete network observability at normal operating mode

This section presents the OPP formulation results for the selected benchmarked distribution feeders at normal operating mode with and without considering the presence of the passive measurements or ZIB. Table 5.9 provides the required number of PMUs, their locations, installation costs, and the needed/obtained number of measurements without considering the presence of ZIB. In contrast, Table 5.10 presents similar observations considering the presence of ZIB. It can be observed from Tables 5.9 and 5.10 that the GWO approach achieved more measurements than those of the MILP approach for most of the cases with the same number of PMUs deployment by rearranging their locations. For instance, it was required to achieve 13 measurements for the IEEE 13-node test distribution feeder at normal operating mode as given in Table 5.9. The GWO and MILP achieved 23 and 14 measurements, respectively, by installing 6 PMUs at different feeder nodes. Thus, GWO achieved 10 redundant measurements while MILP achieved only one. In addition, the PMU installation cost has been lowered for the GWO method over the MILP method as the measurement redundancy issue has been integrated with the objective function as given in Eq. (5.7). This example demonstrates the efficacy of the metaheuristic approach and its superiority over the MILP approach in obtaining more measurements; hence, increasing measurement redundancy.

However, incorporating the ZIB on the chosen distribution feeder lowered the required number of PMUs significantly for full feeder observability for both solution strategies. For instance, the IEEE 37-node test feeder requires only eight PMUs after ZIB consideration, while it needs 12 PMUs without considering the presence of ZIB. Similar observations can be seen for other test feeders as well.

Table 5.9 Required number of PMUs, their locations, the installation cost for the IEEE benchmark distribution network under normal operating mode without considering ZIB and PMU channel limit.

Test network	Method	Required PMU	Installation cost (unit)	Measurements Achieved	Measurements Required	PMU locations (bus #)
IEEE 13-bus	GWO	6	5.231	23	13	632, 633, 645, 671, 684, 692
	MILP		5.923	14		634, 646, 650, 675, 680, 684
IEEE 34-bus	GWO	12	11.765	42	34	802, 808, 814, 820, 824, 834, 836, 846, 854, 858, 862, 888
	MILP		11.941	36		800, 808, 820, 824, 836, 838, 842, 848, 850, 854, 864, 888
IEEE 37-bus	GWO	12	11.730	47	37	701, 702, 705, 706, 707, 708, 709, 710, 711, 714, 734, 744
	MILP		11.811	44		702, 705, 707, 708, 709, 710, 711, 714, 725, 737, 744, 799

GWO, grey wolf optimization; MILP, mixed-integer linear programming; PMUs, phasor measurement units; ZIB, zero injection bus.

Hence, the presented results endorsed the incorporation of the presence of the passive measurements or ZIB.

5.5.2 Complete network observability under contingency

This section provides the results achieved for OPP formulation developed under contingency cases on the selected distribution feeder. Tables 5.11 and 5.12 present the obtained results under contingency cases without incorporating ZIB. It can be seen that the required number of PMUs has been increased as given in Tables 5.11 and 5.12 compared to the results given in Tables 5.9 and 5.10. For example, the feeder requires seven and 13 PMUs for full grid observability of IEEE 13-node under single PMU loss and single line outage cases, respectively, as given in Table 5.11. In contrast, only six PMUs are required for the same network at normal operating mode without considering ZIB (Table 5.9). However, the required number of PMUs is reduced to six and seven as given in Table 5.12 for single PMU loss and single line outage cases, respectively, which is lower than that of the observations

Table 5.10 Required number of PMUs, their locations, the installation cost for the IEEE benchmark distribution network under normal operating mode considering ZIB without considering PMU channel limit.

Test network	Method	Required PMU	Installation cost (unit)	Measurements Achieved	Measurements Required	PMU locations (bus #)
IEEE 13-bus	GWO	4	3.539	16	10	632, 645, 671, 692
	MILP		3.615	15		632, 645, 684, 692
IEEE 34-bus	GWO	10	9.794	36	29	802, 808, 820, 824, 834, 836, 846, 854, 858, 862
	MILP		9.882	33		800, 808, 820, 824, 836, 838, 844, 846, 854, 858
IEEE 37-bus	GWO	8	7.838	32	26	701, 702, 709, 710, 711, 714, 734, 744
	MILP		7.838	32		701, 702, 709, 710, 711, 714, 734, 744

GWO, grey wolf optimization; MILP, mixed-integer linear programming; PMUs, phasor measurement units; ZIB, zero injection bus.

of Table 5.11, which endorsed the incorporation of passive measurement in the test distribution feeders as well. Similar to previous analysis, both approaches provide same number of PMU deployment for full observability of the test distribution feeders. The GWO method provides more or at least the same number of measurements for all investigated cases. Therefore, the presented results validate the superior performance of the GWO method over the MILP method in obtaining more measurement redundancy.

5.5.3 Complete network observability considering phasor measurement unit channel limit

Like transmission network cases, the efficacy of the considered OPP formulation was also investigated considering PMU channel limitation for the distribution feeders. Therefore, this section provides the obtained results considering one to five channels of the deployed PMUs and the presence of ZIB for the test distribution feeders. For the sake of brevity and to avoid repetition of similar results and discussions, this section avoids presenting those results in detail; instead, all the cases have been summarized in Table 5.13 for the normal operating mode and Table 5.14 for the contingency

Table 5.11 Required number of PMUs and their installation cost for the IEEE benchmark distribution networks under contingencies without considering ZIB.

Test network	Contingency	Method	Required PMU	Installation cost (Unit)	Measurements Achieved	Measurements Required
IEEE 13-bus	Single line outage	GWO	7	6.615	25	20
		MILP		6.769	23	
	Single PMU loss	GWO	13	12.154	37	26
		MILP		12.154	37	
IEEE 34-bus	Single line outage	GWO	19	18.853	64	59
		MILP		18.912	62	
	Single PMU loss	GWO	27	26.677	79	68
		MILP		26.677	79	
IEEE 37-bus	Single line outage	GWO	18	17.784	67	59
		MILP		17.865	64	
	Single PMU loss	GWO	31	30.568	90	74
		MILP		30.568	90	

GWO, grey wolf optimization; MILP, mixed-integer linear programming; PMUs, phasor measurement units; ZIB, zero injection bus.

cases. Tables 5.13 and 5.14 also present the results of the referenced works for comparison purposes. It can be observed for most of the investigated cases that the GWO provides the same or more number of measurements than that of the reported works in most cases. Therefore, the presented results justified the efficacy of the developed OPP formulation and the employed solution methodology.

After investigating the results of Tables 5.13 and 5.14, it can be seen that the required number of PMUs reduces with the increase of available PMU channels for

Table 5.12 Required number of PMUs and their installation cost for the IEEE benchmark distribution networks under contingencies considering ZIB.

Test network	Contingency	Method	Required PMU	Installation cost (unit)	Measurements Achieved	Measurements Required
IEEE 13-bus	Single line outage	GWO	6	5.692	22	18
		MILP		5.769	21	
	Single PMU loss	GWO	7	6.923	21	20
		MILP		6.923	21	
IEEE 34-bus	Single line outage	GWO	15	14.971	51	50
		MILP		14.971	51	
	Single PMU loss	GWO	23	22.706	68	58
		MILP		22.765	66	
IEEE 37-bus	Single line outage	GWO	14	13.919	52	49
		MILP		13.919	52	
	Single PMU loss	GWO	16	15.919	55	52
		MILP		15.919	55	

GWO, grey wolf optimization; MILP, mixed-integer linear programming; PMUs, phasor measurement units; ZIB, zero injection bus.

Table 5.13 Comparisons among different techniques considering channel limits of PMU under normal operating mode.

Channel limit	Method	13-node RP[a]	13-node AM[b]	34-node RP	34-node AM	37-node RP	37-node AM
1	GWO	5	10	15	30	14	28
	BSA [5]	5	10	15	30	14	28
2	GWO	4	12	11	33	9	27
	BSA [5]	4	12	11	33	9	27
3	GWO	4	14	10	36	8	30
	BSA [5]	4	14	10	36	8	30
4	GWO	4	16	10	36	8	32
	BSA [5]	4	16	10	36	8	32
5	GWO	4	16	10	36	8	32
	BSA [5]	4	16	10	36	8	32

AM, achieved measurement; BSA, backtracking search algorithm; GWO, grey wolf optimization; PMU, phasor measurement unit; RP, required number of PMU.

Table 5.14 Comparisons among different techniques considering channel limits of PMU under contingency cases.

Channel Limit	Operating Mode	Method	13-node RP[a]	13-node AM[b]	34-node RP	34-node AM	37-node RP	37-node AM
1	Single line outage	GWO	9	18	26	52	25	50
		BSA [5]	9	18	26	52	25	50
	Single PMU loss	GWO	10	20	30	60	27	54
		BSA [5]	10	20	30	60	27	54
2	Single line outage	GWO	6	18	17	50	17	50
		BSA [5]	6	18	17	50	17	50
	Single PMU loss	GWO	8	22	23	65	19	55
		BSA [5]	8	22	23	65	19	55
3	Single line outage	GWO	5	18	15	51	14	50
		BSA [5]	5	18	15	51	14	50
	Single PMU loss	GWO	7	22	23	68	16	53
		BSA [5]	7	22	23	68	16	53
4	Single line outage	GWO	6	22	15	51	14	52
		BSA [5]	6	22	15	51	14	52
	Single PMU loss	GWO	7	21	23	68	16	55
		BSA [5]	7	21	23	68	16	55
5	Single line outage	GWO	6	22	15	51	14	52
		BSA [5]	6	22	15	51	14	52
	Single PMU loss	GWO	7	21	23	68	16	55
		BSA [5]	7	21	23	68	16	55

AM, achieved measurement; BSA, backtracking search algorithm; GWO, grey wolf optimization; PMU, phasor measurement unit; RP, required number of PMU.

the chosen distribution test feeders. Besides, it should be noted the required number of PMUs was not reduced; instead, PMU installation cost was increased while channels were increased from four to five. Therefore, installing PMUs with more than four channels for the investigated distribution networks is neither cost-effective nor necessary. The power system decision-makers and PMU producers may find this observation meaningful while developing the wide-area measurement and monitoring scheme considering PMU channel availability.

5.6 Summary

This chapter presented an improved and effective OPP formulation considering PMU channel limitations and incorporating network passive measurements, aiming to minimize PMU requirements and maximize the number of measurements subjected to full network observability. It also extended the developed formulation to incorporate different contingency cases associated with electricity grids for example, single PMU loss and single line outage cases. Then, it employed the GWO algorithm to solve the developed OPP formulation on several IEEE standard transmission networks and distribution feeders and compared the obtained results with the MILP method and the existing works. The presented results exhibited the GWO technique superiority over other techniques as it ensured entire grid observability with smaller or at most an equal number of PMU installation. The results also showed the superior performance of the GWO method over the MILP method for obtaining more measurements for most of the investigated cases. Finally, it is worth noting that deployment of PMU with more than four channels does not minimize the number of PMU requirements for the investigated transmission networks and distribution feeders; instead, it increased the investment costs.

References

[1] H. Zaheb, et al., A contemporary novel classification of voltage stability indices, Appl. Sci. 10 (5) (2020) 1639, doi:10.3390/app10051639.

[2] S.D. Ahmed, F.S.M. Al-Ismail, M. Shafiullah, F.A. Al-Sulaiman, I.M. El-Amin, Grid integration challenges of wind energy: a review, IEEE Access, 8, Institute of Electrical and Electronics Engineers Inc., New Jersey, USA, 2020, pp. 10857–10878, doi:10.1109/ACCESS.2020.2964896.

[3] M. Ilius Hasan Pathan, M. Juel Rana, M. Shoaib Shahriar, M. Shafiullah, M. Hasan Zahir, A. Ali, Real-time LFO damping enhancement in electric networks employing PSO optimized ANFIS, Invent 5 (4) (2020) 61, doi:10.3390/inventions5040061.

[4] X. Ji, et al., Real-time robust forecasting-aided state estimation of power system based on data-driven models, Int. J. Electr. Power Energy Syst. 125 (Feb) (2021) 106412, doi:10.1016/j.ijepes.2020.106412.

[5] M. Shafiullah, M. Abido, M. Hossain, A. Mantawy, An improved OPP problem formulation for distribution grid observability, Energies 11 (11) (2018) 3069, doi:10.3390/en11113069.

[6] M. Shafiullah, S.M. Rahman, M.G. Mortoja, B. Al-Ramadan, Role of spatial analysis technology in power system industry: an overview, Renew. Sustain. Energy Rev. 66 (Dec) (2016) 584–595, doi:10.1016/j.rser.2016.08.017.
[7] F. Yang, Z. Ling, M. Wei, T. Mi, H. Yang, R.C. Qiu, Real-time static voltage stability assessment in large-scale power systems based on spectrum estimation of phasor measurement unit data, Int. J. Electr. Power Energy Syst. 124 (Jan) (2021) 106196, doi:10.1016/j.ijepes.2020.106196.
[8] G.L. Kusic, W.E. McGahey, M. Lehtonen, Measurement of power system phase differences by means of GPS timing, 2016. In: Electric Power Quality and Supply Reliability (PQ), IEEE, New Jersey, USA, 2016, pp. 297–300, doi:10.1109/PQ.2016.7724130.
[9] P. Nanda, C.K. Panigrahi, A. Dasgupta, Phasor estimation and modelling techniques of PMU- a review, Energy Procedia 109 (Mar) (2017) 64–77, doi:10.1016/j.egypro.2017.03.052.
[10] D. Georges, Optimal PMU-based monitoring architecture design for power systems, Control Eng. Pract. 30 (Sept) (2014) 150–159, doi:10.1016/j.conengprac.2013.11.019.
[11] A.E. Labrador Rivas, T. Abrão, Faults in smart grid systems: Monitoring, detection and classification, Electr. Power Syst. Res. 189 (Dec) (2020) 106602, doi:10.1016/j.epsr.2020.106602.
[12] C. Qin, Z. Nie, P. Banerjee, A.K. Srivastava, End-to-end remote field testing of phasor measurement units using phasor measurement unit performance analyzer test suite, IEEE Trans. Ind. Appl. 56 (6) (2020) 7067–7076, doi:10.1109/TIA.2020.3019994.
[13] W. Yao, et al., A novel method for phasor measurement unit sampling time error compensation, IEEE Trans. Smart Grid 9 (2) (2018) 1063–1072, doi:10.1109/TSG.2016.2574946.
[14] CRE, Phasor measurement unit for monitoring power systems, 2020. https://phoenix-h2020.eu/phasor-measurement-unit-for-monitoring-power-systems/. (Accessed 24 Feb, 2021).
[15] General Protection, Phasor measurement unit (PMU), 2021. https://new.siemens.com/global/en/products/energy/energy-automation-and-smart-grid/protection-relays-and-control/general-protection/phasor-measurement-unit-pmu.html. (Accessed 24 February 2021).
[16] J. Chen, A. Abur, Improved bad data processing via strategic placement of PMUs, IEEE Power Engineering Society General Meeting (2005) 2759–2763, doi:10.1109/PES.2005.1489694.
[17] A. Enshaee, R.A. Hooshmand, F.H. Fesharaki, A new method for optimal placement of phasor measurement units to maintain full network observability under various contingencies, Electr. Power Syst. Res. 89 (Aug) (2012) 1–10, doi:10.1016/j.epsr.2012.01.020.
[18] E. Abiri, F. Rashidi, T. Niknam, M.R. Salehi, Optimal PMU placement method for complete topological observability of power system under various contingencies, Int. J. Electr. Power Energy Syst. 61 (Oct) (2014) 585–593, doi:10.1016/j.ijepes.2014.03.068.
[19] E. Abiri, F. Rashidi, T. Niknam, An optimal PMU placement method for power system observability under various contingencies, Int. Trans. Electr. Energy Syst. 25 (4) (2015) 589–606, doi:10.1002/etep.1848.
[20] F. Rashidi, E. Abiri, T. Niknam, M.R. Salehi, Optimal placement of PMUs with limited number of channels for complete topological observability of power systems under various contingencies, Int. J. Electr. Power Energy Syst. 67 (May) (2015) 125–137, doi:10.1016/j.ijepes.2014.11.015.
[21] F. Aminifar, A. Khodaei, M. Fotuhi-Firuzabad, M. Shahidehpour, Contingency-constrained PMU placement in power networks, IEEE Trans. Power Syst. 25 (1) (2010) 516–523, doi:10.1109/TPWRS.2009.2036570.

[22] R. Kumar, V.S. Rao, Optimal placement of PMUs with limited number of channels. In: 2011 North American Power Symposium, IEEE, New Jersey, USA, 2011, pp. 1–7, doi:10.1109/NAPS.2011.6024852.
[23] A.H. Al-Mohammed, M.A. Abido, M.M. Mansour, Optimal PMU placement for power system observability using differential evolution, 11th International Conference on Intelligent Systems Design and Applications, IEEE, New Jersey, USA, 2011, pp. 277–282.
[24] M. Shafiullah, M.J. Rana, M.S. Alam, M.A. Uddin, Optimal placement of phasor measurement units for transmission grid observability. In: 2016 International Conference on Innovations in Science, Engineering and Technology (ICISET), IEEE, Chittagong, Bangladesh, 2016, pp. 1–4, doi:10.1109/ICISET.2016.7856492.
[25] M. Dalali, K.H. Kazemi, Optimal PMU placement for full observability of the power network with maximum redundancy using modified binary cuckoo optimisation algorithm, IET Gener. Transm. Distrib. 10 (11) (2016) 2817–2824, doi:10.1049/iet-gtd.2016.0287.
[26] M. Korkali, A. Abur, Placement of PMUs with channel limits. In: 2009 IEEE Power & Energy Society General Meeting, IEEE, New Jersey, USA, 2009, pp. 1–4, doi:10.1109/PES.2009.5275529.
[27] N.M. Manousakis, G.N. Korres, Optimal PMU placement for numerical observability considering fixed channel capacity—a semidefinite programming approach, IEEE Trans. Power Syst. 31 (4) (2016) 3328–3329, doi:10.1109/TPWRS.2015.2490599.
[28] Z. Miljanić, I. Djurović, I. Vujošević, Optimal placement of PMUs with limited number of channels, Electr. Power Syst. Res. 90 (Sept) (2012) 93–98, doi:10.1016/j.epsr.2012.04.010.
[29] M. Shafiullah, M.I. Hossain, M.A. Abido, T. Abdel-Fattah, A.H. Mantawy, A modified optimal PMU placement problem formulation considering channel limits under various contingencies, Measurement 135 (Mar) (2019) 875–885, doi:10.1016/J.MEASUREMENT.2018.12.039.
[30] "Power Systems Test Case Archive - UWEE." https://www2.ee.washington.edu/research/pstca/. (Accessed 19 September 2018).
[31] "Distribution Test Feeders - Distribution test feeder working group - IEEE PES distribution system analysis subcommittee." https://ewh.ieee.org/soc/pes/dsacom/testfeeders/. (Accessed 6 May 2017).

Transmission line parameter and system Thevenin equivalent identification

6.1 Introduction

A reliable power system model can be developed by determining the accurate parameters of the power system network, including transmission line parameters. Such a model is crucial for contingency analysis, power flow analysis, stability analysis, state estimation, network losses analysis, operation and control decision making, protective relay setting calculation, and other purposes [1–6]. In general, the electric utilities compute the transmission line sequence impedances using the electrical and geometric properties of the conductors. The conductor material is usually assumed to be fully paramagnetic. Such approaches can produce considerable errors giving rise to wrong settings on fault locators and distance relays [7–9]. In addition, some lines may involve parts with various conductors having different resistance per unit length. As such, using one resistance value in impedance computation is regarded as the source of errors. Besides, considering the geometric parameters, the arrangement of the conductor can vary from one section to another because of different tower configurations. Moreover, the zero-sequence impedance calculation requires the distance between the conductors to the ground and the ground resistance that is a function of frequency. Accurate estimation of the ground resistivity is difficult due to the nonuniform return path affected by humidity, temperature, and soil composition [7–9].

Like ground resistivity, the transmission line parameters vary with the ambient conditions and operating history [10–13]. Thus, the traditional static line parameter computation methods provide erroneous information [7,8]. Therefore, the researchers explored online computation of the transmission line parameters using the real-time measurement recorded by the advanced metering infrastructures (AMI). The phasor measurement unit (PMU) is the most prominent AMI that provides higher accuracy measurements than other measurement devices. They provide time-synchronized and effective measurements even under various adversities. PMUs are used for various applications in power systems, including parameter identification, state estimation, and fault location [14–21]. Over the past years, many approaches have been reported for transmission line parameter identifications using PMU recorded voltage and current measurements. Reported approaches include Newton-Raphson method [22], two-port ABCD parameter-based method [23], single, double, and multiple measurements based methods [24], least-square method [25–27], concurrent parameter and state estimation method [28], linear estimation theory [29], and nonlinear estimation theory [30]. In [31], the line parameters are estimated using an iterative fault location algorithm within the algorithm itself. However, many of the reported methods can accurately estimate line parameters for the fully transposed transmission lines without measurement noise [29]. Besides, the PMU-based data acquisition process can be contaminated by different noises for various reasons, including data conversion or communication device abnormalities, instrumentation channel anomalies, current transformer saturation errors, system imbalances, etc. Noises severely affect the measurements in short transmission lines compared to the measurements in long transmission lines [5,24,32]. In addition to the issues mentioned earlier, measurement uncertainties also affect the accuracy of the parameter estimation schemes [24,25]. Therefore, all the mentioned issues should be taken into consideration while identifying transmission line parameters.

Like transmission line parameter identification, the PMU measurements are also employed to identify the system Thevenin equivalent (TE) of the electric networks at a particular node. The TE of a specific node simplifies the electric networks for quick and effective analysis of various aspects, including fault diagnosis [33], protection scheme design [34], and voltage stability analysis [35], etc. Different methods were reported for transmission system TE determination, including robust adaptive method [36], lower upper factorization method [37], multiport system based method [38], nonlinear and linear recursive least square method [39], and finite time standard recursive least–fair method [40].

However, the TE at a specific node of a linear system remains constant over time as the changes in the input voltages and currents do not affect it. Though the transmission lines of the electric networks are linear by nature, the node specifications (either P/Q at load nodes or P/V at generation nodes) of the networks are nonlinear. Thus, current and voltage relationships at the system nodes face the nonlinearities that affect the TE of a specific node over time. In response, power flow analysis can handle such issues, but it will become hectic as power flow analysis of the entire electric system should be done each time whenever the system TE is required. For online estimation of the system TE, the supervisory control and data acquisition (SCADA) system are not appropriate as the measurement frequency is very slow [7]. Thus, the PMU becomes very popular in transmission system TE identification as it provides faster and more measurements per cycle than the SCADA system [41]. Nevertheless, the data acquisition or transmission process in PMU can be erroneous due to external interference or internal system faults. Measurement error of the PMU should be taken into consideration for online estimation of the system TE as it severely affects the quality, reliability, and effectiveness of the targeted analysis [7,36,42].

This chapter presents the use of synchronized phasor measurements of voltage and current acquired by the PMUs for online identification of the transmission line parameters and system TE at a particular node. It illustrates three different methods for identifying the transmission line parameters depending on the number of PMU measurement sets available at the two ends of the line. The performances of these approaches are evaluated and quantified without measurement noise and in the presence of random noises and biased errors. Besides, a minimum of three PMU measurement sets at three different loading conditions is used to identify system TE at a particular node. The presented approaches are tested on a 115 kV transmission system chosen from the Saudi Electricity Company (SEC) network and verified with PSCAD/EMTDC and MATLAB/SIMULINK simulations.

The remaining parts of this chapter are organized as follows: Section 6.2 and Section 6.3 provide transmission line parameters and system TE identification methods. Section 6.4 presents the obtained results that confirm the efficacy of the employed methods. Finally, an overall summary of this chapter is presented in Section 6.5.

6.2 Transmission line parameter identification

In this chapter, the line parameters of a two-terminal transmission system are computed using the steady-state amplitudes and phases of the currents and voltages measured from the installed PMUs at the line terminals. There are three approaches for identifying the transmission line parameters based on the available number of PMU measurement sets. For instance, if only one set of measurements is available, the transmission line is represented by the π-type equivalent circuit, as shown in Fig. 6.1.

Where:
Z_{SA}, Z_{SB}: Equivalent source impedances at buses A and B, respectively.
E_A, E_B: Equivalent source voltages at buses A and B, respectively.
Y, Z: Transmission line admittance and impedance, respectively.
I_A, I_B: Normal operating currents at buses A and B, respectively.
V_A, V_B: Normal operating voltages at buses A and B, respectively.

The line parameters of the transmission system network can be calculated using the following equations:

$$Y = \frac{2(I_A + I_B)}{V_A + V_B} \tag{6.1}$$

$$Z = \frac{V_A^2 - V_B^2}{V_B I_A - V_A I_B} \tag{6.2}$$

In parallel lines, the above equations are used to calculate the positive sequence admittance and impedance for each circuit. Then, the mean values are regarded as parameters for parallel lines. The mutual inductance factor is taken into account

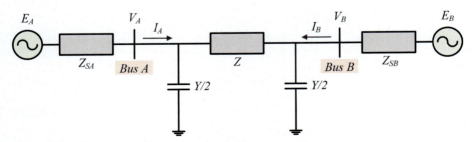

Fig. 6.1 The π-equivalent circuit of a two-terminal line network.

in computing the positive parameters since V_A, I_A, V_B, I_B are influenced by mutual inductance between the lines and phases [24].

However, transmission lines are represented using the two-port ABCD parameters in general. Based on Fig. 6.1, the following equations can be written:

$$V_A = A \cdot V_B + B \cdot I_B \tag{6.3}$$

$$I_A = C \cdot V_B + D \cdot I_B \tag{6.4}$$

If the PMU installed at the line terminals generate two different sets of measurements for different loading conditions, then the following equations can be written from Eqs. (6.3) and (6.4):

$$V_{A1} = A \cdot V_{B1} + B \cdot I_{B1} \tag{6.5}$$

$$I_{A1} = C \cdot V_{B1} + D \cdot I_{B1} \tag{6.6}$$

$$V_{A2} = A \cdot V_{B2} + B \cdot I_{B2} \tag{6.7}$$

$$I_{A2} = C \cdot V_{B2} + D \cdot I_{B2} \tag{6.8}$$

Where, the subscripts (1) and (2) represent the first and the second set of PMU measurements, respectively.

Cramer's Rule yields the following equations for computation of the ABCD parameters from the Eqs. (6.5) to (6.8):

$$A = \frac{I_{B1} \cdot V_{A2} - I_{B2} \cdot V_{A1}}{I_{B1} \cdot V_{B2} - I_{B2} \cdot V_{B1}} \tag{6.9}$$

$$B = \frac{V_{B2} \cdot V_{A1} - V_{B1} \cdot V_{A2}}{I_{B1} \cdot V_{B2} - I_{B2} \cdot V_{B1}} \tag{6.10}$$

$$C = \frac{I_{B1} \cdot I_{A2} - I_{B2} \cdot I_{A1}}{I_{B1} \cdot V_{B2} - I_{B2} \cdot V_{B1}} \tag{6.11}$$

$$D = \frac{I_{A1} \cdot V_{B2} - I_{A2} \cdot V_{B1}}{I_{B1} \cdot V_{B2} - I_{B2} \cdot V_{B1}} \tag{6.12}$$

After computation of the ABCD parameters, the following relationships are used for the identification of transmission line parameters:

$$A = 1 + \frac{YZ}{2} \tag{6.13}$$

$$B = Z \tag{6.14}$$

$$C = Y \cdot \left(1 + \frac{YZ}{4}\right) \tag{6.15}$$

$$D = 1 + \frac{YZ}{2} \tag{6.16}$$

Transmission line parameter and system Thevenin equivalent identification

Since there are only two unknowns, the equations above are redundant. By combining Eqs. (6.13) and (6.14), the following relationships can be deduced for the identification of the transmission line parameters [24]:

$$Z = B \tag{6.17}$$

$$Y = \frac{2 \cdot (A-1)}{B} \tag{6.18}$$

If multiple measurements are available, the following equations can be deduced from the complex Eqs. (6.3) and (6.4) that contain eight variables:

$$\operatorname{Re}[V_A] = \operatorname{Re}[A] \cdot \operatorname{Re}[V_B] - \operatorname{Im}[A] \cdot \operatorname{Im}[V_B] \\ + \operatorname{Re}[B] \cdot \operatorname{Re}[I_B] - \operatorname{Im}[B] \cdot \operatorname{Im}[I_B] \tag{6.19}$$

$$\operatorname{Im}[V_A] = \operatorname{Re}[A] \cdot \operatorname{Im}[V_B] + \operatorname{Im}[A] \cdot \operatorname{Re}[V_B] \\ + \operatorname{Re}[B] \cdot \operatorname{Im}[I_B] + \operatorname{Im}[B] \cdot \operatorname{Re}[I_B] \tag{6.20}$$

$$\operatorname{Re}[I_A] = \operatorname{Re}[C] \cdot \operatorname{Re}[V_B] - \operatorname{Im}[C] \cdot \operatorname{Im}[V_B] \\ + \operatorname{Re}[D] \cdot \operatorname{Re}[I_B] - \operatorname{Im}[D] \cdot \operatorname{Im}[I_B] \tag{6.21}$$

$$\operatorname{Im}[I_A] = \operatorname{Re}[C] \cdot \operatorname{Im}[V_B] + \operatorname{Im}[C] \cdot \operatorname{Re}[V_B] \\ + \operatorname{Re}[D] \cdot \operatorname{Im}[I_B] + \operatorname{Im}[D] \cdot \operatorname{Re}[I_B] \tag{6.22}$$

Eqs. (6.19) to (6.22) can be written in the matrix format as follows:

$$\begin{bmatrix} \operatorname{Re}[V_A] \\ \operatorname{Im}[V_A] \end{bmatrix} = \begin{bmatrix} \operatorname{Re}[V_B] & -\operatorname{Im}[V_B] & \operatorname{Re}[I_B] & -\operatorname{Im}[I_B] \\ \operatorname{Im}[V_B] & \operatorname{Re}[V_B] & \operatorname{Im}[I_B] & \operatorname{Re}[I_B] \end{bmatrix} \cdot \begin{bmatrix} \operatorname{Re}[A] \\ \operatorname{Im}[A] \\ \operatorname{Re}[B] \\ \operatorname{Im}[B] \end{bmatrix} \tag{6.23}$$

$$\begin{bmatrix} \operatorname{Re}[I_A] \\ \operatorname{Im}[I_A] \end{bmatrix} = \begin{bmatrix} \operatorname{Re}[V_B] & -\operatorname{Im}[V_B] & \operatorname{Re}[I_B] & -\operatorname{Im}[I_B] \\ \operatorname{Im}[V_B] & \operatorname{Re}[V_B] & \operatorname{Im}[I_B] & \operatorname{Re}[I_B] \end{bmatrix} \cdot \begin{bmatrix} \operatorname{Re}[C] \\ \operatorname{Im}[C] \\ \operatorname{Re}[D] \\ \operatorname{Im}[D] \end{bmatrix} \tag{6.24}$$

If K number measurements are acquired from the PMU, then the following relationships can be obtained:

$$E = \begin{bmatrix} \operatorname{Re}[V_{A1}] \\ \operatorname{Im}[V_{A1}] \\ \operatorname{Re}[V_{A2}] \\ \operatorname{Im}[V_{A2}] \\ \vdots \end{bmatrix} \tag{6.25}$$

$$H = \begin{bmatrix} \operatorname{Re}[V_{B1}] & -\operatorname{Im}[V_{B1}] & \operatorname{Re}[I_{B1}] & -\operatorname{Im}[I_{B1}] \\ \operatorname{Im}[V_{B1}] & \operatorname{Re}[V_{B1}] & \operatorname{Im}[I_{B1}] & \operatorname{Re}[I_{B1}] \\ \vdots & \vdots & \vdots & \vdots \\ \operatorname{Re}[V_{BK}] & -\operatorname{Im}[V_{BK}] & \operatorname{Re}[I_{BK}] & -\operatorname{Im}[I_{BK}] \\ \operatorname{Im}[V_{BK}] & \operatorname{Re}[V_{BK}] & \operatorname{Im}[I_{BK}] & \operatorname{Re}[I_{BK}] \end{bmatrix} \quad (6.26)$$

$$F = \begin{bmatrix} \operatorname{Re}[A] \\ \operatorname{Im}[A] \\ \operatorname{Re}[B] \\ \operatorname{Im}[B] \end{bmatrix} \quad (6.27)$$

Employing the unbiased least square algorithm, the chain parameters A and B are obtained from the following expression:

$$F = \left(H^T H\right)^{-1} H^T E \quad (6.28)$$

The values of C and D can be obtained using a similar approach. Finally, the Eqs. (6.17) and (6.18) are used for the identification of the transmission line parameters [24].

6.3 Thevenin equivalent identification

The computational procedures for identifying the system TE at a particular node using the basics of a PMU-based algorithm are presented in this section. The following equation represents the node voltage using the TE model [41]:

$$V = E_{th} + Z_{th} \cdot I \quad (6.29)$$

Where:

V, I: Node voltage and current, respectively.
E_{th}, Z_{th}: TE voltage and impedance, respectively.

Without the detailed system representation, all the required variables at a particular node can be reliably estimated if the exact values of the E_{th} and Z_{th} can be obtained. At least two different pairs of current and voltage (V, I) measured at different time instants can be used to determine the TE parameters using the local PMU measurements. Due to variations in system frequency, the measurement reference rotates at the slip frequency between the PMU and the system. The phase angles of I and V at subsequent time steps are measured for various references. Figs. 6.2 and 6.3 present the phasor diagrams for two different sets of measurements. The PMU phase drift has the effect of rotating all phasors by equal angles while the relative angles between individual phasors remain constant.

The TE voltage (E) must be the same in the two cases. Nevertheless, E for the second set of measurements is shifted by an angle equal to the phase drift, but

Transmission line parameter and system Thevenin equivalent identification

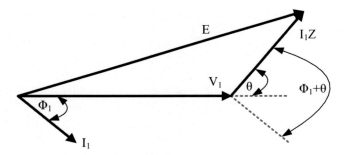

Fig. 6.2 Phasor diagram based on the first measurement set.

its magnitude remains constant. For the first PMU measurement set, E can be written as:

$$E^2 = V_1^2 + I_1^2 Z^2 + 2V_1 I_1 Z \cos(\theta + \varphi) \tag{6.30}$$

Expanding $\cos(\theta + \phi)$, Eq. (6.30) can be rewritten as:

$$E^2 = V_1^2 + I_1^2 Z^2 + 2P_1 r_{th} - 2Q_1 x_{th} \tag{6.31}$$

Where x_{th} and r_{th} denote the reactance and resistance of the Thevenin impedance Z. P_1 and Q_1 represent the real and reactive powers for the first PMU measurement set. Equations of the second PMU measurement set can be written as:

$$E^2 = V_2^2 + I_2^2 Z^2 + 2P_2 r_{th} - 2Q_2 x_{th} \tag{6.32}$$

Subtracting Eq. (6.32) from (6.31) leads to:

$$V_1^2 - V_2^2 + \left(I_1^2 - I_2^2\right)Z^2 + 2\left(P_1 - P_2\right)r_{th} - 2\left(Q_1 - Q_2\right)x_{th} = 0 \tag{6.33}$$

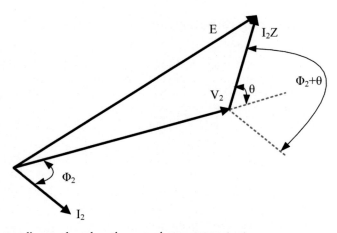

Fig. 6.3 Phasor diagram based on the second measurement set.

The following equation can be reduced by arranging the Eq. (6.33):

$$\left(r_{th}+\frac{P_1-P_2}{I_1^2-I_2^2}\right)^2+\left(x_{th}-\frac{Q_1-Q_2}{I_1^2-I_2^2}\right)^2=\frac{V_2^2-V_1^2}{I_1^2-I_2^2}+\left(\frac{P_1-P_2}{I_1^2-I_2^2}\right)^2+\left(\frac{Q_1-Q_2}{I_1^2-I_2^2}\right)^2 \quad (6.34)$$

Eq. (6.34) represents a circle in the impedance plane that defines the locus for the Thevenin impedance (Z) by satisfying two sets of measurements. However, it does not identify a specific value for Z. Another locus for Z can be determined by taking a third measurement, which can be used with either the first or the second set of PMU measurements in a similar way. The intersection points of the two circles in the Z-plane give the values of the reactance and the resistance of the Thevenin impedance. Detailed procedures of the system TE calculation can be found in [7,41–43].

6.4 Simulation results

A 115 kV transmission system comprised of 38 buses from the SEC, as depicted in Fig. 6.4, was considered for investigating the accuracy of the above-described methods for transmission line parameter and system TE identification. The line connecting bus-30 and bus-38 was chosen, and the nominal-π circuit model was considered. Table 6.1 shows the actual specifications of the selected line, for example, line length L, reactance x, resistance r, and shunt susceptance, B_C where $Y = jB_C$.

The system was modeled in PSCAD/EMTDC platform. Two sets of current and voltage transformers were deployed at both ends of the selected transmission line for simultaneous measurements of currents and voltages. The phasor components of the measured data were processed in real-time through discrete Fourier transform (DFT). These components were further processed to determine the sequence components to identify the positive sequence impedance parameters of the transmission line. It is worth noting that different sets of measurements were recorded at different loading conditions. The following parts of this section present the identified transmission line parameters and system TE.

6.4.1 Identified line parameters under noise-free measurement

This chapter employed three different methods for the identification of the line parameters under noise-free measurement conditions. The first method was designated to use one set of PMU measurements only and termed method # 1. Two sets and multiple sets of PMU measurements were used for the second and the third methods, respectively. They were termed method # 2 and method # 3, respectively. The following equation was employed for accurate calculation of the identified line parameters:

$$\text{error } (\%) = \frac{|\text{Actual value} - \text{Calculated value}|}{\text{Actual value}} \times 100 \quad (6.35)$$

Table 6.2 presents the identified line parameters values and calculated percentage errors. All the employed methods identified the line parameters efficiently with higher accuracy under the noise-free environment as can be seen from the obtained results.

Transmission line parameter and system Thevenin equivalent identification

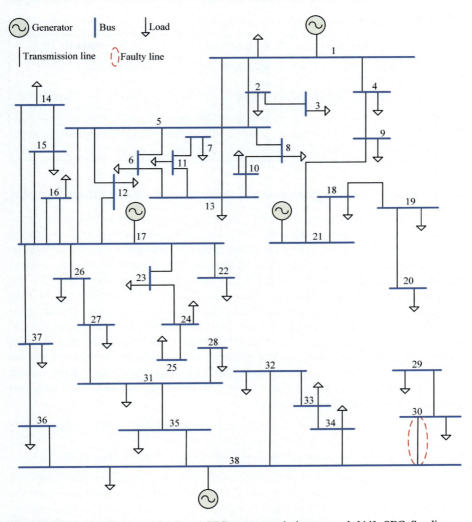

Fig. 6.4 Single line diagram of selected SEC test transmission network [44]. *SEC*, Saudi Electricity Company.

Table 6.1 Selected line parameters of the Saudi Electricity Company transmission network.

Parameter	Value
L	26 kilometers (km)
r	1.9044 Ω /km
x	9.85262 Ω /km
B_C	9.10397e-5 S/km

Table 6.2 Identified line parameters and percentage of errors.

Method	Resistance (r) Calculated value (Ω/km)	Error (%)	Reactance (x) Calculated value (Ω/km)	Error (%)	Susceptance (B_c) Calculated value (S/km)	Error (%)
# 1	1.904405	2.625E-04	9.85284	2.233E-03	9.10416E-05	2.08700E-03
# 2	1.904411	5.776E-04	9.85286	2.436E-03	9.10482E-05	9.33659E-03
# 3	1.904412	6.301E-04	9.85288	2.639E-03	9.10483E-05	9.44643E-03

6.4.2 Identified line parameters under biased and nonbiased noise

The PMU provides measurements with higher accuracy compared to other measurement devices. However, even the PMU based data acquisition process can be contaminated by different noises for various reasons, including data conversion or communication device abnormalities, instrumentation channel anomalies, current transformer saturation errors, system imbalances, etc. Noises severely affect the measurements in short transmission lines compared to the measurements in long transmission lines [5,24,32]. Contemplating the mentioned notes, this chapter considered two types of measurement noises: the bias errors and random noises to the noise-free PMU voltage and current measurements. Considered random noise was normally distributed with 1% standard deviation and zero mean. On the contrary, the bias errors were assumed to have a magnitude of 1% of the mean value of the current or voltage experienced over a 24-hour cycle. The following sources of measurement noises were considered in this chapter:

- Case-1: Noise injection to the time domain samples during PSCAD/EMTDC simulation.
- Case-2: Noise injection to the DFT obtained phasor values.
- Case-3: Noise injection to the symmetrical transformation obtained components.

Table 6.3 presents the simulation results considering the presence of bias errors and random noises in the measurement processes. The results were acceptable if the estimated parameter is within a reasonable range, for example, ±20% for r and ±10% for x and B_C. The satisfactory results are denoted by "Y." Otherwise, the results were unacceptable and denoted by "N." The following deductions can be made from the obtained results:

- A few identified parameters were acceptable for certain types of noise and unacceptable for others. For instance, the rows # 1 and # 5 of method # 1, where the measurements of sending end voltage were contaminated with noise, the computed shunt susceptance was acceptable while the series reactance and resistance were unacceptable. The results indicated that the series reactance and resistance were sensitive to the bias errors and the random noises, whereas the shunt susceptance was not. Similar comments can be made for the other rows in the table.
- Method # 2 outperformed method # 1 for the bias errors even though it was very sensitive to the random noises. This observation indicated that the parameter estimation could be improved with redundancy in the measurements.
- The bias errors and the random noises introduced in the measurements did not affect the calculations since Eqs. (6.9) and (6.10) pertained to method # 2 were not the functions of the sending end current measurements.

Transmission line parameter and system Thevenin equivalent identification

Table 6.3 Simulation results considering the presence of bias errors and random noises in the measurement processes.

Method	Noise type		Case-1 R	X	B_C	Case-2 R	X	B_C	Case-3 R	X	B_C
# 1	Bias errors	V_A	N	N	Y	N	N	Y	N	N	Y
		V_B	N	N	Y	N	N	Y	N	N	Y
		I_A	Y	Y	N	Y	Y	N	Y	Y	N
		I_B	Y	Y	N	Y	Y	N	Y	Y	N
	Random noises	V_A	N	N	Y	N	N	Y	N	N	Y
		V_B	N	N	Y	N	N	Y	N	N	Y
		I_A	Y	Y	Y	Y	Y	Y	Y	Y	Y
		I_B	Y	Y	Y	Y	Y	Y	Y	Y	Y
# 2	Bias errors	V_A	Y	Y	N	Y	Y	N	Y	Y	N
		V_B	Y	Y	N	Y	Y	N	Y	Y	N
		I_A	Y	Y	Y	Y	Y	Y	Y	Y	Y
		I_B	Y	Y	Y	Y	Y	Y	Y	Y	Y
	Random noises	V_A	N	N	N	Y	Y	N	Y	Y	N
		V_B	N	N	N	Y	Y	N	Y	Y	N
		I_A	Y	Y	Y	Y	Y	Y	Y	Y	Y
		I_B	N	N	N	Y	Y	Y	Y	Y	Y
# 3	Bias errors	V_A	Y	Y	N	Y	Y	N	Y	Y	N
		V_B	Y	Y	N	Y	Y	N	Y	Y	N
		I_A	Y	Y	Y	Y	Y	Y	Y	Y	Y
		I_B	Y	Y	Y	Y	Y	Y	Y	Y	Y
	Random noises	V_A	Y	Y	N	Y	Y	N	Y	Y	N
		V_B	N	N	N	Y	Y	N	Y	Y	N
		I_A	Y	Y	Y	Y	Y	Y	Y	Y	Y
		I_B	Y	Y	N	Y	Y	Y	Y	Y	Y

N, unacceptable; Y, acceptable.

- Despite the presence of the measurement noises, the impedance parameters obtained with method # 3 were more accurate than method # 1 and method # 2.
- Method # 3 performed excellently in computing the series reactance and resistance in the presence of bias errors and random noises. However, the calculated shunt susceptance was sensitive to the bias errors and the random noises.

6.4.3 Identified Thevenin equivalent

The PMUs were installed at bus-30 and bus-38 to obtain the current and voltage measurements to determine the system TE of the 115 kV SEC network at buses mentioned above. Three sets of PMU measurements corresponding to three different loading cases were simulated and recorded in the PSCAD/EMTDC platform. To generate the PMU measurements, the loads at bus-30 and bus-38 were varied while representing the generators as voltage sources by setting their voltage and phase angle values to their base case values. The PMU measurements were taken at small intervals, e.g., one measurement per cycle, and loads were varied until attaining the convergence. The Z circles in the impedance plane formed using different combinations of the three sets of measurements recorded at bus 30 and 38 are shown in Figs. 6.5A and 6.6A, respectively.

As can be seen, the circles intersected at the common point for different combinations of the measurement sets. The intersection points provide the values of the TE impedances. However, only two circles are needed to determine the Z_{th} values as shown in Figs. 6.5B and 6.6B. Finally, the calculated TE impedances for bus-30 and bus-38 were 0.9734+ j 4.56 Ω and 6.003+ j 4.835 Ω, respectively.

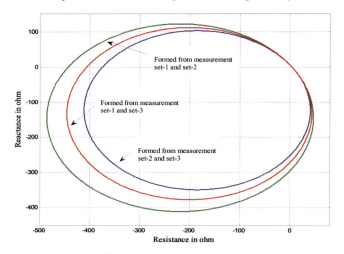

A Z circles in the impedance plane

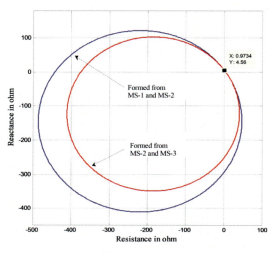

B Thevenin equivalent (Z_{th})

Fig. 6.5 Z circles in the impedance plane and Thevenin equivalent impedance (Z_{th}) identification at bus-30.

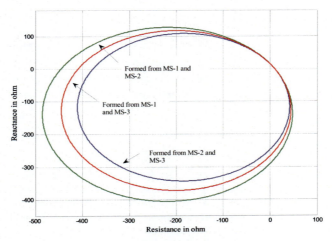

A Z circles in the impedance plane

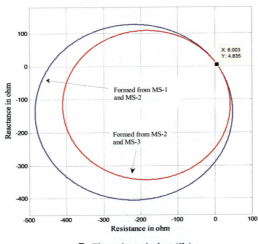

B Thevenin equivalent (Z_{th})

Fig. 6.6 Z circles in the impedance plane and Thevenin equivalent impedance (Z_{th}) identification at bus-38.

6.5 Summary

This chapter described how the synchronized current and voltage measurements obtained from the PMU at both ends of a transmission line were used to identify the transmission line parameters and system TE at a specific end. All the mentioned items were computed online. Three different methods were employed to calculate the transmission line parameters depending on the number of independent sets of synchronized phasor measurements. Besides, a minimum of three PMU measurement sets at three different loading conditions was utilized to determine the TE at a particular node. This chapter investigated the effectiveness of the discussed approaches on a 115 kV transmission system chosen from the SEC and verified through PSCAD/EMTDC and MATLAB/SIMULINK simulations. Obtained results confirmed the accuracy of the transmission line parameters and system TE identification methods even in the presence of biased errors and measurement noises.

References

[1] A.S. Dobakhshari, M. Abdolmaleki, V. Terzija, S. Azizi, Online non-iterative estimation of transmission line and transformer parameters using SCADA data, IEEE Trans. Power Syst. 36 (3) (2021) 2632–2641, doi:10.1109/TPWRS.2020.3037997.

[2] Y. Zhang and Y. Liao, Optimal line parameter estimation method for mid-compensated transmission lines, 2019, doi:10.1109/NAPS46351.2019.9000274.

[3] C. Li, Y. Zhang, H. Zhang, Q. Wu, V. Terzija, Measurement-based transmission line parameter estimation with adaptive data selection scheme, IEEE Trans. Smart Grid 9 (6) (2018) 5764–5773, doi:10.1109/TSG.2017.2696619.

[4] Y. Liao, M. Kezunovic, Online optimal transmission line parameter estimation for relaying applications, IEEE Trans. Power Deliv. 24 (1) (2009) 96–102, doi:10.1109/TPWRD.2008.2002875.

[5] Y. Du, Y. Liao, Online estimation of transmission line parameters, temperature and sag using PMU measurements, Electr. Power Syst. Res. 93 (2012) 39–45, doi:10.1016/j.epsr.2012.07.007.

[6] A. Xue, et al., Robust identification method for transmission line parameters that considers PMU phase angle error, IEEE Access 8 (1) (2020) 86962–86971, doi:10.1109/ACCESS.2020.2992247.

[7] A.H. Al-Mohammed, Adaptive fault location in power system networks based on synchronized phasor measurements, King Fahd University of Petroleum and Minerals, Dhahran, Saudi Arabia, 2012.

[8] S. López, J. Gómez, R. Cimadevilla, O. Bolado, Synchrophasor applications of the National Electric System Operator of Spain, 2008 61st Annual Conference for Protective Relay Engineers, 2008, pp. 436–456, doi:10.1109/CPRE.2008.4515070.

[9] H. Ahmadi, M. Armstrong, Transmission line impedance calculation using detailed line geometry and HEM soil resistivity measurements, Canadian Conference on Electrical and Computer Engineering, 2016, (Oct 2016), doi:10.1109/CCECE.2016.7726808 Oct2016.

[10] M. Reta-Hernández, Transmission line parameters, Electric Power Generation, Transmission, and Distribution, CRC Press, Florida, USA, 2018, pp. 14-1–14-28.

[11] M. Bartos, et al., Impacts of rising air temperatures on electric transmission ampacity and peak electricity load in the United States, Environ. Res. Lett. 11 (11) (2016) 114008, doi:10.1088/1748-9326/11/11/114008.

[12] K. Yu, et al., A novel traveling wave fault location method for transmission network based on time linear dependence, Int. J. Electr. Power Energy Syst. 126 (Mar) (2021) 106608, doi:10.1016/j.ijepes.2020.106608.

[13] M. Shafiullah, M.A. Abido, Z. Al-Hamouz, Wavelet-based extreme learning machine for distribution grid fault location, IET Gener. Transm. Distrib. 11 (17) (2017) 4256–4263, doi:10.1049/iet-gtd.2017.0656.

[14] M. Shafiullah, M.J. Rana, M.S. Alam, M.A. Uddin, Optimal placement of Phasor Measurement Units for transmission grid observability, 2016 International Conference on Innovations in Science, Engineering and Technology (ICISET), IEEE, Chittagong, Bangladesh, 2016, pp. 1–4, doi:10.1109/ICISET.2016.7856492.

[15] M. Shafiullah, M. Abido, M. Hossain, A. Mantawy, An improved OPP problem formulation for distribution grid observability, Energies 11 (11) (2018) 3069, doi:10.3390/en11113069.

[16] M. Shafiullah, M.I. Hossain, M.A. Abido, T. Abdel-Fattah, A.H. Mantawy, A modified optimal PMU placement problem formulation considering channel limits under various contingencies, Meas. J. Int. Meas. Confed. 135 (Mar) (2019), doi:10.1016/j.measurement.2018.12.039.

[17] R.M. Arias Velásquez, Failure analysis and dispatch optimization using phasor measurement units, Eng. Fail. Anal. 121 (Mar) (2021) 105157, doi:10.1016/j.engfailanal.2020.105157.

[18] F.V. Lopes, A. Mouco, R.O. Fernandes, F.C. Neto, Real-world case studies on transmission line fault location feasibility by using M-Class phasor measurement units, Electr. Power Syst. Res. 196 (JUL) (2021) 107261, doi:10.1016/j.epsr.2021.107261.

[19] V. Terzija, Z.M. Radojević, G. Preston, Flexible synchronized measurement technology-based fault locator, IEEE Trans. Smart Grid 6 (2) (2015) 866–873, doi:10.1109/TSG.2014.2367820.

[20] M. Shafiullah, M.A. Abido, A review on distribution grid fault location techniques, Electr. Power Components Syst. 45 (8) (2017) 807–824, doi:10.1080/15325008.2017.1310772.

[21] M. Shafiullah, S.M. Rahman, M.G. Mortoja, B. Al-Ramadan, Role of spatial analysis technology in power system industry: An overview, Renew. Sustain. Energy Rev. 66 (Dec) (2016) 584–595, doi:10.1016/j.rser.2016.08.017.

[22] C.S. Indulkar, K. Ramalingam, Estimation of transmission line parameters from measurements, Int. J. Electr. Power Energy Syst. 30 (5) (2008) 337–342, doi:10.1016/j.ijepes.2007.08.003.

[23] R.E. Wilson, G.A. Zevenbergen, D.L. Mah, A.J. Murphy, Calculation of transmission line parameters from synchronized measurements, Electr. Mach. Power Syst. 27 (12) (1999) 1269–1278, doi:10.1080/073135699268560.

[24] D. Shi, D.J. Tylavsky, N. Logic, and K.M. Koellner, Identification of short transmission-line parameters from synchrophasor measurements, 2008, doi:10.1109/NAPS.2008.5307354.

[25] Y. Liao, Some algorithms for transmission line parameter estimation, 2009 IEEE International Symposium on Sustainable Systems and Technology, ISSST 2009, 2009, pp. 127–132, doi:10.1109/SSST.2009.4806781.

[26] T. Bi, J. Chen, J. Wu, Q. Yang, Synchronized phasor based online parameter identification of overhead transmission line, 3rd International Conference on Deregulation and Restructuring and Power Technologies, DRPT 2008, 2008, pp. 1657–1662, doi:10.1109/DRPT.2008.4523671.

[27] Y. Liao, Algorithms for fault location and line parameter estimation utilizing voltage and current data during the fault, Proceedings of the Annual Southeastern Symposium on System Theory, 2008, pp. 183–187, doi:10.1109/SSST.2008.4480216.

[28] C. Borda, A. Olarte, and H. Diaz, PMU-based line and transformer parameter estimation, 2009, doi:10.1109/PSCE.2009.4840079.

[29] D. Shi, D.J. Tylavsky, K.M. Koellner, N. Logic, D.E. Wheeler, Transmission line parameter identification using PMU measurements, Eur. Trans. Electr. Power 21 (4) (2011) 1574–1588, doi:10.1002/etep.522.

[30] Y. Liao, M. Kezunovic, Optimal estimate of transmission line fault location considering measurement errors, IEEE Trans. Power Deliv. 22 (3) (2007) 1335–1341, doi:10.1109/TPWRD.2007.899554.

[31] R. Che and J. Liang, An accurate fault location algorithm for two-terminal transmission lines combined with parameter estimation, 2009, doi:10.1109/APPEEC.2009.4918128.

[32] Y. Liao, Algorithms for Power System Fault Location and Line Parameter Estimation, 2007 Thirty-Ninth Southeastern Symposium on System Theory. IEEE, 2007, pp. 189–193, doi:10.1109/SSST.2007.352346.

[33] A. Al-Mohammed, M. Abido, A fully adaptive PMU-based fault location algorithm for series-compensated lines, IEEE Trans. Power Syst. 29 (5) (2014) 2129–2137, http://ieeexplore.ieee.org/xpls/abs_all.jsp?arnumber=6736121. (Accessed 19 August 2018).

[34] S. Shen, et al., An adaptive protection scheme for distribution systems with DGs based on optimized thevenin equivalent parameters estimation, IEEE Trans. Power Deliv. 32 (1) (2017) 411–419, doi:10.1109/TPWRD.2015.2506155.

[35] H.Y. Su, C.W. Liu, Estimating the voltage stability margin using PMU measurements, IEEE Trans. Power Syst. 31 (4) (2016) 3221–3229, doi:10.1109/TPWRS.2015.2477426.

[36] A. Zhang, W. Tan, M. Cheng, W. Yang, Thévenin equivalent parameter adaptive robust estimation considering the erroneous measurements of PMU, Energies 13 (18) (2020) 4865, doi:10.3390/en13184865.

[37] Z. Yun, X. Cui, K. Ma, Online Thevenin equivalent parameter identification method of large power grids using LU factorization, IEEE Trans. Power Syst. 34 (6) (2019) 4464–4475, doi:10.1109/TPWRS.2019.2920994.

[38] J.A.V. Vásquez, A.R.R. Matavalam, V. Ajjarapu, Fast calculation of Thévenin equivalents for real-time steady state voltage stability estimation, 2016, doi:10.1109/NAPS.2016.7747849.

[39] M.U. Hashmi, R. Choudhary, J.G. Priolkar, Online thevenin equivalent parameter estimation using nonlinear and linear recursive least square algorithm, 2015, doi:10.1109/ICECCT.2015.7225946.

[40] A. Arancibia, C.A. Soriano-Rangel, F. Mancilla-David, R. Ortega, K. Strunz, Finite–time identification of the Thévenin equivalent parameters in power grids, Int. J. Electr. Power Energy Syst. 116 (Mar) (2020) 105534, doi:10.1016/j.ijepes.2019.105534.

[41] S. Abdelkader, Online Thevenin's equivalent using local PMU measurements, International Conference on Renewable Energies and Power Quality (ICREPQ'11), 2011, pp. 1229–1232, doi:10.24084/repqj09.604.

[42] S.M. Abdelkader, D.J. Morrow, Online Thévenin equivalent determination considering system side changes and measurement errors, IEEE Trans. Power Syst. 30 (5) (2015) 2716–2725, doi:10.1109/TPWRS.2014.2365114.

[43] S.M. Abdelkader, D.J. Morrow, Online tracking of Thévenin equivalent parameters using PMU measurements, IEEE Trans. Power Syst. 27 (2) (2012) 975–983, doi:10.1109/TPWRS.2011.2178868.

[44] A.H. Al-Mohammed, M.A. Abido, An adaptive fault location algorithm for power system networks based on synchrophasor measurements, Electr. Power Syst. Res. 108 (Mar) (2014) 153–163, doi:10.1016/j.epsr.2013.10.013.

Fault diagnosis in two-terminal power transmission lines

7

7.1 Introduction

Electric transmission lines are susceptible to a wide range of transient and permanent faults as they are primarily overhead type and exposed to birds, trees, storms, etc. Power system faults cause power outages, revenue loss, asset damage, and sometimes deaths of birds, animals, and humans. Accurate and rapid identification of the fault locations on a power system network expedites the power restoration and repair of faulty components. This reduces power outages and enhances system reliability. Rapid service restoration also lessens consumer dissatisfaction, crew repair expense, and revenue loss [1–8]. In response, different fault diagnosis schemes for electric power system networks were reported in the literature and books for the last couple of decades [9–15].

However, the ambient condition and operation history of the power system networks affect the power system transmission line parameters [16–19]. In general, such cases are not taken into consideration while developing the classical fault diagnosis algorithms for electric power system networks. Line parameters are usually assumed, in these classical algorithms, unchanged regardless of the change in the system operating conditions [20–22]. In response, the adaptive fault location algorithms estimate the system parameters and system impedance online using the recent measurements obtained through advanced metering infrastructure. The recent proliferation of advanced measurement infrastructure, including the phasor measurement unit (PMU), provides time-synchronized and effective measurements even under various adversities [23–27]. Such accurate measurements paved the way for efficient fault location algorithm development for power system transmission lines [28–30]. Several adaptive fault location algorithms based on PMU measurements were reported to enhance the fault location accuracy of the classical algorithms [31–44]. Some of these algorithms are developed using synchronized voltage and current phasors [38–44]. Other algorithms use only voltage phasor measurements to circumvent the effects of improper operation of current transformers caused by overvoltage and transient state of power network during fault [33–36].

This chapter presents two adaptive fault location algorithms based on PMU measurements for determining fault location on a two-terminal transmission system as an improvement of the works proposed in [31] and [34]. Both algorithms recorded three sets of pre-fault current and voltage phasor measurements at the line terminals for online computation of the Thevenin equivalent (TE) and line parameters. Online determination of the system parameters tackles the discrepancies between the utility provided and real operating information regarding the parameters due to environmental conditions and operation history. The adaptive fault location algorithms are tested on a transmission line of a 115 kV transmission network of the Saudi Electricity

Power System Fault Diagnosis. DOI: https://doi.org/10.1016/B978-0-323-88429-7.00010-2
Copyright © 2022 Elsevier Inc. All rights reserved.

Company (SEC). Simulations are carried out on PSCAD/EMTDC and MATLAB/SIMULINK platforms. Obtained results confirm the efficacy of the presented algorithms and their independence on fault information (type, location, resistance, and inception angle), prefault loading conditions, and measurement uncertainty.

The remaining parts of this chapter are arranged as follows: Section 7.2 provides a detailed explanation of the adaptive fault location algorithms. Section 7.3 presents the obtained results that confirm the efficacy of the algorithms and their independence on system parameter variation, measurement errors, fault information, and prefault loading uncertainty. Finally, an overall summary of this chapter is presented in Section 7.5.

7.2 Fault location algorithms

The essence of developing adaptive fault location algorithms is to locate faults with higher accuracy. In the transmission system, such schemes are mainly concern about the online estimation of the line parameters and system impedance as they can be changed due to the environmental conditions and operation history of the transmission lines. For instance, over current and overload lead to a rise to the conductor temperature that increases the length of the transmission lines; thus, the sag of the transmission lines is also increased. Besides, the resistance of the transmission lines also varies with the change of the temperature. The reactance of the lines also changes when the sag of the line is changed as it is related to the distance between the phase lines. Such uncertainty in the line parameters affects the conventional fault location algorithms [31]. Considering the mentioned notes, this chapter employs two adaptive fault location algorithms to estimate the fault distance in the two-terminal transmission lines, as discussed in the following parts of this section.

7.2.1 First fault location algorithm (FL-1)

The first adaptive fault location algorithm (FL-1) applies the principle of superposition on the power system network to divide the postfault network into a superimposed phase network and a prefault network. It uses prefault PMU measurements obtained from the two terminals of the transmission line to represent the prefault network. Besides, the superimposed electrical measurements (one prefault and one postfault measurements) are used to represent the superimposed network that reduces the impact of prefault load current on the accuracy of the fault location scheme. Then, the superimposed phase network is transformed into the sequence electrical measurement network, as shown in Fig. 7.1 [20].
Where,

i: i^{th} sequence (zero, positive, negative), $i = 0, 1, 2$.
$\Delta V_{Ai}, \Delta V_{Bi}$: i^{th} sequence superimposed voltage of terminals A and B, respectively.
$\Delta I_{Ai}, \Delta I_{Bi}$: i^{th} sequence superimposed current of terminals A and B, respectively.
Z_i, Y_i : i^{th} sequence impedance and admittance between the terminals A and B, respectively.

Fault diagnosis in two-terminal power transmission lines

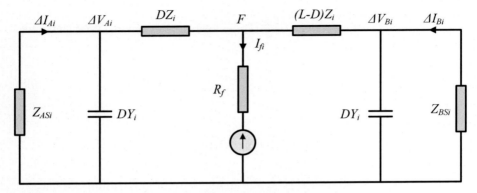

Fig. 7.1 Superimposed network of a two-terminal transmission line.

Z_{ASi}, Z_{BSi}: i^{th} sequence of equivalent source impedance at terminals A and B, respectively.
I_{fi}: i^{th} sequence of the fault current.
R_f: Fault resistance.
L: Transmission line length.
D: Fault point (F) distance from terminal A.

If the superimposed phase measurements are dV_{Aj}, dI_{Aj}, dV_{Bj}, dI_{Bj} (where j denotes the phases a, b, and c), then the symmetrical components can be obtained using the following equations:

$$\Delta V_{Ai} = M^{-1} \times dV_{Aj} \qquad (7.1)$$

$$\Delta I_{Ai} = M^{-1} \times dI_{Aj} \qquad (7.2)$$

$$\Delta V_{Bi} = M^{-1} \times dV_{Bj} \qquad (7.3)$$

$$\Delta I_{Bi} = M^{-1} \times dI_{Bj} \qquad (7.4)$$

Where,

$$M^{-1} = \frac{1}{3}\begin{bmatrix} 1 & 1 & 1 \\ 1 & e^{j120°} & e^{j240°} \\ 1 & e^{j240°} & e^{j120°} \end{bmatrix} \qquad (7.5)$$

The equivalent source impedances at the line terminals are varying depending on the change of generation mode of the system. The source impedances are computed online to mimic real operation mode and subsequently improve the accuracy of the fault location. Referring to Fig. 7.1, the source impedances are calculated as follows:

$$Z_{ASi} = \Delta V_{Ai} / \Delta I_{Ai} \qquad (7.6)$$

$$Z_{BSi} = \Delta V_{Bi} / \Delta I_{Bi} \qquad (7.7)$$

Then, the sequence voltage can be obtained at the fault point from the abruptly changed sequence voltages at the terminals and the currents flowing through the line using the following equations:

$$V_{AFi} = \left(\Delta I_{Ai} - \Delta V_{Ai} DY_i\right) \times \left(\left(Z_{ASi} \parallel \frac{1}{DY_i}\right) + DZ_i\right) \quad (7.8)$$

$$V_{BFi} = \left(\Delta I_{Bi} - \Delta V_{Bi}(L-D)Y_i\right) \times \left(\left(Z_{BSi} \parallel \frac{1}{(L-D)Y_i}\right) + (L-D)Z_i\right) \quad (7.9)$$

If $|V_{AFi}|$ and $|V_{BFi}|$ are plotted along the entire length of the transmission line, the fault distance (D) can be obtained as the point of intersection from bus A of terminal A. For instance, Figs. 7.2, 7.3, 7.4, 7.5, 7.6, 7.7, 7.8, 7.9, 7.10 and 7.11 show the fault point sequence voltages for different types applied on various locations on a 200 km long and 230 kV two-terminal transmission system. As can be seen, the intersection points of the fault point sequence voltages were representing the fault distances from bus A of terminal A. The flowchart of the first adaptive fault location algorithm for the two-terminal transmission line is illustrated in Fig. 7.12.

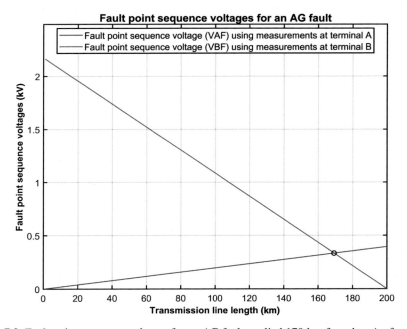

Fig. 7.2 Fault point sequence voltages for an AG fault applied 170 km from bus A of terminal A.

Fault diagnosis in two-terminal power transmission lines 163

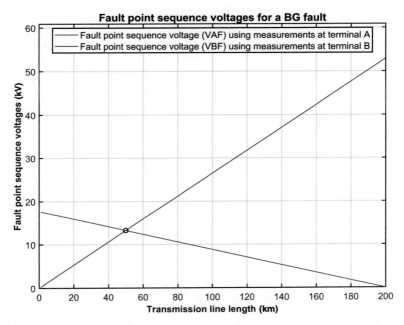

Fig. 7.3 Fault point sequence voltages for a BG fault applied 50 km from bus A of terminal A.

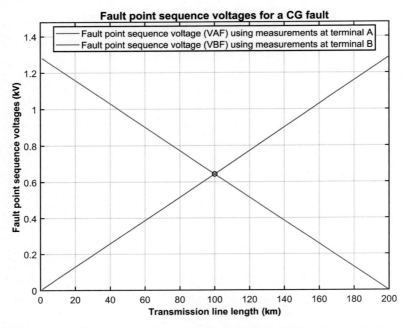

Fig. 7.4 Fault point sequence voltages for a CG fault applied 100 km from bus A of terminal A.

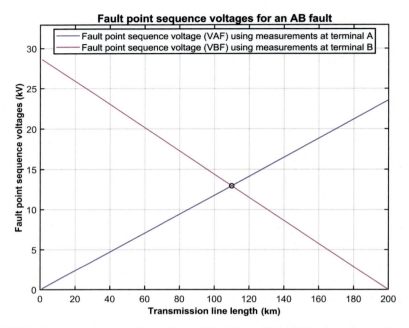

Fig. 7.5 Fault point sequence voltages for an AB fault applied 110 km from bus A of terminal A.

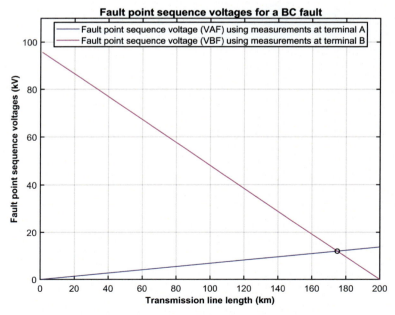

Fig. 7.6 Fault point sequence voltages for a BC fault applied 175 km from bus A of terminal A.

Fault diagnosis in two-terminal power transmission lines 165

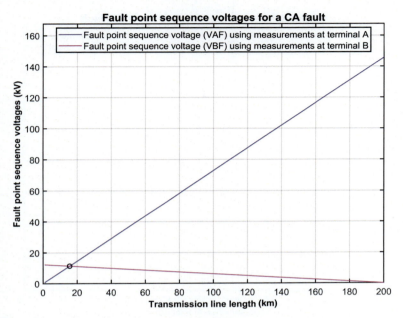

Fig. 7.7 Fault point sequence voltages for a CA fault applied 15 km from bus A of terminal A.

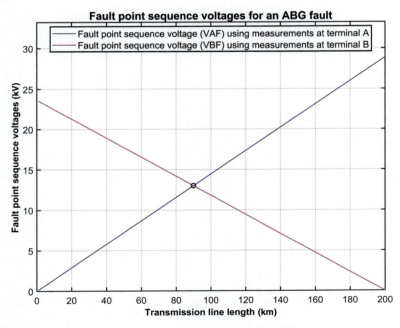

Fig. 7.8 Fault point sequence voltages for an ABG fault applied 90 km from bus A of terminal A.

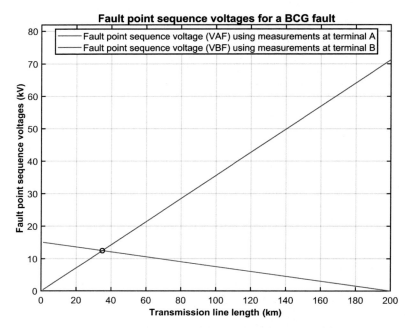

Fig. 7.9 Fault point sequence voltages for a BCG fault applied 35 km from bus A of terminal A.

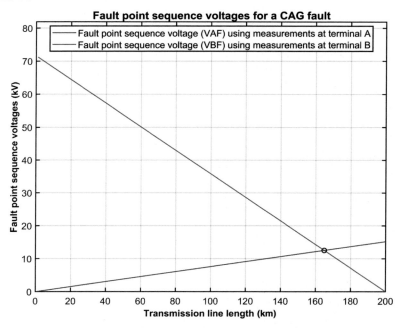

Fig. 7.10 Fault point sequence voltages for a CAG fault applied 165 km from bus A of terminal A.

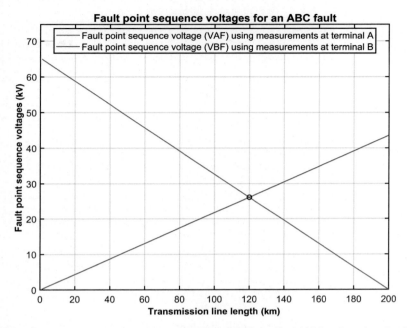

Fig. 7.11 Fault point sequence voltages for an ABC fault applied 120 km from bus A of terminal A.

7.2.2 Second fault location algorithm (FL-2)

The second adaptive fault location algorithm (FL-2) is developed to be independent of fault current to avoid possible errors related to the improper functioning of the current transformers. The faulted single line diagram of the two-terminal transmission system with its π-equivalent model is depicted in Fig. 7.13. Assuming that the line has a length L and an unknown fault occurs at a distance of D_1 from bus A of terminal A and D_2 from bus B of terminal B. Then, the postfault circuit of the system can be represented with the model shown in Fig. 7.14 with all parameters in the three-phase form (abc) [34].

Since the transmission line is divided into two parts with distances of D_1 and D_2 from the buses of the terminals A and B, respectively, thereby:

$$k = \frac{D_1}{L}, (1-k) = \frac{D_2}{L} \tag{7.10}$$

TE model of the faulted system is depicted in Fig. 7.15, where *YSA* and *YSB* represent the Thevenin admittances of the terminals A and B, respectively. Bus voltages at terminals A and B are obtained by subtracting the corresponding prefault voltages

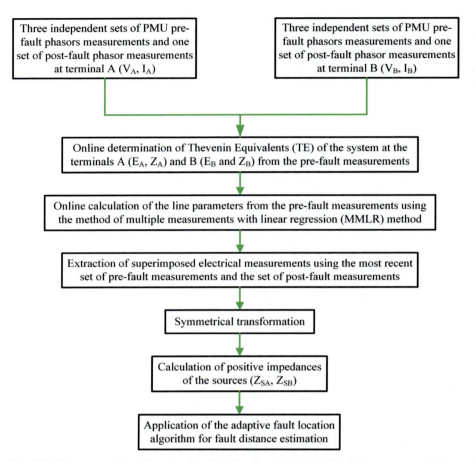

Fig. 7.12 Flowchart of the first adaptive fault location algorithm (FL-1) for a two-terminal transmission network. *PMU*, phasor measurement unit.

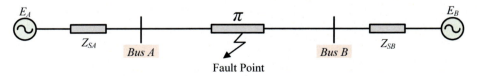

Fig. 7.13 A faulted two-terminal transmission network system represented with its π equivalent model.

Fault diagnosis in two-terminal power transmission lines

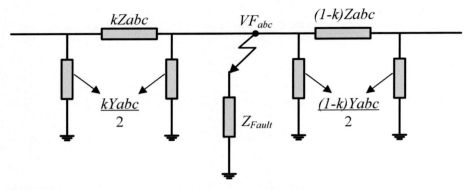

Fig. 7.14 Postfault model of the faulted two-terminal transmission line.

from the postfault voltages. Having (ΔVA) and (ΔVB), the system equations can be expressed as:

$$IA_{abc} = \begin{bmatrix} IA_a \\ IA_b \\ IA_c \end{bmatrix}, IB_{abc} = \begin{bmatrix} IB_a \\ IB_b \\ IB_c \end{bmatrix} \quad (7.11)$$

$$\begin{aligned} IA_{abc} &= \left[YSA_{abc} + \frac{Y_{abc}}{2} k \right] \Delta VA_{abc} \\ IB_{abc} &= \left[YSB_{abc} + \frac{Y_{abc}}{2}(1-k) \right] \Delta VB_{abc} \end{aligned} \quad (7.12)$$

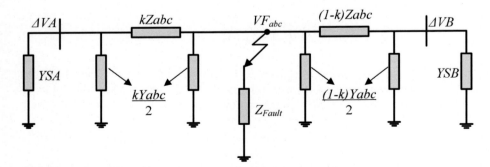

Fig. 7.15 Thevenin equivalent model of the faulted two-terminal transmission system.

Using symmetrical component transformation, the symmetrical components of the phasor currents are expressed using the following equations:

$$IA_{012} = \left[YSA_{012} + \frac{Y_{012}}{2}k \right] \Delta VA_{012}$$
$$IB_{012} = \left[YSB_{012} + \frac{Y_{012}}{2}(1-k) \right] \Delta VB_{012} \tag{7.13}$$

Fault point voltage (VF_{abc}) can be written in terms of ΔVA_{abc} and ΔVB_{abc}, and by applying the symmetrical transformation the following equations can be obtained:

$$VF_{012} = Z_{012}IA_{012}k + \Delta VA_{012}$$
$$VF_{012} = Z_{012}IB_{012}(1-k) + \Delta VB_{012} \tag{7.14}$$

To get the postfault current free formulation, Eq. (7.13) is substituted into Eq. (7.14) as follows:

$$\left[I_{3\times3} + kZ_{012}\left[YSA_{012} + \frac{Y_{012}}{2}k \right] \right] \Delta VA_{012}$$
$$= \left[I_{3\times3} + (1-k)Z_{012}\left[YSB_{012} + \frac{Y_{012}}{2}(1-k) \right] \right] \Delta VB_{012} \tag{7.15}$$

Then, the Eq. (7.15) can be written in the form of Eqs. (7.16) and (7.17):

$$ak^2 + bk + c = 0$$
$$[a]_{3\times1} = Z_{012}\frac{Y_{012}}{2}[\Delta VA_{012} - \Delta VB_{012}] \tag{7.16}$$

$$[b]_{3\times1} = Z_{012}[YSA_{012}\Delta VA_{012} + YSB_{012}\Delta VB_{012}] + Z_{012}Y_{012}\Delta VB_{012}$$
$$[c]_{3\times1} = [\Delta VA_{012} - \Delta VB_{012}] - Z_{012}\left[YSB_{012} + \frac{Y_{012}}{2} \right]\Delta VB_{012} \tag{7.17}$$

Eq. (7.16) can be divided into three equations for zero, positive, and negative sequences as shown in Eq. (7.18), since the system (e.g., generators, parallel admittance, and series impedance of the transmission line) is considered to be symmetric:

$$a_i k^2 + b_i k + c_i; \quad i = 0, 1, 2 \tag{7.18}$$

Eventually, fault-point distance is estimated by substituting k in Eq. (7.10) using the second adaptive fault location algorithm (FL-2). The flowchart of the FL-2 for the two-terminal transmission line is illustrated in Fig. 7.16.

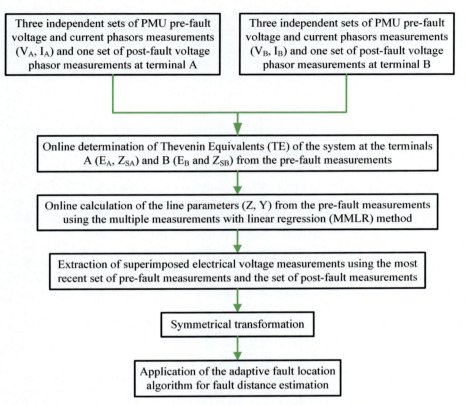

Fig. 7.16 Flowchart of the second adaptive fault location algorithm (FL-2) for a two-terminal transmission network. *PMU*, phasor measurement unit.

7.3 Simulation results

Both fault location algorithms (FL-1 and FL-2) were applied on the transmission line connecting bus-38 (terminal A) and bus-30 (terminal B) of the 115 kV transmission system selected from the SEC network, as depicted in Fig. 6.4 (Chapter 6). The simulations were carried out in PSCAD/EMTDC and MATLAB/SIMULINK platforms. The robustness of the algorithms was evaluated under prefault loading conditions, current and voltage transformers measurement, and fault information (location, type, resistance, and inception angle) uncertainty.

7.3.1 Data generation and conditioning

Prefault and postfault data were generated in PSCAD/EMTDC platform for faults that occurred on the line connecting bus-30 and bus-38. Both FL-1 and FL-2 were

Table 7.1 Transmission line (bus-30 to bus-38) parameters of the SEC network [20].

Line parameter	Value
L	26.0 km
Z	0.0144 + j0.075 p.u. in 100 MVA base
Y	0.01204 p.u. in 100 MVA base
E_A	112.280 kV
E_B	112.400 kV

pu, per unit.

implemented in MATLAB/SIMULINK environment to estimate the fault locations. Table 7.1 presents the values of the line parameters [20]. It is worth noting that the Thevenin equivalent impedances of the system at bus-30 and bus-38 were calculated online, as discussed in detail in Chapter 6. The current and voltage transformers placed at each line terminal were assumed to be ideal devices to show the errors of FL-1 and FL-2 themselves. The three-phase current and voltage signals were sampled at a frequency of 240 Hz (four samples per cycle) and stored for postprocessing. The discrete Fourier transform was applied to extract the voltage and current phasors using the following equation:

$$X = \left(\sqrt{2}/N_s\right)\sum_{k=1}^{N_s} x[k]e^{-j2\pi k/N_s} \tag{7.19}$$

Where, X, $x[k]$, and N_s denote the phasor, the waveform samples, and the total number of samples in one period, respectively. The percentage error of FL-1 and FL-2 were evaluated using the following equation:

$$\%\text{Error} = \frac{|\text{Actual location} - \text{Estimated location}|}{\text{Total line length}} \times 100 \tag{7.20}$$

7.3.2 Accuracy analysis

In this section, various types of faults with different fault locations and resistances were simulated to evaluate the accuracy of FL-1 and FL-2. Tables 7.2, 7.3, 7.4 and 7.5 provide the estimated fault locations for the single line to ground (LG), line to line to ground (LLG), line to line (LL), and three-phase (LLL) faults. The provided fault distances were measured from the bus-38 (terminal A). Obtained results proved the efficacy of the employed algorithms as they calculated the investigated fault locations with less than 1% error. Therefore, it can be concluded that both FL-1 and FL-2 located the faults accurately irrespective of the fault type.

7.3.3 Effect of fault resistance

Fault resistance uncertainty on the accuracy of the employed algorithms was investigated considering the occurrence of different types of faults at 0.80 pu distance from

Table 7.2 LG fault location estimation.

Fault type	Fault resistance (Ω)	Actual fault location (pu)	FL-1 Calculated fault location (pu)	FL-1 Error (%)	FL-2 Calculated fault location (pu)	FL-2 Error (%)
AG	10.0	0.20	0.2007	0.070	0.2000	0.004
		0.40	0.3990	0.100	0.3999	0.011
		0.60	0.5974	0.260	0.5999	0.015
		0.80	0.7958	0.420	0.7998	0.017
	100.0	0.20	0.2007	0.070	0.2000	0.003
		0.40	0.3990	0.100	0.3999	0.007
		0.60	0.5974	0.260	0.5999	0.014
		0.80	0.7958	0.420	0.7998	0.022
BG	10.0	0.20	0.2007	0.070	0.1955	0.447
		0.40	0.3990	0.100	0.3961	0.388
		0.60	0.5974	0.260	0.5970	0.298
		0.80	0.7958	0.420	0.7982	0.177
	100.0	0.20	0.2007	0.070	0.1956	0.445
		0.40	0.3990	0.100	0.3962	0.381
		0.60	0.5974	0.260	0.5971	0.286
		0.80	0.7958	0.420	0.7983	0.171
CG	10.0	0.20	0.2007	0.070	0.1913	0.875
		0.40	0.3990	0.100	0.3926	0.740
		0.60	0.5974	0.260	0.5945	0.548
		0.80	0.7958	0.420	0.7970	0.303
	100.0	0.20	0.2007	0.070	0.1913	0.875
		0.40	0.3990	0.100	0.3926	0.742
		0.60	0.5974	0.260	0.5945	0.546
		0.80	0.7958	0.420	0.7970	0.301

Table 7.3 LL fault location estimation.

Fault type	Fault resistance (Ω)	Actual fault location (pu)	FL-1 Calculated fault location (pu)	FL-1 Error (%)	FL-2 Calculated fault location (pu)	FL-2 Error (%)
AB	1.0	0.20	0.2007	0.070	0.1976	0.245
		0.40	0.3990	0.100	0.3979	0.215
		0.60	0.5974	0.260	0.5985	0.155
		0.80	0.7958	0.420	0.7993	0.071
	10.0	0.20	0.2007	0.070	0.1974	0.257
		0.40	0.3990	0.100	0.3976	0.244
		0.60	0.5974	0.260	0.5980	0.202
		0.80	0.7958	0.420	0.7987	0.126
BC	1.0	0.20	0.2007	0.070	0.1931	0.688
		0.40	0.3990	0.100	0.3940	0.597
		0.60	0.5974	0.260	0.5955	0.447
		0.80	0.7958	0.420	0.7976	0.239
	10.0	0.20	0.2007	0.070	0.1931	0.689
		0.40	0.3990	0.100	0.3939	0.607
		0.60	0.5974	0.260	0.5953	0.468
		0.80	0.7958	0.420	0.7973	0.267
CA	1.0	0.20	0.2007	0.070	0.1973	0.267
		0.40	0.3990	0.100	0.3980	0.202
		0.60	0.5974	0.260	0.5986	0.141
		0.80	0.7958	0.420	0.7992	0.076
	10.0	0.20	0.2007	0.070	0.1973	0.268
		0.40	0.3990	0.100	0.3980	0.200
		0.60	0.5974	0.260	0.5987	0.133
		0.80	0.7958	0.420	0.7993	0.066

Table 7.4 LLG fault location estimation.

Fault type	Fault resistance (Ω)	Actual fault location (pu)	FL-1 Calculated fault location (pu)	FL-1 Error (%)	FL-2 Calculated fault location (pu)	FL-2 Error (%)
ABG	5.0	0.20	0.2007	0.070	0.1981	0.195
		0.40	0.3990	0.100	0.3985	0.146
		0.60	0.5974	0.260	0.5990	0.101
		0.80	0.7958	0.420	0.7996	0.041
	50.0	0.20	0.2007	0.070	0.1976	0.236
		0.40	0.3990	0.100	0.3980	0.204
		0.60	0.5974	0.260	0.5985	0.147
		0.80	0.7958	0.420	0.7993	0.067
BCG	5.0	0.20	0.2007	0.070	0.1936	0.638
		0.40	0.3990	0.100	0.3945	0.547
		0.60	0.5974	0.260	0.5959	0.409
		0.80	0.7958	0.420	0.7978	0.220
	50.0	0.20	0.2007	0.070	0.1932	0.679
		0.40	0.3990	0.100	0.3941	0.588
		0.60	0.5974	0.260	0.5956	0.442
		0.80	0.7958	0.420	0.7976	0.236
CAG	5.0	0.20	0.2007	0.070	0.1959	0.410
		0.40	0.3990	0.100	0.3966	0.341
		0.60	0.5974	0.260	0.5975	0.251
		0.80	0.7958	0.420	0.7986	0.137
	50.0	0.20	0.2007	0.070	0.1971	0.291
		0.40	0.3990	0.100	0.3978	0.223
		0.60	0.5974	0.260	0.5984	0.156
		0.80	0.7958	0.420	0.7992	0.084

Table 7.5 LLL fault location estimation.

Fault type	Fault resistance (Ω)	Actual fault location (pu)	FL-1 Calculated fault location (pu)	FL-1 Error (%)	FL-2 Calculated fault location (pu)	FL-2 Error (%)
ABC	1.0	0.20	0.2007	0.070	0.1960	0.401
		0.40	0.3990	0.100	0.3966	0.341
		0.60	0.5974	0.260	0.5975	0.255
		0.80	0.7958	0.420	0.7986	0.137
	10.0	0.20	0.2007	0.070	0.1960	0.403
		0.40	0.3990	0.100	0.3965	0.349
		0.60	0.5974	0.260	0.5974	0.265
		0.80	0.7958	0.420	0.7985	0.151

the bus-38 (terminal A). This study simulated both low- and high-resistance faults (involving the ground) by varying the fault resistance from 0 to 500 Ω. Similarly, it also simulated faults without involving the ground where the fault resistance was picked randomly from 0 to 30 Ω. In all the investigated cases, the local and remote source impedances were chosen to be the same as the system values. Obtained results for different types of faults are presented in Tables 7.6, 7.7, 7.8 and 7.9. As can be seen, both algorithms estimated the investigated faults with excellent accuracy where the estimation errors were less than one percent.

7.3.4 Effect of fault inception angle

Likewise, fault inception angle uncertainty was investigated for three different types (AG, BC, and CAG) at terminal A (bus-38). The faults were applied 0.60 pu distance from the bus-38 by varying the inception angle from 0 to 150 degrees (°). Table 7.10 presents the obtained results, for example, calculated fault distance and percentage of error. As can be seen, both algorithms (FL-1 and FL-2) estimated the fault locations with satisfactory accuracy that confirmed the effectiveness of the employed algorithms under fault inception angle uncertainty.

7.3.5 Effect of prefault loading

In this section, the pre-fault loading condition uncertainty effect on the accuracy of the fault location algorithm was investigated from three different types of faults (AG, BC, and CAG). The pre-fault loading at terminal A (bus-38)was varied from 0.5 to 3.0 times of the original value, and the faults were applied on 0.60 pu distance from terminal A (bus-38). Table 7.11 presents the obtained results, for example, calculated fault distance and percentage of error. As can be seen, both algorithms (FL-1 and FL-2) estimated fault location with satisfactory accuracy that signaled the robustness of the algorithms even under pre-fault loading condition uncertainty. However, the

Table 7.6 Fault resistance uncertainty effect on the accuracy of fault location algorithms (LG faults occurred 0.80 pu distance from the bus-38).

Fault resistance (Ω)	Algorithm	AG Calculated fault location (pu)	AG Error (%)	BG Calculated fault location (pu)	BG Error (%)	CG Calculated fault location (pu)	CG Error (%)
0.0	FL-1	0.7958	0.420	0.7958	0.420	0.7958	0.420
	FL-2	0.8002	0.015	0.7991	0.086	0.7972	0.283
1.0	FL-1	0.7958	0.420	0.7958	0.420	0.7958	0.420
	FL-2	0.8002	0.022	0.7987	0.129	0.7969	0.306
5.0	FL-1	0.7958	0.420	0.7958	0.420	0.7958	0.420
	FL-2	0.8000	0.001	0.7983	0.167	0.7971	0.292
10.0	FL-1	0.7958	0.420	0.7958	0.420	0.7958	0.420
	FL-2	0.7998	0.017	0.7982	0.177	0.7970	0.303
20.0	FL-1	0.7958	0.420	0.7958	0.420	0.7958	0.420
	FL-2	0.8000	0.004	0.7984	0.158	0.7971	0.289
50.0	FL-1	0.7958	0.420	0.7958	0.420	0.7958	0.420
	FL-2	0.8000	0.004	0.7984	0.156	0.7971	0.290
100.0	FL-1	0.7958	0.420	0.7958	0.420	0.7958	0.420
	FL-2	0.7998	0.022	0.7983	0.171	0.7970	0.301
200.0	FL-1	0.7958	0.420	0.7958	0.420	0.7957	0.430
	FL-2	0.7999	0.007	0.7985	0.155	0.7970	0.297
400.0	FL-1	0.7958	0.420	0.7957	0.430	0.7956	0.440
	FL-2	0.7999	0.007	0.7984	0.163	0.7971	0.294
500.0	FL-1	0.7958	0.420	0.7957	0.430	0.7956	0.440
	FL-2	0.7998	0.017	0.7984	0.163	0.7970	0.302

Table 7.7 Fault resistance uncertainty effect on the accuracy of fault location algorithms (LL faults occurred 0.80 pu distance from the bus-38).

Fault resistance (Ω)	Algorithm	Fault type					
		AB		BC		CA	
		Calculated fault location (pu)	Error (%)	Calculated fault location (pu)	Error (%)	Calculated fault location (pu)	Error (%)
0.0	FL-1	0.7958	0.420	0.7958	0.420	0.7958	0.420
	FL-2	0.7994	0.065	0.7979	0.212	0.7992	0.077
0.50	FL-1	0.7958	0.420	0.7958	0.420	0.7958	0.420
	FL-2	0.7993	0.067	0.7977	0.226	0.7992	0.077
1.0	FL-1	0.7958	0.420	0.7958	0.420	0.7958	0.420
	FL-2	0.7993	0.071	0.7976	0.239	0.7992	0.076
2.50	FL-1	0.7958	0.420	0.7958	0.420	0.7958	0.420
	FL-2	0.7991	0.088	0.7973	0.273	0.7993	0.071
5.0	FL-1	0.7958	0.420	0.7958	0.420	0.7958	0.420
	FL-2	0.7989	0.112	0.7973	0.275	0.7994	0.064
7.50	FL-1	0.7958	0.420	0.7958	0.420	0.7958	0.420
	FL-2	0.7988	0.123	0.7973	0.272	0.7994	0.064
10.0	FL-1	0.7958	0.420	0.7958	0.420	0.7958	0.420
	FL-2	0.7987	0.126	0.7973	0.267	0.7993	0.066
15.0	FL-1	0.7958	0.420	0.7958	0.420	0.7958	0.420
	FL-2	0.7987	0.126	0.7974	0.262	0.7993	0.068
20.0	FL-1	0.7958	0.420	0.7958	0.420	0.7958	0.420
	FL-2	0.7988	0.125	0.7974	0.260	0.7993	0.069
30.0	FL-1	0.7958	0.420	0.7958	0.420	0.7958	0.420
	FL-2	0.7988	0.124	0.7974	0.258	0.7993	0.069

Table 7.8 Fault resistance uncertainty effect on the accuracy of fault location algorithms (LLG faults occurred 0.80 pu distance from the bus-38).

Fault resistance (Ω)	Algorithm	Fault type ABG Calculated fault location (pu)	ABG Error (%)	BCG Calculated fault location (pu)	BCG Error (%)	CAG Calculated fault location (pu)	CAG Error (%)
0.0	FL-1	0.7958	0.420	0.7958	0.420	0.7958	0.420
	FL-2	0.7996	0.041	0.7982	0.184	0.7988	0.124
1.0	FL-1	0.7958	0.420	0.7958	0.420	0.7958	0.420
	FL-2	0.7996	0.041	0.7982	0.183	0.7986	0.138
5.0	FL-1	0.7958	0.420	0.7958	0.420	0.7958	0.420
	FL-2	0.7996	0.041	0.7978	0.220	0.7986	0.137
10.0	FL-1	0.7958	0.420	0.7958	0.420	0.7958	0.420
	FL-2	0.7994	0.057	0.7978	0.220	0.7990	0.102
25.0	FL-1	0.7958	0.420	0.7958	0.420	0.7958	0.420
	FL-2	0.7994	0.063	0.7977	0.231	0.7991	0.088
50.0	FL-1	0.7958	0.420	0.7958	0.420	0.7958	0.420
	FL-2	0.7993	0.067	0.7976	0.236	0.7992	0.084
100.0	FL-1	0.7958	0.420	0.7958	0.420	0.7958	0.420
	FL-2	0.7993	0.069	0.7976	0.237	0.7992	0.079
150.0	FL-1	0.7958	0.420	0.7958	0.420	0.7958	0.420
	FL-2	0.7993	0.070	0.7976	0.238	0.7992	0.078
200.0	FL-1	0.7958	0.420	0.7958	0.420	0.7958	0.420
	FL-2	0.7993	0.070	0.7976	0.238	0.7992	0.078
250.0	FL-1	0.7958	0.420	0.7958	0.420	0.7958	0.420
	FL-2	0.7993	0.070	0.7976	0.238	0.7992	0.078

Table 7.9 Fault resistance uncertainty effect on the accuracy of fault location algorithms (LLL faults occurred 0.80 pu distance from the bus-38).

Fault resistance (Ω)	Algorithm	Calculated fault location (pu)	Error (%)
0.0	FL-1	0.7958	0.420
	FL-2	0.7988	0.118
0.50	FL-1	0.7958	0.420
	FL-2	0.7987	0.127
1.0	FL-1	0.7958	0.420
	FL-2	0.7986	0.137
2.50	FL-1	0.7958	0.420
	FL-2	0.7985	0.150
5.0	FL-1	0.7958	0.420
	FL-2	0.7985	0.152
7.50	FL-1	0.7958	0.420
	FL-2	0.7985	0.152
10.0	FL-1	0.7958	0.420
	FL-2	0.7985	0.151
15.0	FL-1	0.7958	0.420
	FL-2	0.7985	0.150
20.0	FL-1	0.7958	0.420
	FL-2	0.7985	0.149
30.0	FL-1	0.7958	0.420
	FL-2	0.7985	0.149

accuracy of both algorithms degraded with the percentage increase of the prefault loading condition (Fig. 7.17). It is worth noting the FL-1 algorithm outperformed the FL-2 algorithm under prefault load variation.

7.3.6 Effect of measurement errors

The effect of measurement errors on the accuracy of the fault location algorithms (FL-1 and FL-2) was investigated as in reality both current and voltage transformers introduce such errors. This chapter considered 2% errors in the magnitudes and 2° errors in the phase angles for both voltage and currents measured at the line terminals. In addition, a situation where both sending and receiving ends (terminals A and B) measurements had the maximum errors with the opposite signs in the angles (+2° for I_A and −2° for I_B or +2° for V_A and −2° for V_B) and magnitudes (+2% for I_A and −2% for I_B or +2% for V_A and −2% for V_B) were considered. Tables 7.12, 7.13, 7.14 and 7.15 present the results obtained for LG faults. Provided fault distances were measured from the bus-38 (terminal A). As can be seen, both fault location algorithms estimated faults with satisfactory accuracy even under the mentioned measurement errors. However, the FL-1 outperformed the FL-2 in terms of overall accuracy under measurement errors.

Table 7.10 Fault inception angle uncertainty effect on the accuracy of fault location algorithms (investigated faults occurred 0.60 pu distance from the bus-38).

Inception angle (°)	Algorithm	Fault type						
		AG		BC		CAG	
		Calculated fault location (pu)	Error (%)	Calculated fault location (pu)	Error (%)	Calculated fault location (pu)	Error (%)
0.0	FL-1	0.6023	0.230	0.6011	0.110	0.6016	0.160
	FL-2	0.6001	0.010	0.5955	0.450	0.5973	0.270
30.0	FL-1	0.6023	0.230	0.6011	0.110	0.6016	0.160
	FL-2	0.5984	0.160	0.5954	0.460	0.5965	0.350
45.0	FL-1	0.6022	0.220	0.6011	0.110	0.6016	0.160
	FL-2	0.6041	0.410	0.5955	0.450	0.5993	0.070
60.0	FL-1	0.6022	0.220	0.6011	0.110	0.6015	0.150
	FL-2	0.6009	0.090	0.5955	0.450	0.5978	0.220
90.0	FL-1	0.6021	0.210	0.6010	0.100	0.6014	0.140
	FL-2	0.6003	0.030	0.5944	0.560	0.5970	0.300
120.0	FL-1	0.6023	0.230	0.6008	0.080	0.6014	0.140
	FL-2	0.5957	0.430	0.5963	0.370	0.5951	0.490
135.0	FL-1	0.6024	0.240	0.6007	0.070	0.6015	0.150
	FL-2	0.6029	0.290	0.5955	0.450	0.5986	0.140
150.0	FL-1	0.6026	0.260	0.6008	0.080	0.6017	0.170
	FL-2	0.5921	0.790	0.5945	0.550	0.5932	0.680

Table 7.11 Prefault loading condition effect on the accuracy of the fault location algorithms (faults occurred 0.60 pu distance from the bus-38).

Prefault loading factor	Algorithm	Fault type AG Calculated fault location (pu)	Error (%)	BC Calculated fault location (pu)	Error (%)	CAG Calculated fault location (pu)	Error (%)
0.50	FL-1	0.6016	0.160	0.6004	0.040	0.6009	0.090
	FL-2	0.5986	0.140	0.5988	0.120	0.5986	0.140
0.80	FL-1	0.6020	0.200	0.6008	0.080	0.6013	0.130
	FL-2	0.5977	0.230	0.5978	0.220	0.5976	0.240
1.20	FL-1	0.6025	0.250	0.6014	0.140	0.6019	0.190
	FL-2	0.5969	0.310	0.5970	0.300	0.5969	0.310
1.50	FL-1	0.6030	0.300	0.6018	0.180	0.6023	0.230
	FL-2	0.5960	0.400	0.5962	0.380	0.5960	0.400
2.00	FL-1	0.6036	0.360	0.6024	0.240	0.6029	0.290
	FL-2	0.5946	0.540	0.5948	0.520	0.5946	0.540
3.00	FL-1	0.6049	0.490	0.6038	0.380	0.6043	0.430
	FL-2	0.5919	0.810	0.5921	0.790	0.5919	0.810

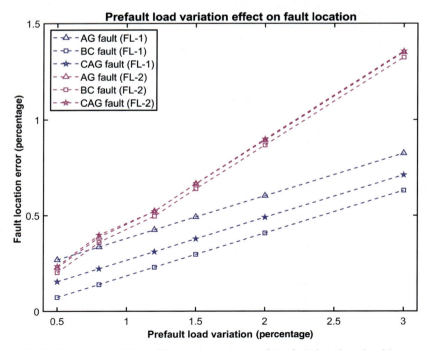

Fig. 7.17 Prefault load variation effect on the accuracy of the fault location algorithms.

Table 7.12 Effect of measurement errors (2% voltage magnitude) on the accuracy of the adaptive fault location algorithms.

Fault type	Fault resistance (Ω)	Actual fault location (pu)	FL-1 Calculated fault location (pu)	FL-1 Error (%)	FL-2 Calculated fault location (pu)	FL-2 Error (%)
AG	10.0	0.20	0.2007	0.070	0.1878	1.218
		0.40	0.3990	0.100	0.3821	1.794
		0.60	0.5974	0.260	0.5774	2.264
		0.80	0.7958	0.420	0.7747	2.530
	100.0	0.20	0.2007	0.070	0.1879	1.213
		0.40	0.3990	0.100	0.3821	1.789
		0.60	0.5974	0.260	0.5774	2.262
		0.80	0.7958	0.420	0.7747	2.532
BG	10.0	0.20	0.2007	0.070	0.1835	1.648
		0.40	0.3990	0.100	0.3784	2.160
		0.60	0.5974	0.260	0.5746	2.541
		0.80	0.7958	0.420	0.7731	2.686
	100.0	0.20	0.2007	0.070	0.1836	1.644
		0.40	0.3990	0.100	0.3784	2.157
		0.60	0.5974	0.260	0.5747	2.532
		0.80	0.7958	0.420	0.7732	2.681
CG	10.0	0.20	0.2007	0.070	0.1794	2.064
		0.40	0.3990	0.100	0.3750	2.504
		0.60	0.5974	0.260	0.5721	2.786
		0.80	0.7958	0.420	0.7719	2.813
	100.0	0.20	0.2007	0.070	0.1794	2.063
		0.40	0.3990	0.100	0.3750	2.503
		0.60	0.5974	0.260	0.5722	2.783
		0.80	0.7958	0.420	0.7719	2.814

Table 7.13 Effect of measurement errors (2° voltage phase angle) on the accuracy of the adaptive fault location algorithms.

Fault type	Fault resistance (Ω)	Actual fault location (pu)	FL-1 Calculated fault location (pu)	FL-1 Error (%)	FL-2 Calculated fault location (pu)	FL-2 Error (%)
AG	10.0	0.20	0.2007	0.070	0.1937	0.630
		0.40	0.3990	0.100	0.3931	0.695
		0.60	0.5974	0.260	0.5925	0.749
		0.80	0.7958	0.420	0.7926	0.742
	100.0	0.20	0.2007	0.070	0.1937	0.629
		0.40	0.3990	0.100	0.3931	0.689
		0.60	0.5974	0.260	0.5925	0.747
		0.80	0.7958	0.420	0.7926	0.744
BG	10.0	0.20	0.2007	0.070	0.1820	1.804
		0.40	0.3990	0.100	0.3892	1.082
		0.60	0.5974	0.260	0.5896	1.038
		0.80	0.7958	0.420	0.7910	0.900
	100.0	0.20	0.2007	0.070	0.1892	1.080
		0.40	0.3990	0.100	0.3892	1.078
		0.60	0.5974	0.260	0.5897	1.027
		0.80	0.7958	0.420	0.7910	0.897
CG	10.0	0.20	0.2007	0.070	0.1848	1.523
		0.40	0.3990	0.100	0.3856	1.445
		0.60	0.5974	0.260	0.5871	1.295
		0.80	0.7958	0.420	0.7897	1.030
	100.0	0.20	0.2007	0.070	0.1848	1.522
		0.40	0.3990	0.100	0.3856	1.445
		0.60	0.5974	0.260	0.5871	1.293
		0.80	0.7958	0.420	0.7897	1.030

Table 7.14 Effect of measurement errors (2% current magnitude) on the accuracy of the adaptive fault location algorithms.

Fault type	Fault resistance (Ω)	Actual fault location (pu)	FL-1 Calculated fault location (pu)	FL-1 Error (%)	FL-2 Calculated fault location (pu)	FL-2 Error (%)
AG	10.0	0.20	0.2007	0.070	0.2043	0.431
		0.40	0.3990	0.100	0.4021	0.209
		0.60	0.5974	0.260	0.5999	0.011
		0.80	0.7958	0.420	0.7977	0.230
	100.0	0.20	0.2007	0.070	0.2044	0.435
		0.40	0.3990	0.100	0.4021	0.211
		0.60	0.5974	0.260	0.5999	0.012
		0.80	0.7958	0.420	0.7977	0.232
BG	10.0	0.20	0.2007	0.070	0.1999	0.006
		0.40	0.3990	0.100	0.3984	0.164
		0.60	0.5974	0.260	0.5971	0.291
		0.80	0.7958	0.420	0.7962	0.385
	100.0	0.20	0.2007	0.070	0.2000	0.001
		0.40	0.3990	0.100	0.3984	0.159
		0.60	0.5974	0.260	0.5972	0.284
		0.80	0.7958	0.420	0.7962	0.380
CG	10.0	0.20	0.2007	0.070	0.1957	0.429
		0.40	0.3990	0.100	0.3949	0.512
		0.60	0.5974	0.260	0.5946	0.538
		0.80	0.7958	0.420	0.7949	0.511
	100.0	0.20	0.2007	0.070	0.1957	0.426
		0.40	0.3990	0.100	0.3949	0.510
		0.60	0.5974	0.260	0.5946	0.536
		0.80	0.7958	0.420	0.7949	0.515

Table 7.15 Effect of measurement errors (2° current phase angle) on the accuracy of the adaptive fault location algorithms.

Fault type	Fault resistance (Ω)	Actual fault location (pu)	FL-1 Calculated fault location (pu)	FL-1 Error (%)	FL-2 Calculated fault location (pu)	FL-2 Error (%)
AG	10.0	0.20	0.2007	0.070	0.2049	0.494
		0.40	0.3990	0.100	0.4040	0.402
		0.60	0.5974	0.260	0.6031	0.313
		0.80	0.7958	0.420	0.8022	0.224
	100.0	0.20	0.2007	0.070	0.2050	0.500
		0.40	0.3990	0.100	0.4041	0.405
		0.60	0.5974	0.260	0.6031	0.314
		0.80	0.7958	0.420	0.8022	0.220
BG	10.0	0.20	0.2007	0.070	0.2005	0.049
		0.40	0.3990	0.100	0.4002	0.022
		0.60	0.5974	0.260	0.6003	0.027
		0.80	0.7958	0.420	0.8007	0.066
	100.0	0.20	0.2007	0.070	0.2006	0.055
		0.40	0.3990	0.100	0.4003	0.029
		0.60	0.5974	0.260	0.6004	0.036
		0.80	0.7958	0.420	0.8007	0.071
CG	10.0	0.20	0.2007	0.070	0.1962	0.381
		0.40	0.3990	0.100	0.3967	0.334
		0.60	0.5974	0.260	0.5977	0.226
		0.80	0.7958	0.420	0.7994	0.064
	100.0	0.20	0.2007	0.070	0.1962	0.379
		0.40	0.3990	0.100	0.3967	0.333
		0.60	0.5974	0.260	0.5978	0.223
		0.80	0.7958	0.420	0.7994	0.063

7.3.7 Effect of number of measurement sets

As presented in Chapter 6, transmission line parameters can be calculated using either one set or multiple sets of PMU current and voltage measurements recorded at the line terminals. This section investigated the effects of calculating the transmission line parameters using single and multiple sets of measurements on the accuracy of the FL-2 algorithm in the presence of current and voltage transformers measurement errors. It considered 2% errors in the magnitudes and 2° errors in the phase angles for both voltage and currents measured at the line terminals. Tables 7.16, 7.17, 7.18 and 7.19 present the obtained results for different types of LG faults. Provided fault distances were measured from the bus-38 (terminal A). As can be seen from the presented results, the performance of the FL-2 was satisfactory for most of the investigated cases except for the single set measurement and 2% voltage magnitude error. Overall, the multiple sets measurement scheme outperformed the single set measurement scheme in terms of accuracy and consistency.

Table 7.16 Effect of the number of measurement sets and measurement errors (2% voltage magnitude) on the accuracy of the FL-2 algorithm.

Fault type	Fault resistance (Ω)	Actual fault location (pu)	Single set measurement Calculated fault location (pu)	Error (%)	Multiple sets of measurement Calculated fault location (pu)	Error (%)
AG	10.0	0.20	0.2105	1.049	0.1878	1.218
		0.40	0.3791	2.091	0.3821	1.794
		0.60	0.5477	5.230	0.5774	2.264
		0.80	0.7163	8.369	0.7747	2.530
	100.0	0.20	0.2106	1.055	0.1879	1.213
		0.40	0.3791	2.090	0.3821	1.789
		0.60	0.5477	5.229	0.5774	2.262
		0.80	0.7163	8.374	0.7747	2.532
BG	10.0	0.20	0.2075	0.752	0.1835	1.648
		0.40	0.3767	2.331	0.3784	2.160
		0.60	0.5460	5.404	0.5746	2.541
		0.80	0.7154	8.463	0.7731	2.686
	100.0	0.20	0.2076	0.755	0.1836	1.644
		0.40	0.3767	2.327	0.3784	2.157
		0.60	0.5460	5.399	0.5747	2.532
		0.80	0.7154	8.462	0.7732	2.681
CG	10.0	0.20	0.2047	0.471	0.1794	2.064
		0.40	0.3746	2.544	0.3750	2.504
		0.60	0.5446	5.542	0.5721	2.786
		0.80	0.7147	8.526	0.7719	2.813
	100.0	0.20	0.2047	0.471	0.1794	2.063
		0.40	0.3746	2.543	0.3750	2.503
		0.60	0.5446	5.539	0.5722	2.783
		0.80	0.7147	8.530	0.7719	2.814

Table 7.17 Effect of the number of measurement sets and measurement errors (2° voltage phase angle) on the accuracy of the FL-2 algorithm.

Fault type	Fault resistance (Ω)	Actual fault location (pu)	Single set measurement Calculated fault location (pu)	Error (%)	Multiple sets of measurement Calculated fault location (pu)	Error (%)
AG	10.0	0.20	0.1874	1.259	0.1937	0.630
		0.40	0.3911	0.888	0.3931	0.695
		0.60	0.5942	0.580	0.5925	0.749
		0.80	0.7966	0.339	0.7926	0.742
	100.0	0.20	0.1874	1.258	0.1937	0.629
		0.40	0.3912	0.883	0.3931	0.689
		0.60	0.5942	0.579	0.5925	0.747
		0.80	0.7966	0.340	0.7926	0.744
BG	10.0	0.20	0.1829	1.714	0.1820	1.804
		0.40	0.3873	1.272	0.3892	1.082
		0.60	0.5914	0.864	0.5896	1.038
		0.80	0.7951	0.495	0.7910	0.900
	100.0	0.20	0.1829	1.710	0.1892	1.080
		0.40	0.3873	1.267	0.3892	1.078
		0.60	0.5915	0.854	0.5897	1.027
		0.80	0.7950	0.496	0.7910	0.897
CG	10.0	0.20	0.1785	2.154	0.1848	1.523
		0.40	0.3837	1.631	0.3856	1.445
		0.60	0.5889	1.115	0.5871	1.295
		0.80	0.7938	0.621	0.7897	1.030
	100.0	0.20	0.1785	2.154	0.1848	1.522
		0.40	0.3837	1.632	0.3856	1.445
		0.60	0.5889	1.112	0.5871	1.293
		0.80	0.7938	0.621	0.7897	1.030

Table 7.18 Effect of the number of measurement sets and measurement errors (2% current magnitude) on the accuracy of the FL-2 algorithm.

Fault type	Fault resistance (Ω)	Actual fault location (pu)	Single set measurement Calculated fault location (pu)	Error (%)	Multiple sets of measurement Calculated fault location (pu)	Error (%)
AG	10.0	0.20	0.2004	0.035	0.2043	0.431
		0.40	0.4002	0.020	0.4021	0.209
		0.60	0.6001	0.006	0.5999	0.011
		0.80	0.8000	0.005	0.7977	0.230
	100.0	0.20	0.2004	0.040	0.2044	0.435
		0.40	0.4003	0.025	0.4021	0.211
		0.60	0.6001	0.008	0.5999	0.012
		0.80	0.7999	0.009	0.7977	0.232
BG	10.0	0.20	0.1959	0.407	0.1999	0.006
		0.40	0.3964	0.358	0.3984	0.164
		0.60	0.5972	0.277	0.5971	0.291
		0.80	0.7984	0.162	0.7962	0.385
	100.0	0.20	0.1960	0.404	0.2000	0.001
		0.40	0.3965	0.354	0.3984	0.159
		0.60	0.5973	0.268	0.5972	0.284
		0.80	0.7984	0.157	0.7962	0.380
CG	10.0	0.20	0.1917	0.835	0.1957	0.429
		0.40	0.3929	0.711	0.3949	0.512
		0.60	0.5947	0.527	0.5946	0.538
		0.80	0.7971	0.291	0.7949	0.511
	100.0	0.20	0.1917	0.835	0.1957	0.426
		0.40	0.3929	0.712	0.3949	0.510
		0.60	0.5948	0.524	0.5946	0.536
		0.80	0.7971	0.293	0.7949	0.515

Table 7.19 Effect of the number of measurement sets and measurement errors (2° current phase angle) on the accuracy of the FL-2 algorithm.

Fault type	Fault resistance (Ω)	Actual fault location (pu)	Single set measurement Calculated fault location (pu)	Error (%)	Multiple sets of measurement Calculated fault location (pu)	Error (%)
AG	10.0	0.20	0.2003	0.033	0.2049	0.494
		0.40	0.4002	0.018	0.4040	0.402
		0.60	0.6001	0.007	0.6031	0.313
		0.80	0.8000	0.004	0.8022	0.224
	100.0	0.20	0.2004	0.037	0.2050	0.500
		0.40	0.4002	0.019	0.4041	0.405
		0.60	0.6001	0.008	0.6031	0.314
		0.80	0.8000	0.004	0.8022	0.220
BG	10.0	0.20	0.1959	0.410	0.2005	0.049
		0.40	0.3964	0.359	0.4002	0.022
		0.60	0.5972	0.276	0.6003	0.027
		0.80	0.7984	0.161	0.8007	0.066
	100.0	0.20	0.1960	0.405	0.2006	0.055
		0.40	0.3965	0.353	0.4003	0.029
		0.60	0.5973	0.267	0.6004	0.036
		0.80	0.7984	0.158	0.8007	0.071
CG	10.0	0.20	0.1916	0.838	0.1962	0.381
		0.40	0.3929	0.712	0.3967	0.334
		0.60	0.5947	0.527	0.5977	0.226
		0.80	0.7971	0.289	0.7994	0.064
	100.0	0.20	0.1917	0.835	0.1962	0.379
		0.40	0.3929	0.711	0.3967	0.333
		0.60	0.5948	0.523	0.5978	0.223
		0.80	0.7971	0.289	0.7994	0.063

7.4 Summary

In this chapter, two fault location algorithms for the determination of the location faults in two-terminal transmission lines using synchronized post and prefault measurements recorded by the PMU. The adaptive algorithms used PMU current and voltage measurements for online computations of line parameters and TE that reduced parameter uncertainty due to the operational and environmental conditions. The first algorithm (FL-1) used the abruptly changed currents and voltages to get abruptly changed positive current and voltage components to solve the system impedance during the fault time. However, the transient state and overvoltage of the power network during fault occurrence leads to improper functioning of current transformers that affect the accuracy of the algorithm. In response, this chapter employed the second algorithm (FL-2) that was independent of the fault currents. However, from the obtained results, it was evident that the presented fault location algorithms estimated the fault distances for the investigated faults with satisfactory accuracy. Besides, both algorithms showed their independence on fault information (location, type, resistance, and inception angle), measurement, and prefault loading uncertainty. In terms of overall accuracy and consistency, the second algorithm outperformed the first one. Furthermore, the multiple measurement-based approach estimated fault distances more accurately than the single measurement-based approach.

References

[1] J. Doria-García, C. Orozco-Henao, R. Leborgne, O.D. Montoya, W. Gil-González, High impedance fault modeling and location for transmission line, Electr. Power Syst. Res. 196 (Jul) (2021) 107202, doi:10.1016/j.epsr.2021.107202.

[2] A. Mukherjee, P.K. Kundu, A. Das, A supervised principal component analysis-based approach of fault localization in transmission lines for single line to ground faults, Electr. Eng. 1 (4) (2021) 3, doi:10.1007/s00202-021-01221-9.

[3] M. Shafiullah, M.A. Abido, S-transform based FFNN approach for distribution grids fault detection and classification, IEEE Access 6 (1) (2018) 8080–8088, doi:10.1109/ACCESS.2018.2809045.

[4] J. Sadeh, A. Adinehzadeh, Accurate fault location algorithm for transmission line in the presence of series connected FACTS devices, Int. J. Electr. Power Energy Syst. 32 (4) (2010) 323–328, doi:10.1016/j.ijepes.2009.09.001.

[5] W. Bo, Q. Jiang, and Y. Cao, Transmission network fault location using sparse PMU measurements, 2009, doi:10.1109/SUPERGEN.2009.5348286.

[6] C.A. Apostolopoulos, G.N. Korres, A novel algorithm for locating faults on transposed/untransposed transmission lines without utilizing line parameters, IEEE Trans. Power Deliv. 25 (4) (2010) 2328–2338, doi:10.1109/TPWRD.2010.2053223.

[7] M. Shafiullah, M. Ijaz, M.A. Abido, Z. Al-Hamouz, Optimized support vector machine & wavelet transform for distribution grid fault location, 11th IEEE International Conference on Compatibility, Power Electronics and Power Engineering (CPE-POWERENG), IEEE, 2017, pp. 77–82, doi:10.1109/CPE.2017.7915148.

[8] A. Aljohani, T. Sheikhoon, A. Fataa, M. Shafiullah, and M.A. Abido, Design and implementation of an intelligent single line to ground fault locator for distribution feeders, 2019, doi:10.1109/ICCAD46983.2019.9037950.

[9] M.M. Saha, J.J. Izykowski, E. Rosolowski, Fault Location on Power Networks, Springer Science & Business Media, Berlin, Germany, 2009.

[10] I.A. Farhat, Fault Diagnosis in Electric Power Transmission Systems: Detection, classification and isolation of faults in electric power transmission lines, Noor Publishing, Chişinău, Republic of Moldova, 2017.

[11] S. Belagoune, N. Bali, A. Bakdi, B. Baadji, K. Atif, Deep learning through LSTM classification and regression for transmission line fault detection, diagnosis and location in large-scale multi-machine power systems, Measurement 177 (Jun) (2021) 109330, doi:10.1016/j.measurement.2021.109330.

[12] M. Shafiullah, M.A. Abido, A review on distribution grid fault location techniques, Electr. Power Components Syst. 45 (8) (2017) 807–824, doi:10.1080/15325008.2017.1310772.

[13] A. Al-Mohammed, M. Abido, A fully adaptive PMU-based fault location algorithm for series-compensated lines, IEEE Trans. Power Syst. 29 (5) (2014) 2129–2137, http://ieeexplore.ieee.org/xpls/abs_all.jsp?arnumber=6736121. (Accessed 19 August 2018).

[14] M. Shafiullah, M. Abido, T. Abdel-Fattah, Distribution grids fault location employing ST based optimized machine learning approach, Energies 11 (9) (2018) 2328, doi:10.3390/en11092328.

[15] A. Mukherjee, P.K. Kundu, A. Das, Transmission line faults in power system and the different algorithms for identification, classification and localization: a brief review of methods, J. Inst. Eng. (India): B 102 (4) (2021) 855–877, doi:10.1007/s40031-020-00530-0.

[16] M. Reta-Hernández, Transmission line parameters, In: Electric Power Generation, Transmission, and Distribution, CRC Press, Florida, USA, 2018, pp. 14-1-14-28.

[17] M. Bartos, et al., Impacts of rising air temperatures on electric transmission ampacity and peak electricity load in the United States, Environ. Res. Lett. 11 (11) (2016) 114008, doi:10.1088/1748-9326/11/11/114008.

[18] K. Yu, et al., A novel traveling wave fault location method for transmission network based on time linear dependence, Int. J. Electr. Power Energy Syst. 126 (Mar) (2021) 106608, doi:10.1016/j.ijepes.2020.106608.

[19] M. Shafiullah, M.A. Abido, Z. Al-Hamouz, Wavelet-based extreme learning machine for distribution grid fault location, IET Gener. Transm. Distrib. 11 (17) (2017) 4256–4263, doi:10.1049/iet-gtd.2017.0656.

[20] A.H. Al-Mohammed, M.A. Abido, An adaptive fault location algorithm for power system networks based on synchrophasor measurements, Electr. Power Syst. Res. 108 (Mar) (Mar 2014) 153–163, doi:10.1016/j.epsr.2013.10.013.

[21] H. Sardari, B. Mozafari, H.A. Shayanfar, Fast & adaptive fault location technique for three-terminal lines, Electr. Power Syst. Res. 179 (Feb) (2020) 106084, doi:10.1016/j.epsr.2019.106084.

[22] A.H. Al-Mohammed, M.A. Abido, Adaptive fault location for three-terminal lines using synchrophasors, 2014 IEEE International Workshop on Applied Measurements for Power Systems, AMPS 2014 - Proceedings, 2014 69–74, doi:10.1109/AMPS.2014.6947710.

[23] M. Shafiullah, M. Abido, M. Hossain, A. Mantawy, An improved OPP problem formulation for distribution grid observability, Energies 11 (11) (2018) 3069, doi:10.3390/en11113069.

[24] M. Shafiullah, M.I. Hossain, M.A. Abido, T. Abdel-Fattah, A.H. Mantawy, A modified optimal PMU placement problem formulation considering channel limits under

various contingencies, Meas. J. Int. Meas. Confed. 135 (Mar) (2019), doi:10.1016/j.measurement.2018.12.039.
[25] R.M. Arias Velásquez, Failure analysis and dispatch optimization using phasor measurement units, Eng. Fail. Anal. 121 (Mar) (2021) 105157, doi:10.1016/j.engfailanal.2020.105157.
[26] S. Hussain, A.H. Osman, Fault location scheme for multi-terminal transmission lines using unsynchronized measurements, Int. J. Electr. Power Energy Syst. 78 (Jun) (2016) 277–284, doi:10.1016/j.ijepes.2015.11.060.
[27] M. Shafiullah, S.M. Rahman, M.G. Mortoja, B. Al-Ramadan, Role of spatial analysis technology in power system industry: An overview, Renew. Sustain. Energy Rev. 66 (Dec) (2016) 584–595, doi:10.1016/j.rser.2016.08.017.
[28] F.V. Lopes, A. Mouco, R.O. Fernandes, F.C. Neto, Real-World case studies on transmission line fault location feasibility by using M-Class phasor measurement units, Electr. Power Syst. Res. 196 (Jul) (2021) 107261, doi:10.1016/j.epsr.2021.107261.
[29] P.R. Chegireddy, R. Bhimasingu, Synchrophasor based fault location algorithm for three terminal homogeneous transmission lines, Electr. Power Syst. Res. 191 (Feb) (2021) 106889, doi:10.1016/j.epsr.2020.106889.
[30] V. Terzija, Z.M. Radojević, G. Preston, Flexible synchronized measurement technology-based fault locator, IEEE Trans. Smart Grid 6 (2) (2015) 866–873, doi:10.1109/TSG.2014.2367820.
[31] C. Fan, X. Du, S. Li, and W. Yu, An adaptive fault location technique based on PMU for transmission line, 2007, doi:10.1109/PES.2007.385545.
[32] Y. Liao, N. Kang, Fault-location algorithms without utilizing line parameters based on the distributed parameter line model, IEEE Trans. Power Deliv. 24 (2) (2009) 579–584, doi:10.1109/TPWRD.2008.2002698.
[33] S. Das, S.P. Singh, B.K. Panigrahi, Transmission line fault detection and location using wide area measurements, Electr. Power Syst. Res. 151 (Oct) (2017) 96–105, doi:10.1016/j.epsr.2017.05.025.
[34] K.G. Firouzjah, A. Sheikholeslami, A current independent method based on synchronized voltage measurement for fault location on transmission lines, Simul. Model. Pract. Theory 17 (4) (2009) 692–707, doi:10.1016/j.simpat.2008.12.003.
[35] E. Nashawati, R. Garcia, T. Rosenberger, Using synchrophasor for fault location identification, 2012 65th Annual Conference for Protective Relay Engineers, 2012 14–21, doi:10.1109/CPRE.2012.6201218.
[36] Q. Jiang, X. Li, B. Wang, H. Wang, PMU-based fault location using voltage measurements in large transmission networks, IEEE Trans. Power Deliv. 27 (3) (2012) 1644–1652, doi:10.1109/TPWRD.2012.2199525.
[37] C.S. Chen, C.W. Liu, J.A. Jiang, A new adaptive PMU based protection scheme for transposed/untransposed parallel transmission lines, IEEE Trans. Power Deliv. 17 (2) (2002) 395–404, doi:10.1109/61.997906.
[38] C. Xiaotao, et al., Fault location technology of transmission line based on asynchronous phasor measurement from PMUs, IOP Conf. Ser. Earth Environ. Sci 645 (1) (2021) 012084, doi:10.1088/1755-1315/645/1/012084.
[39] M. Shiroei, S. Daniar, and M. Akhbari, A new algorithm for fault location on transmission lines, 2009, doi:10.1109/PES.2009.5275763.
[40] A.Q. Khan, Q. Ullah, M. Sarwar, S.T. Gul, N. Iqbal, Transmission line fault detection and identification in an interconnected power network using phasor measurement units, IFAC-PapersOnLine 51 (24) (2018) 1356–1363, doi:10.1016/j.ifacol.2018.09.558.

[41] Y. Cai, A.D. Rajapakse, N.M. Haleem, N. Raju, A threshold free synchrophasor measurement based multi-terminal fault location algorithm, Int. J. Electr. Power Energy Syst. 96 (Mar) (2018) 174–184, doi:10.1016/j.ijepes.2017.09.035.

[42] K.P. Lien, C.W. Liu, C.S. Yu, J.A. Jiang, Transmission network fault location observability with minimal PMU placement, IEEE Trans. Power Deliv. 21 (3) (2006) 1128–1136, doi:10.1109/TPWRD.2005.858806.

[43] K. Mazlumi, H.A. Abyaneh, S.H.H. Sadeghi, S.S. Geramian, Determination of optimal PMU placement for fault-location observability, 3rd International Conference on Deregulation and Restructuring and Power Technologies, DRPT 2008, 2008, pp. 1938–1942, doi:10.1109/DRPT.2008.4523724.

[44] A. Saber, A. Emam, H. Elghazaly, A threshold free PMU-based fault location scheme for multi-end lines, Ain Shams Eng. J. 11 (4) (2020) 1113–1121, doi:10.1016/j.asej.2020.03.001.

Fault diagnosis in three-terminal power transmission lines

8.1 Introduction

Multiterminal lines in electric power system networks have at least three terminals with a considerable generation behind each one. The multiterminal lines are classified according to the number of terminals. For instance, the three-terminal transmission lines comprise three terminals or sections, four-terminal lines consisting of four sections or terminals. Amongst different types of multi-terminal transmission lines, the three-terminal is possibly the most prominent one [1]. Tapped lines are those with at least three terminals with considerable power generation behind a maximum of two of them [2]. Both multiterminal and tapped lines are implemented in the electric networks for various advantages, including economic benefits, regulatory approvals, right-of-way availability, system performance, and environmental protection [1,3–5].

Transmission lines are susceptible to a wide range of transient and permanent faults as they are primarily overhead type and exposed to birds, trees, storms, etc. Any faults in electric networks cause power outages, revenue loss, asset damage, and sometimes deaths of birds, animals, and humans. Precise information about the fault locations is crucial for expediting the restoration process by sending the maintenance crews to the sites to reduce the outage duration and revenue loss [6–10]. In response, the researchers and power system decision makers put enormous efforts into developing different fault location techniques to avoid the possible consequences of the fault occurrence. Plentiful fault location algorithms were reported for identifying fault locations in one-terminal and two-terminal transmission lines [1,11,13]. Such readily available fault location algorithms suffer from a higher percentage of errors in locating faults in multi-terminal transmission lines due to their inherent complexities and operating constraints [12]. In response, the adaptive fault location scheme intends to improve the accuracy of the non-adaptive scheme that is based on effective estimation of transmission line parameters and system impedance [14]. Besides, the recent proliferation of advanced measurement infrastructure, including the phasor measurement unit (PMU), provides time synchronized and effective measurements even under adversities, for example, power swing and out-of-step conditions [15–18]. Such accurate measurements paved the way for efficient fault location algorithm development for multiterminal transmission lines [19–21].

Considering the mentioned issues, this chapter develops an adaptive PMU-based fault location algorithm for three-terminal transmission lines. The algorithm is adaptive as it utilizes the PMU recorded current and voltage measurements for the online estimation of the transmission line parameters and Thevenin equivalent (TE). The calculated parameters are used to determine the practical operating parameters of the network during a fault. Online parameter estimation tackles the discrepancies between the utility-provided parameters and the changed practical parameters due to

Power System Fault Diagnosis. DOI: https://doi.org/10.1016/B978-0-323-88429-7.00009-6
Copyright © 2022 Elsevier Inc. All rights reserved.

environmental conditions and operation history. The adaptive fault location algorithm is applied on a 500 kV system and the simulation results achieved using MATLAB and PSCAD/EMTDC are provided. The robustness of the algorithm is investigated considering different parameter variations. Presented results confirm the efficacy of the developed adaptive PMU-based fault location scheme for three-terminal transmission lines.

The remaining parts of this chapter are organized as follows. Section 8.2 illustrates the parameter estimation process of a three-terminal transmission line. Section 8.3 provides the steps of the adaptive fault location algorithm for the three-terminal line. Section 8.4 presents the obtained results that confirm the efficacy of the algorithm and their independence on fault information (location, type, resistance, and inception angle) and prefault loading uncertainty. Finally, an overall summary of this chapter is presented in Section 8.5.

8.2 Parameter estimation of a three-terminal line

This part shows how PMU measurements are employed to identify the parameters of a three-terminal transmission line. Fig. 8.1 shows a three-terminal transmission network comprising three sections where A, B, and C denote the three terminals and M represents the common branching point amongst the sections. The steady-state π equivalent model of the three-terminal network is shown in Fig. 8.2 [22]. The parameters of the three-terminal line are obtained using synchronized current and voltage measurements acquired by the PMUs deployed at the terminals. Line parameters are determined for the most general case where the parameters of each

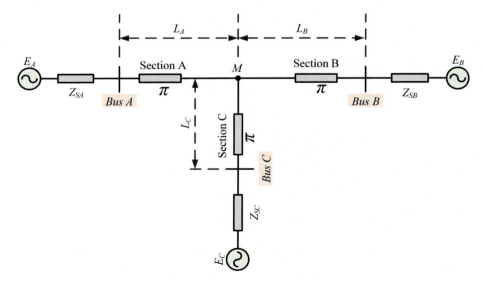

Fig. 8.1 A three-terminal transmission network.

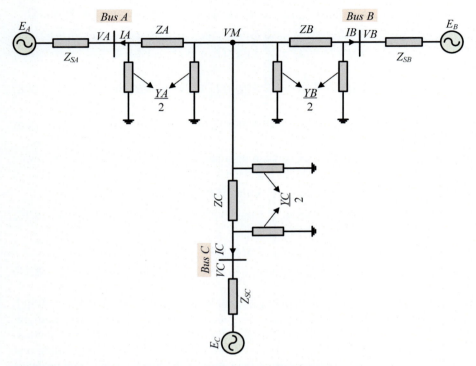

Fig. 8.2 Steady-state π equivalent model of the three-terminal network.

line section are not the same as those of the other two sections and for the special case where the three sections have the same parameters per unit length.

8.2.1 Line sections with identical parameters per unit length

If the three sections have the same admittance (y) and impedance (z) per unit length, then the following equations represent their impedances and admittances:

$$ZA = L_A * z \tag{8.1}$$
$$YA = L_A * y \tag{8.2}$$
$$ZB = L_B * z \tag{8.3}$$
$$YB = L_B * y \tag{8.4}$$
$$ZC = L_C * z \tag{8.5}$$
$$YC = L_C * y \tag{8.6}$$

Where,

L_A, L_B, and L_C: Lengths of the sections A, B, and C, respectively.
ZA, ZB, and ZC: Impedances of the sections A, B, and C, respectively.
YA, YB, and YC: Admittances of the sections A, B, and C, respectively.

From Fig. 8.2, the following Equations can be derived for sections A, B, and C:

$$VM - L_A * IA * z - \frac{1}{2} * VA * (L_A)^2 * y * z - VA = 0 \quad (8.7)$$

$$VM - L_B * IB * z - \frac{1}{2} * VB * (L_B)^2 * y * z - VB = 0 \quad (8.8)$$

$$VM - L_C * IC * z - \frac{1}{2} * VC * (L_C)^2 * y * z - VC = 0 \quad (8.9)$$

Eqs. (8.7)–(8.9) are complex and can be separated into two real parts as follows:

$$\text{Re}[VM] - L_A * \text{Re}[I_A] * \text{Re}[z] + L_A * \text{Im}[I_A] * \text{Im}[z] + 0.5 * (L_A)^2 * \text{Re}[VA] * \text{Im}[z] * \text{Im}[y] + 0.5 * (L_A)^2 * \text{Im}[VA] * \text{Re}[z] * \text{Im}[y] - \text{Re}[VA] = 0 \quad (8.10)$$

$$\text{Im}[VM] - L_A * \text{Re}[I_A] * \text{Im}[z] - L_A * \text{Im}[I_A] * \text{Re}[z] - 0.5 * (L_A)^2 * \text{Re}[VA] * \text{Re}[z] * \text{Im}[y] + 0.5 * (L_A)^2 * \text{Im}[VA] * \text{Im}[z] * \text{Im}[y] - \text{Im}[VA] = 0 \quad (8.11)$$

Eqs. (8.10) and (8.11) are nonlinear with five variables, namely Re[z], Im[z], Im[y], Re[VM], Im[VM]. Therefore, a minimum of five equations are needed to get the values of these unknowns. In addition to Eqs. (8.10) and (8.11), four more equations can be obtained by splitting the Eqs. (8.8) and (8.9) into the corresponding real equations. With a total of six equations, the values of the five unknowns can be determined. For the line sections with the same parameters per unit length, only one set of current and voltage phasor measurements is required to obtain the line parameters for each section. Suppose that:

$$[X_1 \quad X_2 \quad X_3 \quad X_4 \quad X_5]' = [\text{Re}[z] \quad \text{Im}[z] \quad \text{Im}[y] \quad \text{Re}[VM] \quad \text{Im}[VM]]'$$

Then, the line parameters per unit length can be obtained by solving the following nonlinear equations:

$$X_4 - L_A * \text{Re}[I_A] * X_1 + L_A * \text{Im}[I_A] * X_2 + 0.5 * (L_A)^2 * \text{Re}[VA] * X_2 * X_3 \\ + 0.5 * (L_A)^2 * \text{Im}[VA] * X_1 * X_3 - \text{Re}[VA] = 0 \quad (8.12)$$

$$X_5 - L_A * \text{Re}[I_A] * X_2 - L_A * \text{Im}[I_A] * X_1 - 0.5 * (L_A)^2 * \text{Re}[VA] * X_1 * X_3 \\ + 0.5 * (L_A)^2 * \text{Im}[VA] * X_2 * X_3 - \text{Im}[VA] = 0 \quad (8.13)$$

$$X_4 - L_B * \text{Re}[I_B] * X_1 + L_B * \text{Im}[I_B] * X_2 + 0.5 * (L_B)^2 * \text{Re}[VB] * X_2 * X_3 \\ + 0.5 * (L_B)^2 * \text{Im}[VB] * X_1 * X_3 - \text{Re}[VB] = 0 \quad (8.14)$$

$$X_5 - L_B * \text{Re}[I_B] * X_2 - L_B * \text{Im}[I_B] * X_1 - 0.5 * (L_B)^2 * \text{Re}[VB] * X_1 * X_3 \\ + 0.5 * (L_B)^2 * \text{Im}[VB] * X_2 * X_3 - \text{Im}[VB] = 0 \quad (8.15)$$

Fault diagnosis in three-terminal power transmission lines

$$X_4 - L_C * \text{Re}[I_C] * X_1 + L_C * \text{Im}[I_C] * X_2 + 0.5 * (L_C)^2 * \text{Re}[VC] * X_2 * X_3$$
$$+ 0.5 * (L_C)^2 * \text{Im}[VC] * X_1 * X_3 - \text{Re}[VC] = 0 \qquad (8.16)$$

Now, Eqs. (8.1)–(8.6) can be used to obtain the line parameters of sections A, B, and C knowing the length of each section.

8.2.2 Line sections with nonidentical parameters

For the case of nonidentical line sections having different parameters per unit length, more unknowns are involved. Thus, more sets of measurements are required. Now, a total of 11 unknowns should be evaluated namely, Re[ZA], Im[ZA], Im[YA], Re[ZB], Im[ZB], Im[YB], Re[ZC], Im[ZC], Im[YC], Re[VM], Im[VM]. From one set of measurements, six equations can be derived as described in the previous section. With the second set of measurements, six more equations can be derived. However, these new six equations will add two more unknowns about the tap point voltage (VM). Hence, with 12 equations and 13 unknowns, one more set of measurements becomes necessary. For the third set of measurements, a similar procedure can be followed. In that case, an additional six equations can be derived that will introduce two more unknowns. Therefore, obtained 18 equations from three sets of measurements are more than enough to evaluate the fifteen unknowns. Considering the first set of measurements, the following Equations can be formulated from Fig. 8.2:

$$(VM)_1 - ZA * (IA)_1 - \frac{1}{2} * (VA)_1 * ZA * YA - (VA)_1 = 0 \qquad (8.17)$$

$$(VM)_1 - ZB * (IB)_1 - \frac{1}{2} * (VB)_1 * ZB * YB - (VB)_1 = 0 \qquad (8.18)$$

$$(VM)_1 - ZC * (IC)_1 - \frac{1}{2} * (VC)_1 * ZC * YC - (VC)_1 = 0 \qquad (8.19)$$

The subscript one (1) is used to represent the first set of measurements. Likewise, the second and the third sets of measurements are represented using the subscripts two (2) and three (3), respectively:

$$(VM)_2 - ZA * (IA)_2 - \frac{1}{2} * (VA)_2 * ZA * YA - (VA)_2 = 0 \qquad (8.20)$$

$$(VM)_2 - ZB * (IB)_2 - \frac{1}{2} * (VB)_2 * ZB * YB - (VB)_2 = 0 \qquad (8.21)$$

$$(VM)_2 - ZC * (IC)_2 - \frac{1}{2} * (VC)_2 * ZC * YC - (VC)_2 = 0 \qquad (8.22)$$

$$(VM)_3 - ZA * (IA)_3 - \frac{1}{2} * (VA)_3 * ZA * YA - (VA)_3 = 0 \qquad (8.23)$$

$$(VM)_3 - ZB * (IB)_3 - \frac{1}{2} * (VB)_3 * ZB * YB - (VB)_3 = 0 \qquad (8.24)$$

$$(VM)_3 - ZC*(IC)_3 - \frac{1}{2}*(VC)_3 * ZC*YC - (VC)_3 = 0 \quad (8.25)$$

Eq. (8.17) is a complex nonlinear equation that can be separated into two real nonlinear equation as follows:

$$\begin{aligned}&\text{Re}\big[(VM)_1\big] - \text{Re}[ZA]*\text{Re}\big[(IA)_1\big] + \text{Im}[ZA]*\text{Im}\big[(IA)_1\big] + 0.5*\\ &\text{Re}\big[(VA)_1\big]*\text{Im}[ZA]*\text{Im}[YA] + 0.5*\text{Im}\big[(VA)_1\big]*\text{Im}[YA]*\\ &\text{Re}[ZA] - \text{Re}\big[(VA)_1\big] = 0\end{aligned} \quad (8.26)$$

$$\begin{aligned}&\text{Im}\big[(VM)_1\big] - \text{Re}[ZA]*\text{Im}\big[(IA)_1\big] - \text{Im}[ZA]*\text{Re}\big[(IA)_1\big] - 0.5*\text{Re}\big[(VA)_1\big]*\\ &\text{Im}[YA]*\text{Re}[ZA] + 0.5*\text{Im}\big[(VA)_1\big]*\text{Im}[ZA]*\text{Im}[YA] - \text{Im}\big[(VA)_1\big] = 0\end{aligned} \quad (8.27)$$

By proceeding in a similar fashion, 16 more equations can be derived from Eqs. (8.18) to (8.25). Thus, the 15 unknowns can be evaluated from these derived 18 equations. The unknowns are Re[ZA], Im[ZA], Im[YA], Re[ZB], Im[ZB], Im[YB], Re[ZC], Im[ZC], Im[YC], Re[$(VM)_1$], Im[$(VM)_1$], Re[$(VM)_2$], Im[$(VM)_2$], Re[$(VM)_3$], Im[$(VM)_3$].

8.3 Adaptive fault location algorithm for three-terminal transmission line

The adaptive fault location algorithm presented in Chapter 7 for the two-terminal transmission line can be extended for the three-terminal transmission lines. In a three-terminal line, a fault may take place in any one of the sections. Referring to Fig. 8.1; the voltage of the node M can be computed in terms of the voltage of the buses A, B, or C. By taking the line section A into consideration, IA can be expressed as:

$$IA = \left[YSA + \frac{YA}{2}\right]\Delta VA \quad (8.28)$$

and

$$VM = \Delta VA + ZAIA \quad (8.29)$$

If the value of IA is substituted into Eq. (8.29), VM can be obtained as:

$$VM = \left(I_{3\times 3} + ZA\left(YSA + \frac{YA}{2}\right)\right)\Delta VA \quad (8.30)$$

Likewise, taking the sections B and C into consideration, VM can also be expressed as:

$$VM = \left(I_{3\times 3} + ZB\left(YSB + \frac{YB}{2}\right)\right)\Delta VB \quad (8.31)$$

$$VM = \left(I_{3\times 3} + ZC\left(YSC + \frac{YC}{2}\right)\right)\Delta VC \tag{8.32}$$

The fault locations in the three-terminal lines are evaluated in two stages. In the first stage, the faulty section is identified, and in the second stage, fault distance is calculated. Node M voltage (VM) cannot be evaluated accurately using the measured bus voltage of the fault occurring section. However, voltage (VM) can be assessed accurately using the bus voltages of the other two sections (nonfaulted). Thus, the faulty section can be easily identified by evaluating the VM values using Eqs. (8.30) to (8.32) as the evaluated value of the faulty section disagrees with the values of the nonfaulty sections. Fig. 8.3 illustrates the steps of the faulty section identification scheme of a three-terminal transmission network.

After identification of the faulty section, the fault distance from the bus of the faulty section is computed. To do so, it is assumed that a fault occurs in section B, and the distance of the fault from bus B is the $l1$ unit. Fig. 8.4 depicts the TE model of the assumed faulted three-terminal transmission system. For this case, the node M voltage (VM) can be evaluated using the bus voltages of sections A or C. Therefore, the required currents and voltages for fault distance calculation of Fig. 8.4 can be expressed as:

$$IM = IMA + IMC \tag{8.33}$$

$$IMA = IA + \frac{YA}{2}VM \tag{8.34}$$

$$IMC = IC + \frac{YC}{2}VM \tag{8.35}$$

$$IFM = IM + \frac{YB}{2}(1-k)VM \tag{8.36}$$

$$VF = VM + ZB(1-k)IFM \tag{8.37}$$

Besides, the following equations can be written to obtain the bus voltage of section B:

$$IB = \left[YSB + \frac{YB}{2}k\right]\Delta VB \tag{8.38}$$

$$VF = \Delta VB + ZBkIB \tag{8.39}$$

From Eqs. (8.37) and (8.39), the following equations can be written:

$$\Delta VB + ZBkIB = VM + ZB(1-k)IFM \tag{8.40}$$

Therefore, k can be represented as the function of the measured bus voltages of three terminals and the system parameters as:

$$k = f(\Delta VA, \Delta VB, \Delta VC) \rightarrow ak^2 + bk + c = 0 \tag{8.41}$$

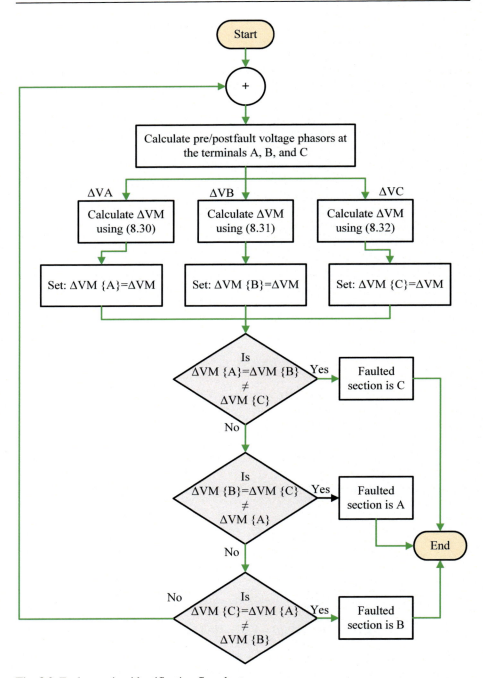

Fig. 8.3 Faulty section identification flowchart.

Fault diagnosis in three-terminal power transmission lines   203

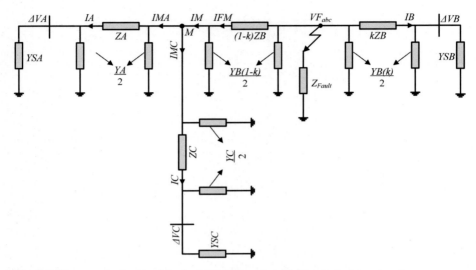

Fig. 8.4 The π-equivalent model of a faulted three-terminal transmission network.

Such that,

$$a = ZB\frac{YB}{2}\Delta VB + ZB\frac{YA}{2}\Delta VA + ZB\frac{YC}{2}\Delta VC \qquad (8.42)$$

$$b = ZBYSB\Delta VB - ZB\frac{YA}{2}\Delta VA - ZB\frac{YC}{2}\Delta VC + ZB\frac{YB}{2}VM + ZBYSA\Delta VA$$
$$+ ZBYSC\Delta VC + ZB\left(\frac{YA}{2} + \frac{YB}{2} + \frac{YC}{2}\right)VM$$

$$c = \Delta VB - VM - ZBYSA\Delta VA - ZBYSC\Delta VC - ZB\left(\frac{YA}{2} + \frac{YB}{2} + \frac{YC}{2}\right)VM$$

The *VM* can be obtained using either Eq. (8.30) or (8.32). Based on the obtained value of VM and other system parameters, the fault point distance from the bus B of section B is computed as:

$$l1 = k \times L_B \qquad (8.43)$$

Likewise, the coefficients (*a*, *b*, and *c*) of the quadratic Eq. (8.41) can be obtained using the following equation if a fault occurred at a distance of *l*1 from bus A of section A:

$$a = ZA\frac{YA}{2}\Delta VA + ZA\frac{YB}{2}\Delta VB + ZA\frac{YC}{2}\Delta VC \qquad (8.44)$$

$$b = ZAYSA\Delta VA - ZA\frac{YB}{2}\Delta VB - ZA\frac{YC}{2}\Delta VC + ZA\frac{YA}{2}VM + ZAYSB\Delta VB$$
$$+ ZAYSC\Delta VC + ZA\left(\frac{YA}{2} + \frac{YB}{2} + \frac{YC}{2}\right)VM$$

$$c = \Delta VA - VM - ZAYSB\Delta VB - ZAYSC\Delta VC - ZA\left(\frac{YA}{2} + \frac{YB}{2} + \frac{YC}{2}\right)VM$$

Then, the *VM* can be obtained from either Eq. (8.31) or (8.32). Eventually, the fault distance from bus A is computed as:

$$l1 = k \times L_A \tag{8.45}$$

Finally, the coefficients (*a*, *b*, and *c*) of the quadratic Eq. (8.41) can be obtained using the following equation if a fault occurred at a distance of *l*1 from bus C of section C:

$$a = ZC\frac{YC}{2}\Delta VC + ZC\frac{YA}{2}\Delta VA + ZC\frac{YB}{2}\Delta VB \tag{8.46}$$

$$b = ZCYSC\Delta VC - ZC\frac{YA}{2}\Delta VA - ZC\frac{YB}{2}\Delta VB + ZC\frac{YC}{2}VM + ZCYSA\Delta VA$$
$$+ ZCYSB\Delta VB + ZC\left(\frac{YA}{2} + \frac{YB}{2} + \frac{YC}{2}\right)VM$$

$$c = \Delta VC - VM - ZCYSA\Delta VA - ZCYSB\Delta VB - ZC\left(\frac{YA}{2} + \frac{YB}{2} + \frac{YC}{2}\right)VM$$

Then, the *VM* can be obtained from either Eq. (8.30) or (8.31). Eventually, the fault distance from bus C is computed as:

$$l1 = k \times L_C \tag{8.47}$$

The steps of the adaptive fault location algorithm for the three-terminal transmission lines are illustrated in Fig. 8.5 and summarized as follows:

1. Three different sets of PMU prefault current and voltage phasor measurements at terminals A, B, and C are measured and stored.
2. One set of postfault voltage phasor measurements at terminals A, B, and C are measured and stored.
3. Utilizing the PMU prefault measurements, the system TE at the terminals A, B, and C are obtained online, as explained earlier in Chapter 6.
4. Besides, the PMU prefault measurements are used for the online parameter identification of sections A, B, and C.
5. The superimposed electrical voltage measurements (ΔVA, ΔVB, and ΔVC) are acquired at each terminal (A, B, and C) using the most recent prefault and the postfault measurements (differences between the two sets of measurements).

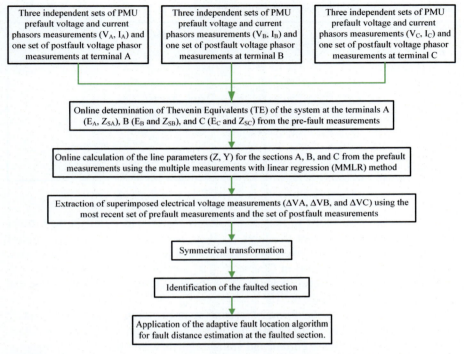

Fig. 8.5 Adaptive fault location algorithm flowchart for a three-terminal transmission network.

6. The phasor quantities are transformed into symmetrical components through the symmetrical transformation.
7. Faulted section is identified by following the steps mentioned above (Fig. 8.3).
8. Finally, the fault location is computed using the relevant Equations as discussed.

8.4 Simulation results

The adaptive fault location algorithm was tested on a 500 kV three-terminal transmission line, as shown in Fig. 8.1. A single section was assumed to be faulted amongst the three sections A, B, and C. The simulation was carried out in MATLAB and PSCAD/EMTDC platform. The robustness of the employed algorithm was evaluated considering the fault information (fault resistance, inception angle, location, and prefault loading condition) uncertainty.

8.4.1 Data generation and conditioning

The adaptive fault location algorithm was implemented in the MATLAB environment and assessed using prefault and postfault data acquired from the PSCAD/

Table 8.1 Three-terminal transmission network parameters.

Parameter	Value
R (Ω/mile)	0.249168
L (mH/mile)	1.556277
C (F/mile)	19.469e-9
E_A (kV)	500 ∠0°
E_B (kV)	475 ∠ − 15°
E_C (kV)	472 ∠ − 10°
Z_{SA} (Ω)	5.7257 + j15.1762
Z_{SB} (Ω)	5.1033 + j15.3082
Z_{SC} (Ω)	5.4145 + j15.2422

EMTDC platform for faults in the three sections (A, B, and C) of the test three-terminal transmission system. Sections A, B, C were 150, 100, and 110 miles, respectively. The Thevenin impedance of the system at terminals A, B, and C were evaluated online, as explained in Chapter 6. The parameters of the test system are tabulated in Table 8.1. It also includes TE voltages that were calculated using the method explained in Chapter 6.

Current and voltage transformers located at each terminal were deliberately assumed as ideal devices to investigate the errors of the adaptive fault location algorithm itself. The three-phase current and voltage signals were sampled at a frequency of 240 Hz, corresponding to 4 samples per cycle, and stored for post processing. Sections A, B, and C were assumed to have the same admittance and the same impedance per unit length. Based on this assumption, only one set of prefault current and voltage phasor measurements at terminal A, B, and C were employed for online computation of the line parameters, as explained earlier. To extract the current and voltage phasors, the discrete Fourier transform was applied as given by the following Equation:

$$X = \left(\sqrt{2}/N_s\right) \sum_{k=1}^{N_s} x[k] e^{-j2\pi k/N_s} \qquad (8.48)$$

Where, $x[k]$ is the waveform samples, X is the phasor, and N_s is the total number of samples in one period. The accuracy in terms of percentage error of the adaptive fault location algorithm is expressed as:

$$\%\text{Error} = \frac{|\text{Actual location} - \text{Estimated location}|}{\text{Total line length}} \times 100 \qquad (8.49)$$

8.4.2 Accuracy analysis

The accuracy of the adaptive fault location algorithm was evaluated by simulating the system considering various faults with different fault resistances and fault locations. Tables 8.2, 8.3, 8.4 and 8.5 present the estimated fault locations determined for

Table 8.2 Fault location results for LG faults on the test three-terminal network.

Fault type	Fault resistance (Ω)	Actual fault location (pu)	Section A Calculated fault location (pu)	Section A Error (%)	Section B Calculated fault location (pu)[a]	Section B Error (%)	Section C Calculated fault location (pu)	Section C Error (%)
AG	10.0	0.20	0.1990	0.100	0.1996	0.040	0.2008	0.080
		0.40	0.4010	0.100	0.4019	0.190	0.4036	0.360
		0.60	0.6039	0.390	0.6056	0.560	0.6107	1.070
		0.80	0.8084	0.840	0.8095	0.950	0.8146	1.460
	100.0	0.20	0.1993	0.070	0.1975	0.250	0.2027	0.270
		0.40	0.4017	0.170	0.4006	0.060	0.4051	0.510
		0.60	0.6064	0.640	0.6061	0.610	0.6206	2.060
		0.80	0.8115	1.150	0.8108	1.080	0.8218	2.180
BG	10.0	0.20	0.1998	0.020	0.2004	0.040	0.2015	0.150
		0.40	0.4018	0.180	0.4037	0.370	0.4045	0.450
		0.60	0.6047	0.470	0.6053	0.530	0.6127	1.270
		0.80	0.8088	0.880	0.8114	1.140	0.8141	1.410
	100.0	0.20	0.2015	0.150	0.2022	0.220	0.2052	0.520
		0.40	0.4050	0.500	0.4070	0.700	0.4090	0.900
		0.60	0.6075	0.750	0.6108	1.080	0.6220	2.200
		0.80	0.8113	1.130	0.8150	1.500	0.8221	2.210
CG	10.0	0.20	0.1994	0.060	0.2011	0.110	0.2015	0.150
		0.40	0.4023	0.230	0.4057	0.570	0.4059	0.590
		0.60	0.6034	0.340	0.6053	0.530	0.6124	1.240
		0.80	0.8086	0.860	0.8109	1.090	0.8139	1.390
	100.0	0.20	0.1991	0.090	0.2004	0.040	0.2029	0.290
		0.40	0.4028	0.280	0.4053	0.530	0.4073	0.730
		0.60	0.6060	0.600	0.6085	0.850	0.6191	1.910
		0.80	0.8110	1.100	0.8121	1.210	0.8204	2.040

[a]pu = per unit.

Table 8.3 Fault location results for LL faults on the test three-terminal network.

Fault type	Fault resistance (Ω)	Actual fault location (pu)	Section A Calculated fault location (pu)	Section A Error (%)	Section B Calculated fault location (pu)	Section B Error (%)	Section C Calculated fault location (pu)	Section C Error (%)
AB	1.0	0.20	0.1994	0.060	0.1998	0.020	0.2007	0.070
		0.40	0.4008	0.080	0.4013	0.130	0.4027	0.270
		0.60	0.6042	0.420	0.6051	0.510	0.6088	0.880
		0.80	0.8083	0.830	0.8098	0.980	0.8126	1.260
	10.0	0.20	0.1994	0.060	0.1997	0.030	0.2008	0.080
		0.40	0.4009	0.090	0.4015	0.150	0.4029	0.290
		0.60	0.6042	0.420	0.6051	0.510	0.6092	0.920
		0.80	0.8083	0.830	0.8099	0.990	0.8128	1.280
BC	1.0	0.20	0.1999	0.010	0.2009	0.090	0.2014	0.140
		0.40	0.4020	0.200	0.4049	0.490	0.4047	0.470
		0.60	0.6038	0.380	0.6041	0.410	0.6110	1.100
		0.80	0.8087	0.870	0.8113	1.130	0.8119	1.190
	10.0	0.20	0.1998	0.020	0.2010	0.100	0.2014	0.140
		0.40	0.4021	0.210	0.4052	0.520	0.4050	0.500
		0.60	0.6038	0.380	0.6045	0.450	0.6112	1.120
		0.80	0.8086	0.860	0.8114	1.140	0.8122	1.220
CA	1.0	0.20	0.1992	0.080	0.2006	0.060	0.2009	0.090
		0.40	0.4015	0.150	0.4042	0.420	0.4046	0.460
		0.60	0.6027	0.270	0.6050	0.500	0.6091	0.910
		0.80	0.8080	0.800	0.8097	0.970	0.8125	1.250
	10.0	0.20	0.1990	0.100	0.2005	0.050	0.2008	0.080
		0.40	0.4015	0.150	0.4039	0.390	0.4045	0.450
		0.60	0.6028	0.280	0.6051	0.510	0.6092	0.920
		0.80	0.8081	0.810	0.8096	0.960	0.8127	1.270

Table 8.4 Fault location results for LLG faults on the test three-terminal network.

Fault type	Fault resistance (Ω)	Actual fault location (pu)	Section A Calculated fault location (pu)	Section A Error (%)	Section B Calculated fault location (pu)	Section B Error (%)	Section C Calculated fault location (pu)	Section C Error (%)
ABG	5.0	0.20	0.1994	0.060	0.2001	0.010	0.2008	0.080
		0.40	0.4011	0.110	0.4023	0.230	0.4032	0.320
		0.60	0.6038	0.380	0.6049	0.490	0.6085	0.850
		0.80	0.8082	0.820	0.8100	1.000	0.8120	1.200
	50.0	0.20	0.1996	0.040	0.2003	0.030	0.2011	0.110
		0.40	0.4012	0.120	0.4022	0.220	0.4033	0.330
		0.60	0.6042	0.420	0.6056	0.560	0.6087	0.870
		0.80	0.8082	0.820	0.8102	1.020	0.8124	1.240
BCG	5.0	0.20	0.1997	0.030	0.2008	0.080	0.2012	0.120
		0.40	0.4018	0.180	0.4043	0.430	0.4043	0.430
		0.60	0.6035	0.350	0.6042	0.420	0.6098	0.980
		0.80	0.8085	0.850	0.8108	1.080	0.8116	1.160
	50.0	0.20	0.1999	0.010	0.2011	0.110	0.2015	0.150
		0.40	0.4020	0.200	0.4051	0.510	0.4048	0.480
		0.60	0.6037	0.370	0.6043	0.430	0.6108	1.080
		0.80	0.8086	0.860	0.8113	1.130	0.8118	1.180
CAG	5.0	0.20	0.1993	0.070	0.2006	0.060	0.2009	0.090
		0.40	0.4015	0.150	0.4038	0.380	0.4043	0.430
		0.60	0.6030	0.300	0.6049	0.490	0.6088	0.880
		0.80	0.8081	0.810	0.8099	0.990	0.8121	1.210
	50.0	0.20	0.1994	0.060	0.2009	0.090	0.2012	0.120
		0.40	0.4018	0.180	0.4044	0.440	0.4048	0.480
		0.60	0.6029	0.290	0.6055	0.550	0.6092	0.920
		0.80	0.8080	0.800	0.8100	1.000	0.8126	1.260

Table 8.5 Fault location results for LLL faults on the test three-terminal network.

Fault type	Fault resistance (Ω)	Actual fault location (pu)	Section A Calculated fault location (pu)	Section A Error (%)	Section B Calculated fault location (pu)	Section B Error (%)	Section C Calculated fault location (pu)	Section C Error (%)
ABC	1.0	0.20	0.1994	0.060	0.2004	0.040	0.2009	0.090
		0.40	0.4014	0.140	0.4033	0.330	0.4037	0.370
		0.60	0.6033	0.330	0.6045	0.450	0.6083	0.830
		0.80	0.8082	0.820	0.8101	1.010	0.8113	1.130
	10.0	0.20	0.1992	0.080	0.2004	0.040	0.2007	0.070
		0.40	0.4014	0.140	0.4035	0.350	0.4038	0.380
		0.60	0.6033	0.330	0.6048	0.480	0.6084	0.840
		0.80	0.8081	0.810	0.8100	1.000	0.8115	1.150

the single line to ground (LG) faults, line to line to ground (LLG) faults, line to line (LL) faults, and three-phase (LLL) faults occurred in sections A, B, and C. As can be observed from the results obtained, the adaptive fault location algorithm proved to be highly accurate irrespective of the fault type.

8.4.3 Effect of fault resistance

The effect of fault resistance uncertainty on the adaptive fault location algorithm was investigated. Two types of faults were studied where the ground was involved in one type to capture both low and high resistance (0 to 500 Ω) faults, and for other kinds, no ground was involved that captured only low resistance (0 to 30 Ω) faults. The local and remote source impedances were set, in all cases, to be equal to the system values. Obtained results for different types of faults are presented in Tables 8.6, 8.7, 8.8, 8.9, 8.10, 8.11, 8.12, 8.13, 8.14 and 8.15, considering the fault occurrence at 0.4 pu distance from terminals A, B, and C. As can be seen from the obtained results, the adaptive fault location algorithm accurately located different types of faults at different terminals that confirmed the independence of algorithm under fault resistance uncertainty.

8.4.4 Effect of fault inception angle

Like the effect of fault resistance uncertainty, the effect of fault inception angle uncertainty on the accuracy of the adaptive fault location algorithm was investigated. The fault inception angles were varied between 0 to 150 degrees (°), where the faults were applied at 0.40 pu distance from the respective buses of the three terminals.

Table 8.6 Fault resistance uncertainty effect on the accuracy of fault location algorithm for LG faults on terminal A (faults were applied 0.40 pu distance from bus A).

Fault resistance (Ω)	Fault type					
	AG		BG		CG	
	Calculated fault location (pu)	Error (%)	Calculated fault location (pu)	Error (%)	Calculated fault location (pu)	Error (%)
0.0	0.3998	0.020	0.4004	0.040	0.4018	0.180
1.0	0.3998	0.020	0.4004	0.040	0.4018	0.180
5.0	0.3997	0.030	0.4005	0.050	0.4018	0.180
15.0	0.3993	0.070	0.4007	0.070	0.4017	0.170
20.0	0.3992	0.080	0.4009	0.090	0.4016	0.160
50.0	0.3983	0.170	0.4015	0.150	0.4011	0.110
150.0	0.3961	0.390	0.4022	0.220	0.3985	0.150
200.0	0.3952	0.480	0.4023	0.230	0.3973	0.270
400.0	0.3928	0.720	0.4026	0.260	0.3931	0.690
500.0	0.3920	0.800	0.4029	0.290	0.3915	0.850

Tables 8.16, 8.17 and 8.18 present the obtained results for three types of faults (AG, BC, and BCG) on the terminals A, B, and C, respectively. The results confirm the efficacy of the adaptive fault location accuracy under fault inception angle uncertainty as it accurately located the investigated faults.

Table 8.7 Fault resistance uncertainty effect on the accuracy of fault location algorithm for LL faults on terminal A (faults were applied 0.40 pu distance from bus A).

Fault resistance (Ω)	Fault type					
	AB		BC		CA	
	Calculated fault location (pu)	Error (%)	Calculated fault location (pu)	Error (%)	Calculated fault location (pu)	Error (%)
0.0	0.4000	0.000	0.4020	0.200	0.4014	0.140
0.50	0.4000	0.000	0.4020	0.200	0.4014	0.140
1.50	0.4000	0.000	0.4020	0.200	0.4014	0.140
2.50	0.4000	0.000	0.4020	0.200	0.4014	0.140
5.0	0.4000	0.000	0.4020	0.200	0.4013	0.130
7.50	0.4000	0.000	0.4020	0.200	0.4013	0.130
15.0	0.4000	0.000	0.4021	0.210	0.4012	0.120
20.0	0.4000	0.000	0.4022	0.220	0.4011	0.110
25.0	0.4000	0.000	0.4022	0.220	0.4010	0.100
30.0	0.3999	0.010	0.4023	0.230	0.4009	0.090

Table 8.8 Fault resistance uncertainty effect on the accuracy of fault location algorithm for LLG faults on terminal A (faults were applied 0.40 pu distance from bus A).

Fault resistance (Ω)	Fault type						
	ABG		BCG		CAG	
	Calculated fault location (pu)	Error (%)	Calculated fault location (pu)	Error (%)	Calculated fault location (pu)	Error (%)
0.0	0.4008	0.080	0.4018	0.180	0.4015	0.150
1.0	0.4008	0.080	0.4018	0.180	0.4015	0.150
2.0	0.4008	0.080	0.4018	0.180	0.4015	0.150
10.0	0.4008	0.080	0.4019	0.190	0.4015	0.150
25.0	0.4007	0.070	0.4020	0.200	0.4016	0.160
75.0	0.4006	0.060	0.4021	0.210	0.4018	0.180
100.0	0.4006	0.060	0.4021	0.210	0.4018	0.180
150.0	0.4005	0.050	0.4021	0.210	0.4018	0.180
200.0	0.4005	0.050	0.4021	0.210	0.4018	0.180
250.0	0.4004	0.040	0.4021	0.210	0.4018	0.180

Table 8.9 Fault resistance uncertainty effect on the accuracy of fault location algorithm for LG faults on terminal B (faults were applied 0.40 pu distance from bus B).

Fault resistance (Ω)	Fault type						
	AG		BG		CG	
	Calculated fault location (pu)	Error (%)	Calculated fault location (pu)	Error (%)	Calculated fault location (pu)	Error (%)
0.0	0.4030	0.300	0.4032	0.320	0.4050	0.500
1.0	0.4030	0.300	0.4032	0.320	0.4050	0.500
5.0	0.4027	0.270	0.4034	0.340	0.4050	0.500
15.0	0.4023	0.230	0.4038	0.380	0.4050	0.500
20.0	0.4020	0.200	0.4040	0.400	0.4050	0.500
50.0	0.4009	0.090	0.4047	0.470	0.4044	0.440
150.0	0.3979	0.210	0.4057	0.570	0.4019	0.190
200.0	0.3968	0.320	0.4061	0.610	0.4011	0.110
400.0	0.3942	0.580	0.4083	0.830	0.3995	0.050
500.0	0.3936	0.640	0.4094	0.940	0.3994	0.060

Table 8.10 Fault resistance uncertainty effect on the accuracy of fault location algorithm for LL faults on terminal B (faults were applied 0.40 pu distance from bus B).

Fault resistance (Ω)	Fault type					
	AB		BC		CA	
	Calculated fault location (pu)	Error (%)	Calculated fault location (pu)	Error (%)	Calculated fault location (pu)	Error (%)
0.0	0.4024	0.240	0.4044	0.440	0.4043	0.430
0.50	0.4024	0.240	0.4045	0.450	0.4043	0.430
1.50	0.4024	0.240	0.4045	0.450	0.4042	0.420
2.50	0.4024	0.240	0.4045	0.450	0.4042	0.420
5.0	0.4024	0.240	0.4046	0.460	0.4042	0.420
7.50	0.4024	0.240	0.4046	0.460	0.4041	0.410
15.0	0.4024	0.240	0.4048	0.480	0.4039	0.390
20.0	0.4024	0.240	0.4049	0.490	0.4038	0.380
25.0	0.4024	0.240	0.4050	0.500	0.4037	0.370
30.0	0.4023	0.230	0.4051	0.510	0.4036	0.360

Table 8.11 Fault resistance uncertainty effect on the accuracy of fault location algorithm for LLG faults on terminal B (faults were applied 0.40 pu distance from bus B).

Fault resistance (Ω)	Fault type					
	ABG		BCG		CAG	
	Calculated fault location (pu)	Error (%)	Calculated fault location (pu)	Error (%)	Calculated fault location (pu)	Error (%)
0.0	0.4031	0.310	0.4041	0.410	0.4040	0.400
1.0	0.4031	0.310	0.4041	0.410	0.4040	0.400
2.0	0.4031	0.310	0.4041	0.410	0.4040	0.400
10.0	0.4031	0.310	0.4043	0.430	0.4041	0.410
25.0	0.4032	0.320	0.4045	0.450	0.4044	0.440
75.0	0.4031	0.310	0.4047	0.470	0.4047	0.470
100.0	0.4031	0.310	0.4047	0.470	0.4047	0.470
150.0	0.4030	0.300	0.4047	0.470	0.4047	0.470
200.0	0.4030	0.300	0.4047	0.470	0.4047	0.470
250.0	0.4030	0.300	0.4047	0.470	0.4047	0.470

Table 8.12 Fault resistance uncertainty effect on the accuracy of fault location algorithm for LG faults on terminal C (faults were applied 0.40 pu distance from bus C).

Fault resistance (Ω)	AG Calculated fault location (pu)	Error (%)	BG Calculated fault location (pu)	Error (%)	CG Calculated fault location (pu)	Error (%)
0.0	0.4038	0.380	0.4039	0.390	0.4056	0.560
1.0	0.4037	0.370	0.4039	0.390	0.4056	0.560
5.0	0.4035	0.350	0.4040	0.400	0.4056	0.560
15.0	0.4031	0.310	0.4043	0.430	0.4056	0.560
20.0	0.4028	0.280	0.4045	0.450	0.4055	0.550
50.0	0.4016	0.160	0.4051	0.510	0.4050	0.500
150.0	0.3985	0.150	0.4059	0.590	0.4026	0.260
200.0	0.3973	0.270	0.4061	0.610	0.4015	0.150
400.0	0.3944	0.560	0.4077	0.770	0.3992	0.080
500.0	0.3936	0.640	0.4085	0.850	0.3987	0.130

Table 8.13 Fault resistance uncertainty effect on the accuracy of fault location algorithm for LL faults on terminal C (faults were applied 0.40 pu distance from bus C).

Fault resistance (Ω)	AB Calculated fault location (pu)	Error (%)	BC Calculated fault location (pu)	Error (%)	CA Calculated fault location (pu)	Error (%)
0.0	0.4035	0.350	0.4051	0.510	0.4049	0.490
0.50	0.4033	0.330	0.4051	0.510	0.4049	0.490
1.50	0.4033	0.330	0.4051	0.510	0.4049	0.490
2.50	0.4033	0.330	0.4051	0.510	0.4049	0.490
5.0	0.4032	0.320	0.4052	0.520	0.4048	0.480
7.50	0.4032	0.320	0.4052	0.520	0.4048	0.480
15.0	0.4032	0.320	0.4053	0.530	0.4046	0.460
20.0	0.4031	0.310	0.4054	0.540	0.4045	0.450
25.0	0.4031	0.310	0.4055	0.550	0.4044	0.440
30.0	0.4031	0.310	0.4056	0.560	0.4043	0.430

Fault diagnosis in three-terminal power transmission lines

Table 8.14 Fault resistance uncertainty effect on the accuracy of fault location algorithm for LLG faults on terminal C (faults were applied 0.40 pu distance from bus C).

	Fault type					
	ABG		BCG		CAG	
Fault resistance (Ω)	Calculated fault location (pu)	Error (%)	Calculated fault location (pu)	Error (%)	Calculated fault location (pu)	Error (%)
0.0	0.4040	0.400	0.4049	0.490	0.4049	0.490
1.0	0.4040	0.400	0.4049	0.490	0.4049	0.490
2.0	0.4041	0.410	0.4050	0.500	0.4049	0.490
10.0	0.4042	0.420	0.4052	0.520	0.4051	0.510
25.0	0.4043	0.430	0.4054	0.540	0.4053	0.530
75.0	0.4042	0.420	0.4056	0.560	0.4056	0.560
100.0	0.4042	0.420	0.4056	0.560	0.4056	0.560
150.0	0.4041	0.410	0.4056	0.560	0.4056	0.560
200.0	0.4041	0.410	0.4056	0.560	0.4056	0.560
250.0	0.4040	0.400	0.4055	0.550	0.4056	0.560

Table 8.15 Fault resistance uncertainty effect on the accuracy of fault location algorithm for LLL faults (faults were applied 0.40 pu distance from the respective buses).

	Section A		Section B		Section C	
Fault resistance (Ω)	Calculated fault location (pu)	Error (%)	Calculated fault location (pu)	Error (%)	Calculated fault location (pu)	Error (%)
0.0	0.4016	0.160	0.4040	0.400	0.4045	0.450
0.50	0.4016	0.160	0.4041	0.410	0.4046	0.460
1.50	0.4016	0.160	0.4041	0.410	0.4046	0.460
2.50	0.4016	0.160	0.4041	0.410	0.4046	0.460
5.0	0.4016	0.160	0.4042	0.420	0.4046	0.460
7.50	0.4016	0.160	0.4042	0.420	0.4047	0.470
15.0	0.4014	0.140	0.4044	0.440	0.4048	0.480
20.0	0.4015	0.150	0.4045	0.450	0.4048	0.480
25.0	0.4014	0.140	0.4046	0.460	0.4049	0.490
30.0	0.4014	0.140	0.4047	0.470	0.4049	0.490

Table 8.16 Fault inception angle uncertainty effect on the accuracy of fault location algorithm (faults were applied 0.40 pu distance from bus A of terminal A).

	Fault type					
	AG		BC		BCG	
Fault inception angle (°)	Calculated fault location (pu)	Error (%)	Calculated fault location (pu)	Error (%)	Calculated fault location (pu)	Error (%)
0.0	0.4009	0.090	0.4030	0.300	0.4026	0.260
30.0	0.4009	0.090	0.4030	0.300	0.4026	0.260
45.0	0.4009	0.090	0.4030	0.300	0.4026	0.260
60.0	0.4009	0.090	0.4030	0.300	0.4026	0.260
90.0	0.4009	0.090	0.4029	0.290	0.4025	0.250
120.0	0.4010	0.100	0.4028	0.280	0.4024	0.240
135.0	0.4010	0.100	0.4031	0.310	0.4026	0.260
150.0	0.4010	0.100	0.4031	0.310	0.4027	0.270

Table 8.17 Fault inception angle uncertainty effect on the accuracy of fault location algorithm (faults were applied 0.40 pu distance from bus B of terminal B).

	Type of Fault					
	AG		BC		BCG	
Fault inception angle (°)	Calculated fault location (pu)	Error (%)	Calculated fault location (pu)	Error (%)	Calculated fault location (pu)	Error (%)
0.0	0.4025	0.250	0.4047	0.470	0.4043	0.430
30.0	0.4025	0.250	0.4047	0.470	0.4043	0.430
45.0	0.4025	0.250	0.4047	0.470	0.4042	0.420
60.0	0.4025	0.250	0.4047	0.470	0.4043	0.430
90.0	0.4025	0.250	0.4047	0.470	0.4042	0.420
120.0	0.4025	0.250	0.4047	0.470	0.4043	0.430
135.0	0.4024	0.240	0.4047	0.470	0.4043	0.430
150.0	0.4025	0.250	0.4046	0.460	0.4041	0.410

8.4.5 Effect of prefault loading

The effect of the prefault loading uncertainty on the accuracy of the adaptive fault location algorithm was also investigated for three different fault types (AG, BC, and BCG). Prefault loading conditions were varied from 0.5 to 3.0 times of the original loading value. Tables 8.19, 8.20 and 8.21 present the obtained results for the faults applied on 0.40 pu distance from the respective buses of the terminals A, B, and

Fault diagnosis in three-terminal power transmission lines 217

Table 8.18 Fault inception angle uncertainty effect on the accuracy of fault location algorithm (faults were applied 0.40 pu distance from bus C of terminal C).

Fault inception angle (°)	Fault type					
	AG		BC		BCG	
	Calculated fault location (pu)	Error (%)	Calculated fault location (pu)	Error (%)	Calculated fault location (pu)	Error (%)
0.0	0.4033	0.330	0.4052	0.520	0.4049	0.490
30.0	0.4033	0.330	0.4052	0.520	0.4049	0.490
45.0	0.4033	0.330	0.4052	0.520	0.4049	0.490
60.0	0.4033	0.330	0.4052	0.520	0.4049	0.490
90.0	0.4033	0.330	0.4053	0.530	0.4050	0.500
120.0	0.4032	0.320	0.4054	0.540	0.4050	0.500
135.0	0.4034	0.340	0.4052	0.520	0.4048	0.480
150.0	0.4032	0.320	0.4051	0.510	0.4047	0.470

Table 8.19 Prefault loading condition uncertainty effect on the accuracy of fault location algorithm (faults were applied 0.40 pu distance from bus A of the terminal A).

Prefault loading factor	Fault type					
	AG		BC		BCG	
	Calculated fault location (pu)	Error (%)	Calculated fault location (pu)	Error (%)	Calculated fault location (pu)	Error (%)
0.50	0.3989	0.110	0.4011	0.110	0.4007	0.070
0.80	0.4001	0.010	0.4022	0.220	0.4018	0.180
1.20	0.4017	0.170	0.4038	0.380	0.4034	0.340
1.50	0.4028	0.280	0.4049	0.490	0.4045	0.450
2.00	0.4048	0.480	0.4068	0.680	0.4064	0.640
3.00	0.4087	0.870	0.4107	1.070	0.4103	1.030

Table 8.20 Prefault loading condition uncertainty effect on the accuracy of fault location algorithm (faults were applied 0.40 pu distance from bus B of the terminal B).

Prefault loading factor	Fault type					
	AG		BC		BCG	
	Calculated fault location (pu)	Error (%)	Calculated fault location (pu)	Error (%)	Calculated fault location (pu)	Error (%)
0.50	0.4003	0.030	0.4025	0.250	0.4021	0.210
0.80	0.4016	0.160	0.4038	0.380	0.4034	0.340
1.20	0.4034	0.340	0.4056	0.560	0.4051	0.510
1.50	0.4047	0.470	0.4069	0.690	0.4064	0.640
2.00	0.4069	0.690	0.4090	0.900	0.4086	0.860
3.00	0.4114	1.140	0.4135	1.350	0.4130	1.300

Table 8.21 Prefault loading condition uncertainty effect on the accuracy of fault location algorithm (faults were applied 0.40 pu distance from bus C of the terminal C).

Prefault loading factor	Fault type					
	AG		BC		BCG	
	Calculated fault location (pu)	Error (%)	Calculated fault location (pu)	Error (%)	Calculated fault location (pu)	Error (%)
0.50	0.4008	0.080	0.4028	0.280	0.4025	0.250
0.80	0.4023	0.230	0.4043	0.430	0.4039	0.390
1.20	0.4043	0.430	0.4062	0.620	0.4058	0.580
1.50	0.4058	0.580	0.4077	0.770	0.4073	0.730
2.00	0.4083	0.830	0.4101	1.010	0.4097	0.970
3.00	0.4133	1.330	0.4150	1.500	0.4147	1.470

C, respectively. As can be observed from the obtained results, the developed fault location algorithm is practically independent of the prefault loading conditions as it located the faults with satisfactory accuracy.

8.4.6 Comparison with nonadaptive algorithm for three-terminal lines

The nonadaptive fault location algorithm does not consider the variations of the system parameters. It neglects the effects of the system operating and surrounding environment conditions on the system parameters. Nevertheless, such assumptions are regarded as the source of errors impacting the accuracy of the fault location. On the contrary, the effects of the operation history and the surrounding environment conditions on the line parameters and system impedance are nullified using the adaptive fault location algorithm to identify the parameters online from the PMU measurements. Such action assists in mimicking the real operating conditions of the system before and after the fault occurrence. Considering the mentioned notes, the impact of system line parameters and system impedance uncertainty on the accuracy of the adaptive fault location algorithm was investigated in this chapter. A ±25% variations of the system parameters were assumed from their nominal values and applied different types of faults (AG, CAG, BC, and ABC) on 0.40 pu distance from the bus A of terminal A. Table 8.22 presents the obtained results where the fault location accuracy was degraded with the increase of the system parameter variation and can be degraded up to around 6 percent for 25 percent parameter variation. It was also observed that the fault location accuracy degradation was proportional to the parameter variation for the investigated faults on the three-terminal system (Fig. 8.6). Therefore, it can be concluded that the lower the parameter variation, the lower the degradation of the accuracy.

Fault diagnosis in three-terminal power transmission lines

Table 8.22 Line parameter and system impedance variation effect on the accuracy of the adaptive fault location algorithm (faults were applied 0.40 pu distance from bus A of terminal A).

Parameter variation (%)	Fault type							
	AG		CAG		BC		ABC	
	Calculated fault location (pu)	Error (%)	Calculated fault location (pu)	Error (%)	Calculated fault location (pu)	Error (%)	Calculated fault location (pu)	Error (%)
−25.0	0.3423	5.770	0.3446	5.540	0.3446	5.540	0.3442	5.580
−20.0	0.3557	4.430	0.3563	4.370	0.3572	4.280	0.3563	4.370
−15.0	0.3671	3.290	0.3685	3.150	0.3681	3.190	0.3681	3.190
−10.0	0.3782	2.180	0.3809	1.910	0.3809	1.910	0.3798	2.020
−5.0	0.3903	0.970	0.3912	0.880	0.3920	0.800	0.3913	0.870
0.0	0.4015	0.150	0.4021	0.210	0.4034	0.340	0.4028	0.280
5.0	0.4129	1.290	0.4130	1.300	0.4146	1.460	0.4139	1.390
10.0	0.4232	2.320	0.4243	2.430	0.4257	2.570	0.4243	2.430
15.0	0.4330	3.300	0.4345	3.450	0.4353	3.530	0.4345	3.450
20.0	0.4431	4.310	0.4453	4.530	0.4454	4.540	0.4451	4.510
25.0	0.4544	5.440	0.4557	5.570	0.4561	5.610	0.4550	5.500

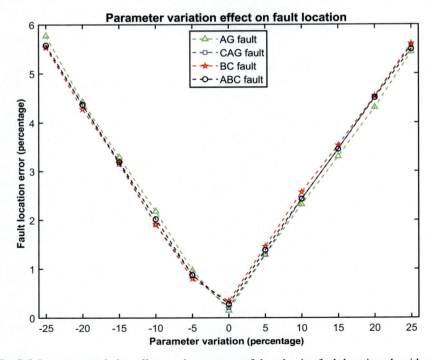

Fig. 8.6 Parameter variation effect on the accuracy of the adaptive fault location algorithm for the investigated three-terminal transmission system.

8.5 Summary

This chapter presented an adaptive algorithm for obtaining the fault locations in a three-terminal transmission system using the synchronized prefault and postfault measurements captured by the PMUs. The algorithm consists of two stages where the first stage detects the faulty section, and the second stage calculates the fault distances from the bus of the faulty section. The presented algorithm uses the PMU current and voltage measurements for the online determination of line parameters and the system TE. Online parameter determination helps to curb parameter uncertainty that occurs due to operating and environmental conditions. The adaptive fault location algorithm was applied on a 500 kV three-terminal transmission system, and the study was carried out on MATLAB and PSCAD/EMTDC platform. Presented results verified the effectiveness of the adaptive fault location algorithm as it located the faults with high accuracy. Furthermore, the algorithm proved its robustness by exhibiting its independence on the fault information (resistance, inception angle, type, and location), prefault loading condition, and system parameter uncertainty.

References

[1] S.R. Punam, A.V Satpute, Review on multiterminal transmission line protection techniques, Int. J. Eng. Res. Technol. 4 (30) (2018). www.ijert.org.

[2] M.M. Saha, J.J. Izykowski, E. Rosolowski, Fault Location on Power Networks, Springer Science & Business Media. Berlin, Germany, 2009.

[3] A. Saber, B.R. Bhalja, Phasor-based fault location algorithm for three-end multi-section nonhomogeneous parallel transmission lines, Int. J. Electr. Power Energy Syst. 130 (2021) 106958, doi:10.1016/j.ijepes.2021.106958.

[4] A.H. Al-Mohammed, M.A. Abido, Adaptive fault location for three-terminal lines using synchrophasors, 2014 IEEE International Workshop on Applied Measurements for Power Systems, AMPS 2014 - Proceedings, 2014, pp. 69–74, doi:10.1109/AMPS.2014.6947710.

[5] H.A. Jimenez, D. Guillen, R. Tapia-Olvera, G. Escobar, F. Beltran-Carbajal, An improved algorithm for fault detection and location in multi-terminal transmission lines based on wavelet correlation modes, Electr. Power Syst. Res. 192 (2021) 106953, doi:10.1016/j.epsr.2020.106953.

[6] A. Saber, New fault location algorithm for four-circuit overhead lines using unsynchronized current measurements, Int. J. Electr. Power Energy Syst. 120 (2020) 106037, doi:10.1016/j.ijepes.2020.106037.

[7] M. Shafiullah, M.A. Abido, S-transform based FFNN approach for distribution grids fault detection and classification, IEEE Access 6 (2018) 8080–8088, doi:10.1109/ACCESS.2018.2809045.

[8] S. Azizi, M. Sanaye-Pasand, M. Paolone, Locating faults on untransposed, meshed transmission networks using a limited number of synchrophasor measurements, IEEE Trans. Power Syst. 31 (6) (2016) 4462–4472, doi:10.1109/TPWRS.2016.2517185.

[9] M. Shafiullah, M.A. Abido, A review on distribution grid fault location techniques, Electr. Power Components Syst. 45 (8) (2017) 807–824, doi:10.1080/15325008.2017.1310772.

[10] A. Mukherjee, P.K. Kundu, A. Das, A supervised principal component analysis-based approach of fault localization in transmission lines for single line to ground faults, Electr. Eng. 1 (2021) 3, doi:10.1007/s00202-021-01221-9.

[11] B. Mahamedi, M. Sanaye-Pasand, S. Azizi, J.G Zhu, Unsynchronised fault-location technique for three-terminal lines, IET Gener. Transm. Distrib. 9 (15) (2015) 2099–2107, doi:10.1049/iet-gtd.2015.0062.

[12] M.M. Devi, M. Geethanjali, A.R. Devi, Fault localization for transmission lines with optimal phasor measurement units, Comput. Electr. Eng. (2018), doi:10.1016/J.COMPELECENG.2018.01.043.

[13] T.C. Lin, Z.R. Xu, F.B. Ouedraogo, Y.J. Lee, A new fault location technique for three-terminal transmission grids using unsynchronized sampling, Int. J. Electr. Power Energy Syst. 123 (2020) 106229, doi:10.1016/j.ijepes.2020.106229.

[14] A. Al-Mohammed, M. Abido, A fully adaptive PMU-based fault location algorithm for series-compensated lines, IEEE Trans. Power Syst. 29 (5) (2014) 2129–2137, http://ieeexplore.ieee.org/xpls/abs_all.jsp?arnumber=6736121. (Accessed 19 August 2018).

[15] S. Hussain, A.H. Osman, Fault location scheme for multi-terminal transmission lines using unsynchronized measurements, Int. J. Electr. Power Energy Syst. 78 (2016) 277–284, doi:10.1016/j.ijepes.2015.11.060.

[16] M. Shafiullah, M.I. Hossain, M.A. Abido, T. Abdel-Fattah, A.H. Mantawy, A modified optimal PMU placement problem formulation considering channel limits under various contingencies, Meas. J. Int. Meas. Confed. 135 (2019), doi:10.1016/j.measurement.2018.12.039.

[17] A. Al-Mohammed, M. Abido, Fault location based on synchronized measurements: a comprehensive survey, Sci. World Journal (2014), http://www.hindawi.com/journals/tswj/2014/845307/abs/. (Accessed 23 May 2016).

[18] M. Shafiullah, S.M. Rahman, M.G. Mortoja, B. Al-Ramadan, Role of spatial analysis technology in power system industry: an overview, Renew. Sustain. Energy Rev. 66 (2016) 584–595, doi:10.1016/j.rser.2016.08.017.

[19] N. Zhang, M. Kezunovic, Improving real-time fault analysis and validating relay operations to prevent or mitigate cascading blackouts, Proceedings of the IEEE Power Engineering Society Transmission and Distribution Conference, 2006, pp. 847–852, doi:10.1109/TDC.2006.1668608.

[20] N. Zhang and M. Kezunovic, A study of synchronized sampling based fault location algorithm performance under power swing and out-of-step conditions, 2005, doi:10.1109/PTC.2005.4524595.

[21] P.R. Chegireddy, R. Bhimasingu, Synchrophasor based fault location algorithm for three terminal homogeneous transmission lines, Electr. Power Syst. Res. 191 (2021) 106889, doi:10.1016/j.epsr.2020.106889.

[22] K.G. Firouzjah, A. Sheikholeslami, A current independent method based on synchronized voltage measurement for fault location on transmission lines, Simul. Model. Pract. Theory 17 (4) (2009) 692–707, doi:10.1016/j.simpat.2008.12.003.

Fault diagnosis in series compensated power transmission lines

9.1 Introduction

Most power plants generate electricity in the form of alternating current (AC) power. Besides, most of the electricity-consuming devices connected to the power grids operate on AC. Thus, AC power transmission is considered as the most cost-effective and reliable option for bulk power transmission over long distances in most cases [1–3]. However, the electric power system operators run the transmission grids at their near-maximum capacity and thermal limit due to surging growth of energy demand, increased transmission lines building expenditure, availability of transmission corridors, and environmental and regulatory concerns. Such issues lead the system operators to explore alternatives, including enhancing the power transmission capacity of the existing transmission lines. Like other flexible AC transmission devices, series compensation in transmission lines improves the power transfer capability in addition to other benefits, for example, low-frequency oscillation damping, reduction of system losses, improvement of load division between the parallel lines, and reactive power control for voltage stability. With series compensation, the problem of very long distance as the drawback for the AC transmission can be eliminated in most cases [4–10].

The series compensators are equipped with metal oxide varistors (MOVs) that provide overvoltage protection to the compensators. However, the MOV introduces nonlinearity in the systems causing high-frequency transients, direct current (DC) decaying, and nonfundamental decaying [9–12]. Besides, the series compensated lines (SCLs) introduce multiple difficulties such as an abrupt change in impedance at compensation point, significant increase in steady-state currents than the fault currents, inversion of the current and voltage signals, subsynchronous oscillations, odd-harmonic components, etc. Thus, the distance protection relays and fault locators start to either lose their accuracy or malfunction [9–14]. The mentioned issues make the fault diagnosis for the series compensated transmission systems a challenging task. However, precise information about power system faults is crucial for expediting the restoration process by sending the maintenance crews to the sites to reduce the outage duration and revenue loss, ensure quality power supply, and enhance system reliability [15–19]. In response, the researchers explored different techniques for fault diagnosis in SCL, including impedance estimation, traveling wave, synchronized and unsynchronized measurements, differential protection, distributed line model, artificial intelligence, and hybrid techniques [20–24].

This chapter presents an adaptive fault location algorithm for SCL using synchronized phasor measurements collected from the phasor measurement units (PMUs).

The algorithm is model-free, and as such, it can be applied to transmission lines regardless of the type of series compensation. Moreover, the algorithm utilizes PMU measurements for online computation of the system Thevenin equivalent (TE) and line parameters to tackle the parameter uncertainty that occurs due to the operating history and environmental conditions. The adaptive fault location algorithm is implemented on a 400 kV transmission network with an SCL, and the study is carried out on PSCAD/EMTDC and MATLAB/SIMULINK platforms. Simulation results confirm the satisfactory accuracy and reliability of the adaptive fault location algorithm. The efficacy of the adaptive algorithm is investigated under fault information (resistance, inception angle, location, and type), prefault loading condition, and line compensation degree uncertainty.

The remaining parts of this chapter are organized as follows: Section 9.2 presents series capacitor locations while Section 9.3 discusses series capacitor schemes for the transmission lines. Section 9.4 illustrates the PMU-based parameter calculation scheme of an SCL, whereas Section 9.5 demonstrates the adaptive fault location subroutines. Section 9.6 presents the obtained results that confirm the efficacy of the algorithm and their independence on fault information (location, type, resistance, and inception angle), prefault loading, system parameter, and line compensation degree uncertainty. Finally, an overall summary of this chapter is presented in Section 9.7.

9.2 Series capacitor locations

In transmission lines, the series capacitors are installed either at both ends of the line (Fig. 9.1) or at the middle of the line (Fig. 9.2) to lower the line impedance for enhancing the power transfer capacity. Usually, the series capacitors located along the transmission line are unmanned. However, such installation reduces the worst-case fault current; thus, it requires a comparatively lower MOV rating. On the other hand, the line ends series capacitors are close to the substations and therefore attended. For this case, the fault current is high; thus, it requires a comparatively higher MOV rating. The locations of series capacitor banks are crucial due to several reasons. To start with, the compensation effectiveness is a function of the locations of the capacitors. Moreover, the voltage profiles along the line are influenced by the capacitor locations. Also, their locations affect the main circuit equipment and the protection schemes of the transmission lines. In addition, their maintenance behaviors are impacted by their locations. Last but not the least, MOV ratings for the series capacitors depend upon their locations [25].

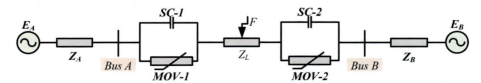

Fig. 9.1 MOV and SC installed at both ends of a transmission line. *MOV*, metal oxide varistor.

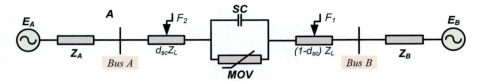

Fig. 9.2 MOV and SC installed at midpoint of a transmission line. *MOV*, metal oxide varistor.

9.3 Series capacitor schemes

A series capacitor needs control, protection, and supervision to function properly as a critical component of the power system networks. The capacitor should also be fully insulated to the ground since it is operated at the system voltage level. Among different schemes, Figs. 9.3 and 9.4 show the circuit diagrams of the series capacitor with a gapped scheme and a gapless scheme, respectively.

The MOV serves as the primary protective device for a series capacitor. It is a nonlinear resistor that limits the voltage across the capacitor for safety purposes. During a fault, the MOV protects the capacitor bank against the unexpected high voltage by bypassing the fault current. In many cases, if the MOV is not sufficient to absorb the fault current, a spark gap is used to bypass the fault current. A bypass switch is included in the approach to bypass and inserts the series capacitor when needed. In addition, the bypass switch is necessary to extinguish the spark gap or bypass the MOV together with faults near the series capacitor. Finally, a current limiting damping circuit (CLDC) is implemented to limit and damp the high-frequency discharge current resulting from the spark gap operation or closing of the bypass switch. Typically, the CLDC comprises an air-core reactor, which gives an LC discharge circuit and the capacitor. When high damping of the capacitor discharge current is needed, a parallel damping resistor is connected across the reactor [25,26].

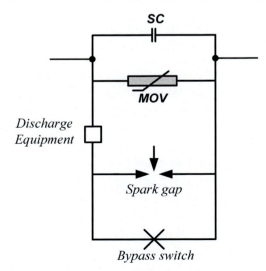

Fig. 9.3 Circuit diagram for a gapped series capacitor scheme.

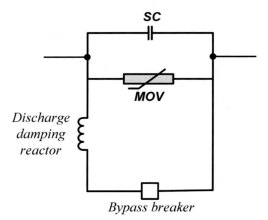

Fig. 9.4 Circuit diagram for a gapless series capacitor scheme.

9.3.1 Metal oxide varistor voltage-current relationship

The following nonlinear equation can approximate the relationship between MOV's current and voltage:

$$i_v = I_c \left(\frac{V_v}{V_p} \right)^{\alpha} \quad (9.1)$$

Where,

I_c, i_v: MOV coordination and instantaneous currents, respectively.
V_v, V_p: MOV instantaneous voltage and protective level voltage defined at I_c, respectively.
α: MOV material manufacturing process-related constant.

The value of α typically lies in the range 20–30. With a close look at Eq. (9.1), one may note that, as the applied voltage starts increasing beyond a specific value, the MOV material will start to conduct a rising amount of current while the voltage across it is almost fixed. This sharp MOV voltage-current characteristic provides direct overvoltage protection to the series capacitor during anomalies in the electric networks. The automatic and instantaneous restoration and reinsertion offered by the MOV can improve the line transfer restriction but introduce other problems associated with transient instability [27–32].

9.3.2 Internal and external faults

Different requirements of the series capacitors for different types (internal and external) faults should be considered while designing the SC for the transmission lines. Concerning a series capacitor, internal faults are failures of the power system occurring in the same part of the transmission line that contains the capacitor. However, external faults are those occurring outside of the line segment containing the series capacitor. The series capacitor shall remain in the circuit during and after the external faults. It shall also resist the fault current passing through it. The protecting MOV of the series capacitor must have the adequate thermal capacity to withstand the fault current flowing through the circuit. After clearing the fault, the MOV voltage drops and, as a result, it stops conducting, and the series capacitor returns into operation instantaneously and automatically. Such action is referred to as the instantaneous insertion of the series capacitor, a crucial system behavior after being subjected to any disturbance. Fast reinsertion of the series capacitors is often required for system postfault stability [25].

The functions of the series capacitors for internal faults are a bit different. They are taken out while the faulted line sections are disconnected as they do not need to take any fault current at that time. However, they must carry the fault currents before complete disconnection that requires 60 to 100 milliseconds (ms). Generally, it is allowed to bypass the series capacitor during an internal fault to reduce MOV capacity. In the gapped schemes, bypassing is accomplished through the spark gap and the bypass switch. On the contrary, it is achieved through the bypass switch alone for the gapless schemes. The gapped and gapless schemes are used when the fault duties for the internal faults are high and low, respectively. The fault duties for the internal faults depend both on series capacitor location and system short circuit capacity. The bypass switch is opened, and the series capacitor is reinserted back into the system operation with reclosing of the disconnected line section [25].

9.4 Phasor measurement unit-based parameter calculation of series-compensated line

Before building the adaptive fault location algorithm for an SCL, its parameters should be computed online considering the operating history and environmental conditions. Thus, this section presents the calculation procedures of the positive-sequence parameters using the synchronized current and voltage measurements captured from the installed PMU at both ends of the SCL [33].

In Fig. 9.5, a transposed SCL between terminal P and Q is considered where E_G and E_H denote the TEs at the terminals P and Q, respectively. The series compensation device is placed in between the terminals (at location R). It is aimed to determine the SCL positive sequence series reactance, series resistance, and shunt susceptance from

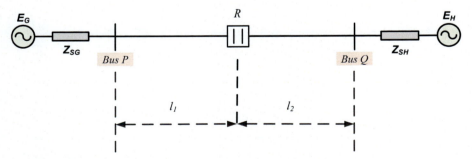

Fig. 9.5 Single line diagram of an SCL. *SCL*, series-compensated line.

the steady-state PMU current and voltage measurements recorded from the terminals P and Q. Usually, the zero sequence components of the SCL are negligible during the normal operating situations. However, the algorithm can be used for computation of the SCL zero sequence components if such components emerge due to external faults or unbalanced loading conditions. System positive sequence network under normal operating conditions is shown in Fig. 9.6.

Where,

V_{pi}, V_{qi}: The ith moment positive sequence voltage phasors at terminal P and Q, respectively.
I_{pi}, I_{qi}: The ith moment positive sequence current phasors at terminal P and Q, respectively.
$i=1, 2, ..., N$; N is the total number of measurements sets where each set consists of V_{pi}, I_{pi}, V_{qi}, and I_{qi}.
V_{si}, I_{ri}: Voltage-drop across and current flow through the SC, respectively.
V_{ri}: Location R left side voltage.
Z_c, γ: Characteristic impedance and propagation constant of the line.
Z_{pr}, Z_{qr}: Equivalent series impedances of the line segment PR and QR, respectively.
Y_{pr}, Y_{qr}: Equivalent shunt admittances of the line segment PR and QR, respectively.
l_1, l_2: Lengths of the line segments PR and QR, respectively.

Equivalent line parameters can be expressed as:

$$Z_{pr} = Z_c \sinh(\gamma l_1) \tag{9.2}$$

$$Z_{qr} = Z_c \sinh(\gamma l_2) \tag{9.3}$$

$$Y_{pr} = \frac{2}{Z_c} \tanh\left(\frac{\gamma l_1}{2}\right) \tag{9.4}$$

$$Y_{qr} = \frac{2}{Z_c} \tanh\left(\frac{\gamma l_2}{2}\right) \tag{9.5}$$

$$Z_c = \sqrt{\frac{z_1}{y_1}} \tag{9.6}$$

$$\gamma = \sqrt{z_1 y_1} \tag{9.7}$$

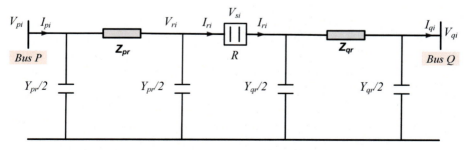

Fig. 9.6 Positive sequence network of an SCL during normal operation. *SCL*, series-compensated line.

Fault diagnosis in series compensated power transmission lines

Where,

z_1, y_1: Positive sequence series impedance and shunt admittance per unit length, respectively.

It is required to determine the real and imaginary components of z_1 and the imaginary component of y_1. Concerning Fig. 9.6 and considering that the direction of current from P to Q, the following Equations can be written:

$$V_{pi} = V_{ri} \cosh(\gamma l_1) + I_{ri} Z_c \sinh(\gamma l_1) \tag{9.8}$$

$$I_{pi} = I_{ri} \cosh(\gamma l_1) + \frac{V_{ri}}{Z_c} \sinh(\gamma l_1) \tag{9.9}$$

$$V_{ri} - V_{si} = V_{qi} \cosh(\gamma l_2) + I_{qi} Z_c \sinh(\gamma l_2) \tag{9.10}$$

$$I_{ri} = I_{qi} \cosh(\gamma l_2) + \frac{V_{qi}}{Z_c} \sinh(\gamma l_2) \tag{9.11}$$

By removing I_{ri} and V_{ri} and knowing that $l = l_1 + l_2$, the following equation can be obtained:

$$V_{pi} = V_{si} \cosh(\gamma l_1) + V_{qi} \cosh(\gamma l) + I_{qi} Z_c \sinh(\gamma l) \tag{9.12}$$

$$I_{pi} = V_{si} \frac{\sinh(\gamma l_1)}{Z_c} + I_{qi} \cosh(\gamma l) + \frac{V_{qi}}{Z_c} \sinh(\gamma l) \tag{9.13}$$

The above two complex equations are obtained for one set of measurements and can be split into four real equations with five unknowns, for example, three-line parameters and imaginary and real components of V_{si}. Four more real equations with two real unknowns can be obtained from the second set of measurements for the new V_{si}. The eight equations can be solved to identify seven unknowns. A more robust estimate for the unknowns can be achieved using the classical least-squares approach. Defining the unknown variables as:

$$X = [x_1, x_2, \ldots, x_{2N}, x_{2N+1}, x_{2N+2}, x_{2N+3}]^T \tag{9.14}$$

Where,

N: Number of measurement sets.
x_{2i-1}, x_{2i}: Variables representing voltage across the series compensation device, e.g., $V_{si} = x_{2i-1} e^{jx_{2i}}$ where $I = 1, 2, \ldots, N$.
x_{2N+1}: Positive sequence series resistance per unit length of the line.
x_{2N+2}: Positive sequence series reactance per unit length of the line.
x_{2N+3}: Positive sequence shunt susceptance per unit length of the line.

Using the defined variables, Eqs. (9.12) and (9.13) can be written in the form $f_{2i-1}(X) = 0$ and $f_{2i}(X) = 0$, respectively as shown below:

$$f_{2i-1}(X) = x_{2i-1} e^{jx_{2i}} \cosh(\gamma l_1) + V_{qi} \cosh(\gamma l) + I_{qi} Z_c \sinh(\gamma l) - V_{pi} = 0 \tag{9.15}$$

$$f_{2i}(X) = x_{2i-1}e^{jx_{2i}}\frac{\sinh(\gamma l_1)}{Z_c} + I_{qi}\cosh(\gamma l) + \frac{V_{qi}}{Z_c}\sinh(\gamma l) - I_{pi} = 0 \qquad (9.16)$$

Where,

$$Z_c = \sqrt{(x_{2N+1} + jx_{2N+2})/(jx_{2N+3})} \qquad (9.17)$$

$$\gamma = \sqrt{(x_{2N+1} + jx_{2N+2})(jx_{2N+3})} \qquad (9.18)$$

Defining the function vector, $F(X)$ as:

$$F_{2i-1}(X) = \text{Re}(f_i(X)), i = 1, 2, \ldots, 2N \qquad (9.19)$$

$$F_{2i}(X) = \text{Im}(f_i(X)), i = 1, 2, \ldots, 2N \qquad (9.20)$$

Then, the unknown variable vector, X, is derived as:

$$X_{k+1} = X_k + \Delta X \qquad (9.21)$$

$$\Delta X = -(H^T H)^{-1}\left[H^T F(X_k)\right] \qquad (9.22)$$

$$H = \frac{\partial F(X_k)}{\partial X} \qquad (9.23)$$

Where,

k: Iteration number starting from 1.
X_k: Variable vector at k^{th} iteration.
X_{k+1}: Variable vector next to k^{th} iteration.
ΔX: Variable update.
H: A matrix consisting of the derivatives of the function with respect to the unknown variables.

9.5 Fault location algorithm description

Fig. 9.7 shows a line equipped with a single stack of the fixed series capacitor with MOV for overvoltage protection [32]. A fault may occur at both ends of the capacitor stack; therefore, faults F_A and F_B shall be considered. Consequently, the faults are located using two subroutines S_A and S_B. Moreover, the valid subroutine is indicated using a selection procedure. Then, the actual fault location is obtained.

Figs. 9.8 and 9.9 show how the compensating bank partitions the line into two segments of per-unit length d_{SC} and $(1-d_{SC})$. The relative distances d_{FA} and d_{FB} are related to the per-unit distances d_A and d_B by the following relationships:

$$d_A = d_{FA} \cdot d_{SC} \qquad (9.24)$$

$$d_B = d_{SC} + (1 - d_{FB}) \cdot (1 - d_{SC}) \qquad (9.25)$$

Fault diagnosis in series compensated power transmission lines 231

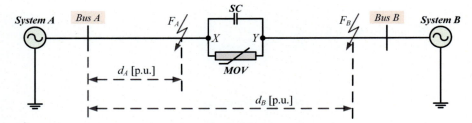

Fig. 9.7 Single line diagram of an SCL equipped with SC and MOV. *MOV*, metal oxide varistor; *SCL*, series-compensated line.

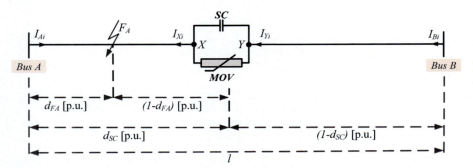

Fig. 9.8 Subroutine S_A – scheme of SCL under fault F_A in section A-X. *SCL*, series-compensated line.

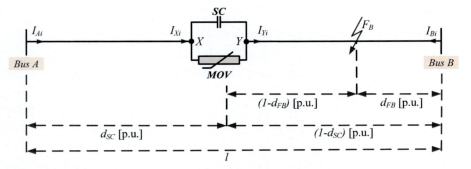

Fig. 9.9 Subroutine S_B – scheme of SCL under fault F_B in section B-Y. *SCL*, series-compensated line.

9.5.1 Fault location subroutine, S_A

The subroutine (S_A) for a fault (F_A) in section A-X is derived by neglecting the effect of line shunt capacitances. Following generalized fault loop model is used for the subroutine:

$$V_{Ap} - d_{FA} Z_{1LA} I_{Ap} - R_{FA} I_{FA} = 0 \tag{9.26}$$

Where,

d_{FA}: Per unit distance to fault, F_A.
R_{FA}: Fault resistance for fault, F_A.
V_{Ap}: Fault loop voltage for fault, F_A.
I_{Ap}: Fault loop current for fault, F_A.
I_{FA}: Total fault current for fault, F_A.
Z_{1LA}: Line section (A-X) positive sequence impedance.

Fault loop current and voltage are composed according to the fault type. They are written as shown below:

$$V_{Ap} = a_1 V_{A1} + a_2 V_{A2} + a_0 V_{A0} \tag{9.27}$$

$$I_{Ap} = a_1 I_{A1} + a_2 I_{A2} + a_0 \frac{Z_{0LA}}{Z_{1LA}} I_{A0} \tag{9.28}$$

Where,

a_1, a_2, a_0: Weighting coefficients as listed in Table 9.1.
V_{A1}, V_{A2}, V_{A0}: Side-A positive, negative, and sequence voltages, respectively.
I_{A1}, I_{A2}, I_{A0}: Side-A positive, negative, and sequence currents, respectively.
Z_{0LA}: Line section (A-X) zero sequence impedance.

The total fault current in Eq. (9.26) is obtained using the generalized fault model as shown below:

$$I_{FA} = a_{F1} I_{FA1} + a_{F2} I_{FA2} + a_{F0} I_{FA0} \tag{9.29}$$

Where,

a_{F1}, a_{F2}, a_{F0}: Share coefficients as listed in Table 9.2.

The ith sequence component of the total fault current is computed by summing up the ith sequence components of the currents of the two ends of faulted section (A-X):

$$I_{FAi} = I_{Ai} + I_{Xi} \tag{9.30}$$

Table 9.1 Weighting coefficients [26].

Fault type	a_1	a_2	a_0
AG	1.0	1.0	1.0
BG	$-0.50 - j0.50\sqrt{3}$	$-0.50 + j0.50\sqrt{3}$	1.0
CG	$-0.50 + j0.50\sqrt{3}$	$-0.50 - j0.50\sqrt{3}$	1.0
AB, ABG, ABC	$1.50 + j0.50\sqrt{3}$	$1.50 - j0.50\sqrt{3}$	0.0
BC, BCG	$-j\sqrt{3}$	$j\sqrt{3}$	0.0
CA, CAG	$-1.50 + j0.50\sqrt{3}$	$-1.50 - j0.50\sqrt{3}$	0.0

Table 9.2 Share coefficients [26].

Fault type	a_{F1}	a_{F2}	a_{F0}
AG	0.0	3.0	0.0
BG	0.0	$-1.50 + j1.50\sqrt{3}$	0.0
CG	0.0	$-1.50 - j1.50\sqrt{3}$	0.0
AB	0.0	$1.50 - j0.50\sqrt{3}$	0.0
BC	0.0	$j\sqrt{3}$	0.0
CA	0.0	$-1.50 - j0.50\sqrt{3}$	0.0
ABG	0.0	$3.0 - j\sqrt{3}$	$j\sqrt{3}$
BCG	0.0	$j2.0\sqrt{3}$	$j\sqrt{3}$
CAG	0.0	$3.0 + j\sqrt{3}$	$j\sqrt{3}$
ABC	$1.50 + j0.50\sqrt{3}$	$1.50 - j0.50\sqrt{3}$	0.0

Where,

i: Index of the symmetrical components ("0," "1," and "2" for zero, positive, and negative sequence components, respectively).

I_{Ai}, I_{Xi}: The ith sequence current components at bus A and point X, respectively.

If the line shunt capacitances are neglected, then I_{Xi} will be equal to the current I_{Bi}. Thus, the following relationship can be obtained:

$$I_{FAi} = I_{Ai} + I_{Bi} \tag{9.31}$$

It is recommended to follow the following guidelines to assure high accuracy of fault location [26]:

- Use negative sequence components for phase-to-ground (LG) and phase-to-phase (LL) faults.
- Use negative and zero sequence components for phase-to-phase-to-ground (LLG) faults.
- Use superimposed positive sequence components for three-phase symmetrical (LLL) faults.

Total fault current is obtained for the case of three-phase balanced faults collecting the superimposed positive sequence currents from both ends (A and B) as follows:

$$I_{FA1} = I_{A1}^{\text{superimp.}} + I_{B1}^{\text{superimp.}} \tag{9.32}$$

Where,

$I_{A1}^{\text{superimp.}}$: Superimposed positive sequence current at end A.
$I_{B1}^{\text{superimp.}}$: Superimposed positive sequence current at end B.

The superimposed positive sequence currents are calculated by subtracting the prefault quantity (superscript: "pre") from the fault quantity:

$$I_{FA1} = \left(I_{A1} - I_{A1}^{pre}\right) + \left(I_{B1} - I_{B1}^{pre}\right) \tag{9.33}$$

Where,

I_{A1}^{pre}, I_{B1}^{pre}: Prefault positive sequence currents at the ends A and B, respectively.
I_{A1}, I_{B1}: Positive sequence fault currents at the ends A and B, respectively.

Referring to Table 9.2, accurate computation of the total fault current can be assured as the positive-sequence components are excluded for all fault types. The fault distance is determined using the following relationship after resolving Eq. (9.26) into the corresponding imaginary and real parts and removing the unknown fault resistance:

$$d_{FA} = \frac{\text{real}(V_{Ap})\text{imag}(I_{FA}) - \text{imag}(V_{Ap})\text{real}(I_{FA})}{\text{real}(Z_{1LA}I_{Ap})\text{imag}(I_{FA}) - \text{imag}(Z_{1LA}I_{Ap})\text{real}(I_{FA})} \quad (9.34)$$

Knowing the fault distance, the fault resistance can be determined.

9.5.2 Fault location subroutine, S_B

For a fault F_B in section B-Y, the distance to the fault be calculated without representing the bank if the fault location function is available at terminal B. Fault location subroutine S_B can be employed for fault location like the subroutine S_A. Based on the comprehensive fault loop model, the equation for fault F_B can be expressed as:

$$V_{Bp} - d_{FB}Z_{1LB}I_{Bp} - R_{FB}I_{FB} = 0 \quad (9.35)$$

Where,

d_{FB}: Per-unit distance to fault, F_B.
R_{FB}: Fault resistance for fault, F_B.
V_{Bp}: Fault loop voltage for fault, F_B.
I_{Bp}: Fault loop current for fault, F_B.
I_{FB}: Total fault current for fault, F_B.
Z_{1LB}: Line section (B-Y) positive sequence impedance.

Fault loop voltage and current are composed according to the type of fault and they can be written as:

$$V_{Bp} = a_1 V_{B1} + a_2 V_{B2} + a_0 V_{B0} \quad (9.36)$$

$$I_{Bp} = a_1 I_{B1} + a_2 I_{B2} + a_0 \frac{Z_{0LB}}{Z_{1LB}} I_{B0} \quad (9.37)$$

Where,

V_{B1}, V_{B2}, V_{B0}: Side-B positive, negative, and zero sequence voltages, respectively.
I_{B1}, I_{B2}, I_{B0}: Side-B positive, negative, and zero sequence currents, respectively.
Z_{0LB}: Zero sequence impedance of the line section B-Y.

Total fault current in Eq. (9.35) can be determined using the generalized fault model as follows:

$$I_{FB} = a_{F1}I_{FB1} + a_{F2}I_{FB2} + a_{F0}I_{FB0} \quad (9.38)$$

By adding the ith sequence components of currents from both ends of the faulted section B-Y, the ith sequence component of the total fault current is determined as:

$$I_{FBi} = I_{Bi} + I_{Yi} \tag{9.39}$$

Where,

I_{Bi}, I_{Yi}: The ith sequence current components at bus B and point Y, respectively.

If the line shunt capacitances are neglected, I_{Yi} equals I_{Ai}, and Eq. (9.39) can be written as:

$$I_{FBi} = I_{Ai} + I_{Bi} \tag{9.40}$$

For three-phase balanced faults, the total fault current is determined by adding the superimposed positive sequence currents from both ends A and B as follows:

$$I_{FB1} = I_{A1}^{\text{superimp.}} + I_{B1}^{\text{superimp.}} \tag{9.41}$$

Like, subroutine S_A, Eq. (9.41) can be put in the following form:

$$I_{FB1} = \left(I_{A1} - I_{A1}^{pre}\right) + \left(I_{B1} - I_{B1}^{pre}\right) \tag{9.42}$$

By splitting Eq. (9.35) into the imaginary and real components and removing the unknown fault resistance, the fault distance can be calculated as:

$$d_{FB} = \frac{\text{real}(V_{Bp})\,\text{imag}(I_{FB}) - \text{imag}(V_{Bp})\,\text{real}(I_{FB})}{\text{real}(Z_{1LB}I_{Bp})\,\text{imag}(I_{FB}) - \text{imag}(Z_{1LB}I_{Bp})\,\text{real}(I_{FB})} \tag{9.43}$$

Knowing the fault distance, the fault resistance can be determined.

9.5.3 Selection procedure

The procedure explained in [32] is adopted for choosing the proper subroutine from S_A and S_B. The subroutine which yields a negative fault resistance or a fault distance outside of the section range is rejected. This allows, in most cases, to choose the proper subroutine. If this is not the case, network circuit diagrams for the negative sequence related to the subroutines (S_A and S_B) are considered. Superimposed positive-sequence components are considered for three-phase balanced faults. The valid subroutine is chosen if the obtained remote source impedance has an R-L character, and the calculated value is close to the actual source impedance.

9.6 Simulation results

The adaptive fault location algorithm was applied on a 400 kV transmission network with a series compensated line, as shown in Fig. 9.10. The line was assumed to be faulted at both sides (A-X and B-Y) of the series compensation device. This part presents the simulation results achieved using the PSCAD/EMTDC and the MATLAB/

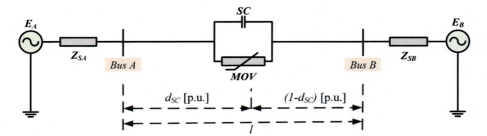

Fig. 9.10 Test series compensated transmission line system.

SIMULINK environment. The efficacy of the adaptive algorithm was investigated under fault information (resistance, inception angle, location, and type), prefault loading condition, and line compensation degree uncertainty.

9.6.1 Data generation and conditioning

The adaptive fault location algorithm was implemented in the MATLAB/SIMULINK environment and assessed using prefault and postfault data captured from the PSCAD/EMTDC simulations for faults that occurred on sections A-X and B-Y. The 400 kV SCL, compensated in the middle with a degree of 70% compensation, was modeled with its distributed parameters. The line parameters and system TEs are listed in Table 9.3. The system TEs at terminals A and B were computed online as explained in Chapter 6 and in [26,34–36].

The current and voltage transformers located at each line terminal were deliberately considered as the ideal devices to investigate the accuracy of the adaptive fault location algorithm. The three-phase current and voltage signals were sampled at a frequency of 240 Hz (four samples per cycle) and stored for postprocessing. The

Table 9.3 Series-compensated line network parameters [26].

Parameter	Value
l (km)	300.0
d_{SC} (p.u.)	0.50
Z_{1L} (Ω)	8.28 + j94.50
Z_{0L} (Ω)	82.50 + j307.90
C_{1L} (nF/km)	13.0
C_{0L} (nF/km)	8.50
Z_{1SA} (Ω)	1.31 + j15.0
Z_{1SB} (Ω)	1.31 + j15.0
Z_{0SA} (Ω)	2.33 + j22.50
Z_{0SB} (Ω)	2.33 + j22.50
E_A (kV)	400.0∠0°
E_B (kV)	390.0 ∠ − 10°
I_c (kA)	1.0
V_p (kV)	150.0
A	23.0

voltage and current phasors were extracted employing the discrete Fourier transform as given in the following equation:

$$X = \left(\sqrt{2}/N_s\right)\sum_{k=1}^{N_s} x[k]e^{-j2\pi k/N_s} \qquad (9.44)$$

Where, N_s is the total number of samples in one period, X is the phasor, and $x[k]$ is the waveform samples. The percentage error of the fault location was calculated using the following equation:

$$\% \text{ Error} = \frac{|\text{Actual fault location} - \text{Estimated fault location}|}{\text{Total line length}} \times 100 \qquad (9.45)$$

9.6.2 Accuracy analysis

Different types of faults at various locations of the series compensated transmission line by varying fault resistance were simulated to verify the effectiveness of the adaptive fault location algorithm. Tables 9.4, 9.5, 9.6 and 9.7 present the calculated fault locations and percentage error for LG, LL, LLG, and LLL faults at various locations

Table 9.4 Fault location results for LG faults on the test SC transmission network.

Fault type	Fault resistance (Ω)	Actual fault location (pu)	Calculated fault location (pu)	Error (%)
AG	10.0	0.20	0.1992	0.080
		0.40	0.4009	0.090
		0.60	0.6047	0.470
		0.80	0.8005	0.050
	100.0	0.20	0.1990	0.100
		0.40	0.4050	0.500
		0.60	0.6075	0.750
		0.80	0.7943	0.570
BG	10.0	0.20	0.2032	0.320
		0.40	0.4094	0.940
		0.60	0.5948	0.520
		0.80	0.7960	0.400
	100.0	0.20	0.2035	0.350
		0.40	0.4083	0.830
		0.60	0.5932	0.680
		0.80	0.7953	0.470
CG	10.0	0.20	0.2013	0.130
		0.40	0.4068	0.680
		0.60	0.5927	0.730
		0.80	0.7997	0.030
	100.0	0.20	0.1989	0.110
		0.40	0.4001	0.010
		0.60	0.5972	0.280
		0.80	0.8026	0.260

Table 9.5 Fault location results for LL faults on the test SC transmission network.

Fault type	Fault resistance (Ω)	Actual fault location (pu)	Calculated fault location (pu)	Error (%)
AB	1.0	0.20	0.2005	0.050
		0.40	0.4032	0.320
		0.60	0.5968	0.320
		0.80	0.7995	0.050
	10.0	0.20	0.2007	0.070
		0.40	0.4043	0.430
		0.60	0.5960	0.400
		0.80	0.7995	0.050
BC	1.0	0.20	0.2016	0.160
		0.40	0.4054	0.540
		0.60	0.5945	0.550
		0.80	0.7984	0.160
	10.0	0.20	0.2021	0.210
		0.40	0.4053	0.530
		0.60	0.5960	0.400
		0.80	0.7989	0.110
CA	1.0	0.20	0.2007	0.070
		0.40	0.4038	0.380
		0.60	0.5955	0.450
		0.80	0.7991	0.090
	10.0	0.20	0.2007	0.070
		0.40	0.4061	0.610
		0.60	0.5940	0.600
		0.80	0.7978	0.220

with varying fault resistances on the line segments *A-X* and *B-Y*. As can be observed from the obtained results, the adaptive fault location algorithm effectively estimated the fault distances of the applied faults with satisfactory accuracy.

9.6.3 Effect of fault resistance

The effect of fault resistance uncertainty on the fault location algorithm was investigated, and obtained results are presented in this section. Two types of faults were studied where the ground was involved in one type to capture both low and high resistance (0 to 500 Ω) faults, and for other kinds, no ground was involved that captured only low resistance (0 to 30 Ω) faults. The local and remote source impedances were set, in all cases, to be equal to the system values. Obtained results for different faults are presented in Tables 9.8, 9.9, 9.10 and 9.11, considering the fault occurrence at 0.6 pu distance from the bus A of the SCL. As can be seen from the obtained results, the adaptive fault location algorithm accurately located different types of faults with satisfactory accuracy that confirmed the independence of the algorithm under fault resistance uncertainty.

Table 9.6 Fault location results for LLG faults on the test SC transmission network.

Fault type	Fault resistance (Ω)	Actual fault location (pu)	Calculated fault location (pu)	Error (%)
ABG	5.0	0.20	0.2005	0.050
		0.40	0.4033	0.330
		0.60	0.5968	0.320
		0.80	0.7995	0.050
	50.0	0.20	0.2005	0.050
		0.40	0.4033	0.330
		0.60	0.5969	0.310
		0.80	0.7996	0.040
BCG	5.0	0.20	0.2014	0.140
		0.40	0.4054	0.540
		0.60	0.5946	0.540
		0.80	0.7986	0.140
	50.0	0.20	0.2014	0.140
		0.40	0.4054	0.540
		0.60	0.5947	0.530
		0.80	0.7985	0.150
CAG	5.0	0.20	0.2009	0.090
		0.40	0.4042	0.420
		0.60	0.5957	0.430
		0.80	0.7988	0.120
	50.0	0.20	0.2008	0.080
		0.40	0.4038	0.380
		0.60	0.5957	0.430
		0.80	0.7991	0.090

Table 9.7 Fault location results for LLL faults on the test SC transmission network.

Fault type	Fault resistance (Ω)	Actual fault location (pu)	Calculated fault location (pu)	Error (%)
ABC	1.0	0.20	0.1995	0.050
		0.40	0.4019	0.190
		0.60	0.5985	0.150
		0.80	0.8007	0.070
	10.0	0.20	0.1957	0.430
		0.40	0.3968	0.320
		0.60	0.6063	0.630
		0.80	0.8052	0.520

Table 9.8 Fault resistance uncertainty effect on the accuracy of fault location algorithm for LG faults on SCL (faults were applied 0.60 pu distance from the bus A of the SCL).

	Fault type					
	AG		BG		CG	
Fault resistance (Ω)	Calculated fault location (pu)	Error (%)	Calculated fault location (pu)	Error (%)	Calculated fault location (pu)	Error (%)
0.00	0.5969	0.310	0.5955	0.450	0.5948	0.520
1.00	0.5972	0.280	0.5946	0.540	0.5945	0.550
5.00	0.5999	0.010	0.5929	0.710	0.5943	0.570
10.0	0.6047	0.470	0.5948	0.520	0.5927	0.730
20.0	0.6055	0.550	0.6028	0.280	0.5822	1.780
50.0	0.5864	1.360	0.6022	0.220	0.6075	0.750
100.0	0.6075	0.750	0.6018	0.180	0.5972	0.280
200.0	0.6049	0.490	0.6013	0.130	0.5849	1.510
400.0	0.5863	1.370	0.5975	0.250	0.5859	1.410
500.0	0.5792	2.080	0.5960	0.400	0.5731	2.690

SCL, series-compensated line.

Table 9.9 Fault resistance uncertainty effect on the accuracy of fault location algorithm for LL faults on SCL (faults were applied 0.60 pu distance from the bus A of the SCL).

	Fault type					
	AB		BC		CA	
Fault resistance (Ω)	Calculated fault location (pu)	Error (%)	Calculated fault location (pu)	Error (%)	Calculated fault location (pu)	Error (%)
0.00	0.5969	0.310	0.5947	0.530	0.5959	0.4100
0.50	0.5969	0.310	0.5946	0.540	0.5957	0.4300
1.00	0.5968	0.320	0.5945	0.550	0.5955	0.4500
2.50	0.5966	0.340	0.5943	0.570	0.5950	0.5000
5.00	0.5964	0.360	0.5945	0.550	0.5943	0.5700
7.50	0.5963	0.370	0.5950	0.500	0.5940	0.6000
10.0	0.5960	0.400	0.5960	0.400	0.5940	0.6000
15.0	0.5953	0.470	0.5983	0.170	0.5947	0.5300
20.0	0.5939	0.610	0.6007	0.070	0.5964	0.3600
30.0	0.5893	1.070	0.6033	0.330	0.6022	0.2200

SCL, series-compensated line.

9.6.4 Effect of fault inception angle

Like the effect of fault resistance uncertainty, the effect of fault inception angle uncertainty on the accuracy of the fault location algorithm was also investigated. The fault inception angles were varied between 0 and 150°, where the faults were applied at

Table 9.10 Fault resistance uncertainty effect on the accuracy of fault location algorithm for LLG faults on SCL (faults were applied 0.60 pu distance from the bus A of the SCL).

Fault resistance (Ω)	Fault type					
	ABG		BCG		CAG	
	Calculated fault location (pu)	Error (%)	Calculated fault location (pu)	Error (%)	Calculated fault location (pu)	Error (%)
0.00	0.5967	0.330	0.5946	0.540	0.5960	0.400
1.00	0.5968	0.320	0.5946	0.540	0.5959	0.410
5.00	0.5968	0.320	0.5946	0.540	0.5957	0.430
10.0	0.5969	0.310	0.5946	0.540	0.5957	0.430
25.0	0.5969	0.310	0.5946	0.540	0.5957	0.430
50.0	0.5969	0.310	0.5947	0.530	0.5957	0.430
100.0	0.5969	0.310	0.5946	0.540	0.5958	0.420
150.0	0.5969	0.310	0.5947	0.530	0.5958	0.420
200.0	0.5969	0.310	0.5947	0.530	0.5958	0.420
250.0	0.5970	0.300	0.5947	0.530	0.5959	0.410

SCL, series-compensated line.

Table 9.11 Fault resistance uncertainty effect on the accuracy of fault location algorithm for LLL faults on SCL (faults were applied 0.60 pu distance from the bus A of the SCL).

Fault resistance (Ω)	Calculated fault location (pu)	Error (%)
0.00	0.5969	0.310
0.50	0.5973	0.270
1.00	0.5985	0.150
2.50	0.5989	0.110
5.00	0.6011	0.110
7.50	0.6036	0.360
10.0	0.6063	0.630
15.0	0.6121	1.210
20.0	0.6179	1.790
30.0	0.6286	2.860

SCL, series-compensated line.

0.60 pu distance from the bus A of the SCL. Table 9.12 presents the obtained results for three types of faults (AG, BC, and BCG). Presented results verified the efficacy of the adaptive fault location accuracy under fault inception angle uncertainty as it accurately located the investigated faults.

9.6.5 Effect of prefault loading

The effect of the prefault loading uncertainty on the accuracy of the adaptive fault location algorithm was also investigated for three different fault types (AG, BC, and

Table 9.12 Fault inception angle uncertainty effect on the accuracy of fault location algorithm (faults were applied 0.60 pu distance from the bus A of the SCL).

Inception angle (°)	Fault type					
	AG		BC		ABG	
	Calculated fault location (pu)	Error (%)	Calculated fault location (pu)	Error (%)	Calculated fault location (pu)	Error (%)
0.0	0.5969	0.310	0.5947	0.530	0.5967	0.330
30.0	0.5969	0.310	0.5950	0.500	0.5968	0.320
45.0	0.5967	0.330	0.5951	0.490	0.5968	0.320
60.0	0.5965	0.350	0.5951	0.490	0.5969	0.310
90.0	0.5960	0.400	0.5936	0.640	0.5969	0.310
120.0	0.5965	0.350	0.5910	0.900	0.5967	0.330
135.0	0.5972	0.280	0.5903	0.970	0.5965	0.350
150.0	0.5981	0.190	0.5906	0.940	0.5963	0.370

SCL, series-compensated line.

BCG). Prefault loading conditions were varied from 0.5 to 3.0 times of the original loading value. Table 9.13 presents the obtained results for the faults applied on 0.60 pu distance from the bus A of the SCL. As can be observed from the obtained results, the presented fault location algorithm was practically independent of the prefault loading conditions as it located the investigated faults with satisfactory accuracy.

9.6.6 Effect of line compensation degree

Effect of the line compensation degree uncertainty on the accuracy of the fault location algorithm was also investigated for three different fault types (AG, BC, and

Table 9.13 Prefault loading condition uncertainty effect on the accuracy of fault location algorithm (faults were applied 0.60 pu distance from the bus A of the SCL).

Prefault loading factor	Fault type					
	AG		BC		ABG	
	Calculated fault location (pu)	Error (%)	Calculated fault location (pu)	Error (%)	Calculated fault location (pu)	Error (%)
0.50	0.5970	0.300	0.5945	0.550	0.5968	0.320
0.80	0.5970	0.300	0.5945	0.550	0.5968	0.320
1.20	0.5970	0.300	0.5945	0.550	0.5968	0.320
1.50	0.5970	0.300	0.5945	0.550	0.5968	0.320
2.00	0.5970	0.300	0.5946	0.540	0.5968	0.320
3.00	0.5970	0.300	0.5946	0.540	0.5968	0.320

SCL, series-compensated line.

Table 9.14 Line compensation degree uncertainty effect on the accuracy of fault location algorithm (faults were applied 0.60 pu distance from the bus A of the SCL).

Compensation degree (%)	Fault type					
	AG		BC		ABG	
	Calculated fault location (pu)	Error (%)	Calculated fault location (pu)	Error (%)	Calculated fault location (pu)	Error (%)
50.0	0.5967	0.330	0.5950	0.500	0.5969	0.310
60.0	0.5968	0.320	0.5947	0.530	0.5969	0.310
70.0	0.5969	0.310	0.5947	0.530	0.5967	0.330
80.0	0.5969	0.310	0.5945	0.550	0.5967	0.330
90.0	0.5970	0.300	0.5945	0.550	0.5968	0.320

SCL, series-compensated line.

BCG). The line compensation degree was varied from 50% to 90%. Table 9.14 presents the obtained results for the faults applied on 0.60 pu distance from the bus A of the SCL. It is evident from the obtained results that the algorithm located all investigated faults accurately. Thus, it can be concluded that the algorithm was independent of the line compensation degree uncertainty.

9.6.7 Effect of system parameter variation

The adaptive fault location algorithm for the SCL estimates the line parameters and system TE online from the PMU recorded synchronized measurements to overcome the effect of the ambient conditions and system operating history on the network parameters. On the other hand, the nonadaptive fault location algorithm does not consider the variations of the system parameters due to the mentioned issues. Thus, the accuracy of the nonadaptive algorithm is impacted as such assumptions are regarded as the source of errors. Considering the mentioned notes, this section investigated the effect of parameter variation on the accuracy of the adaptive fault location algorithm. To do so, the system impedance and line parameters were varied in a range of ±25% from their nominal values. Table 9.15 presents the obtained results after incorporating the variation of the system parameters for four different types of faults (AG, BC, CAG, and ABC). All investigated faults were applied 0.60 pu distance from the bus A of the SCL. As can be observed from the presented results, the fault location accuracy was degraded with the increase of the system parameter variation. The degradation reached up to around 14% for 25% parameter variation. It was also observed that the fault location accuracy degradation was proportional to the parameter variation for the investigated faults on the SCL system (Fig. 9.11). Therefore, it can be concluded that the higher the parameter variation, the more significant the impact (degradation) on the accuracy of the fault location algorithm.

Table 9.15 Line parameter and system impedance variation effect on the accuracy of the fault location algorithm (faults were applied 0.60 pu distance from the bus A of the SCL).

Parameter variation (%)	AG Calculated fault location (pu)	AG Error (%)	CAG Calculated fault location (pu)	CAG Error (%)	BC Calculated fault location (pu)	BC Error (%)	ABC Calculated fault location (pu)	ABC Error (%)
−25.0	0.4631	13.690	0.4617	13.830	0.4592	14.080	0.4634	13.660
−20.0	0.4968	10.320	0.4952	10.480	0.4934	10.660	0.4962	10.380
−15.0	0.5263	7.370	0.5253	7.470	0.5231	7.690	0.5261	7.390
−10.0	0.5520	4.800	0.5512	4.880	0.5550	4.500	0.5527	4.730
−5.0	0.5764	2.360	0.5754	2.460	0.5734	2.660	0.5763	2.370
0.0	0.5974	0.260	0.5962	0.380	0.5952	0.480	0.5974	0.260
5.0	0.6169	1.690	0.6153	1.530	0.6147	1.470	0.6161	1.610
10.0	0.6345	3.450	0.6339	3.390	0.6312	3.120	0.6340	3.400
15.0	0.6492	4.920	0.6491	4.910	0.6472	4.720	0.6495	4.950
20.0	0.6643	6.430	0.6634	6.340	0.6623	6.230	0.6646	6.460
25.0	0.6787	7.870	0.6778	7.780	0.6765	7.650	0.6780	7.800

SCL, series-compensated line.

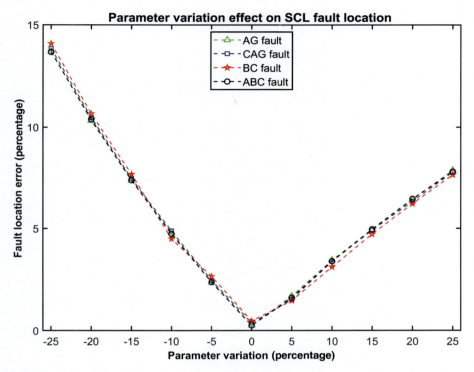

Fig. 9.11 Parameter variation effect on fault location algorithm accuracy in an SCL. *SCL*, series-compensated line.

9.7 Summary

This chapter presented an adaptive fault location algorithm for series compensated transmission line network. The presented algorithm uses the PMU recorded measurements for online computation of the system TE and line parameters. Online parameter determination helps to curb parameter uncertainty that occurs due to the operating history and environmental conditions. The adaptive fault location algorithm was applied on a 400 kV power system network with an SCL assumed to be faulted at both sides. The investigation was carried out on PSCAD/EMTDC and MATLAB/SIMULINK platforms. The adaptive fault location algorithm estimated locations of the investigated faults with high accuracy. Moreover, the presented results confirm the algorithm robustness under fault information (resistance, inception angle, location, and type), prefault loading condition, system parameter, and line compensation degree uncertainty.

References

[1] U.S. Annakkage, N. Perera, V. Liyanage, Accurate fault location estimation in transmission lines under high dc-offset and sub-harmonic conditions, 2013 CIGRÉ Canada Conference, 2013, pp. 1–8, http://www.cigre.org. (Accessed 19 May 2021).

[2] A.R. Adly, S.H.E.A. Aleem, M.A. Algabalawy, F. Jurado, Z.M. Ali, A novel protection scheme for multi-terminal transmission lines based on wavelet transform, Electr. Power Syst. Res. 183 (2020) 106286, doi:10.1016/j.epsr.2020.106286.

[3] G. Dharan et al., Alternating current, 2018. https://energyeducation.ca/encyclopedia/Alternating_current#cite_note-1. (Accessed 18 May 2021).

[4] A. Swetapadma, P. Mishra, A. Yadav, A.Y. Abdelaziz, A non-unit protection scheme for double circuit series capacitor compensated transmission lines, Electr. Power Syst. Res. 148 (2017) 311–325, doi:10.1016/j.epsr.2017.04.002.

[5] M.K. Jena, S.R. Samantaray, Intelligent relaying scheme for series-compensated double circuit lines using phase angle of differential impedance, Int. J. Electr. Power Energy Syst. 70 (2015) 17–26, doi:10.1016/j.ijepes.2015.01.035.

[6] H.A. Abd el-Ghany, M.A. Elsadd, E.S. Ahmed, A faulted side identification scheme-based integrated distance protection for series-compensated transmission lines, Int. J. Electr. Power Energy Syst. 113 (2019) 664–673, doi:10.1016/j.ijepes.2019.06.021.

[7] M.M. Saha, E. Rosolowski, J. Izykowski, P. Pierz, Evaluation of relaying impedance algorithms for series-compensated line, Electr. Power Syst. Res. 138 (2016) 106–112, doi:10.1016/j.epsr.2016.03.046.

[8] Y. Deng, Z. He, R. Mai, S. Lin, L. Fu, Fault location estimator for series compensated transmission line under power oscillation conditions, IET Gener. Transm. Distrib. 10 (13) (2016) 3135–3141, doi:10.1049/iet-gtd.2015.1017.

[9] I.B.M. Taha, A.E. ELGebaly, E.S. Ahmed, H.A. Abd el-Ghany, Generalized voltage estimation of TCSC-compensated transmission lines for adaptive distance protection, Int. J. Electr. Power Energy Syst. 130 (2021) 107018, doi:10.1016/j.ijepes.2021.107018.

[10] B. Vyas, B. Das, R.P. Maheshwari, An improved scheme for identifying fault zone in a series compensated transmission line using undecimated wavelet transform and Chebyshev Neural Network, Int. J. Electr. Power Energy Syst. 63 (2014) 760–768, doi:10.1016/j.ijepes.2014.06.030.

[11] M.T. Hoq, J. Wang, N. Taylor, Review of recent developments in distance protection of series capacitor compensated lines, Electr. Power Syst. Res. 190 (2021) 106831, doi:10.1016/j.epsr.2020.106831.

[12] M. Sahani, P.K. Dash, Fault location estimation for series-compensated double-circuit transmission line using parameter optimized variational mode decomposition and weighted P-norm random vector functional link network, Appl. Soft Comput. J. 85 (2019) 105860, doi:10.1016/j.asoc.2019.105860.

[13] A. Saffarian, M. Abasi, Fault location in series capacitor compensated three-terminal transmission lines based on the analysis of voltage and current phasor equations and asynchronous data transfer, Electr. Power Syst. Res. 187 (2020) 106457, doi:10.1016/j.epsr.2020.106457.

[14] J. Izykowski, E. Rosolowski, P. Balcerek, M. Fulczyk, M.M. Saha, Fault location on double-circuit series-compensated lines using two-end unsynchronized measurements, IEEE Trans. Power Deliv. 26 (4) (2011) 2072–2080, doi:10.1109/TPWRD.2011.2158670.

[15] M. Shafiullah, M.A. Abido, S-transform based FFNN approach for distribution grids fault detection and classification, IEEE Access 6 (2018) 8080–8088, doi:10.1109/ACCESS.2018.2809045.

[16] M. Shafiullah, M.A. Abido, A review on distribution grid fault location techniques, Electr. Power Components Syst. 45 (8) (2017) 807–824, doi:10.1080/15325008.2017.1310772.
[17] A. Mukherjee, P.K. Kundu, A. Das, A supervised principal component analysis-based approach of fault localization in transmission lines for single line to ground faults, Electr. Eng. 1 (2021) 3, doi:10.1007/s00202-021-01221-9.
[18] Y.D. Mamuya, Y.-D. Lee, J.-W. Shen, M. Shafiullah, C.-C. Kuo, Application of machine learning for fault classification and location in a radial distribution grid, Appl. Sci. 10 (14) (2020) 4965, doi:10.3390/app10144965.
[19] X. Xue, M. Cheng, T. Hou, G. Wang, N. Peng, R. Liang, Accurate location of faults in transmission lines by compensating for the electrical distance, Energies 13 (3) (2020) 767, doi:10.3390/en13030767.
[20] A.I. Çapar, A. Basa Arsoy, A performance-oriented impedance based fault location algorithm for series compensated transmission lines, Int. J. Electr. Power Energy Syst. 71 (2015) 209–214, doi:10.1016/j.ijepes.2015.02.020.
[21] H. Livani, C.Y. Evrenosoglu, A machine learning and wavelet-based fault location method for hybrid transmission lines, IEEE Trans. Smart Grid 5 (1) (2014) 51–59, doi:10.1109/TSG.2013.2260421.
[22] G. Manassero Junior, S.G. Di Santo, D.G. Rojas, Fault location in series-compensated transmission lines based on heuristic method, Electr. Power Syst. Res. 140 (2016) 950–957, doi:10.1016/j.epsr.2016.03.049.
[23] M. Mirzaei, B. Vahidi, S.H. Hosseinian, Fault location on a series-compensated three-terminal transmission line using deep neural networks, IET Sci. Meas. Technol. 12 (6) (2018) 746–754, doi:10.1049/iet-smt.2018.0036.
[24] M. Zand, O. Neghabi, M.A. Nasab, M. Eskandari, M. Abedini, A Hybrid Scheme for Fault Locating in Transmission Lines Compensated by the TCSC, 2020 15th International Conference on Protection and Automation of Power Systems, IPAPS 2020, 2020 130–135, doi:10.1109/IPAPS52181.2020.9375626.
[25] R. Grünbaum and J. Samuelsson, Series capacitors facilitate long distance AC power transmission, 2005, doi:10.1109/PTC.2005.4524354.
[26] A.H. Al-Mohammed, Adaptive fault location in power system networks based on synchronized phasor measurements, King Fahd University of Petroleum and Minerals, Dhahran, Saudi Arabia, 2012.
[27] M. Al-Dabbagh, S.K. Kapuduwage, Using instantaneous values for estimating fault locations on series compensated transmission lines, Electr. Power Syst. Res. 76 (1–3) (2005) 25–32, doi:10.1016/j.epsr.2005.03.004.
[28] J. Sadeh, A.M. Ranjbar, N. Hadsaid, R. Feuillet, Accurate fault location algorithm for series compensated transmission lines, 2000 IEEE Power Engineering Society, Conference Proceedings, 4 (2000) 2527–2532, doi:10.1109/PESW.2000.847211.
[29] M. Fulczyk, P. Balcerek, J. Izykowski, E. Rosolowski, M.M. Saha, Fault locator using two-end unsynchronized measurements for UHV series compensated parallel lines, 2008 International Conference on High Voltage Engineering and Application, ICHVE, 2008, (2008) pp. 88–91, doi:10.1109/ICHVE.2008.4773880.
[30] M. Fulczyk, P. Balcerek, J. Izykowski, E. Rosolowski, and M.M. Saha, Two-end unsynchronized fault location algorithm for double-circuit series compensated lines, 2008, doi:10.1109/PES.2008.4596638.
[31] R. Dutra, L. Fabiano, M.M. Saha, S. Lidström, Fault location on parallel transmission lines with series compensation, 2004 IEEE/PES Transmission and Distribution Conference and Exposition: Latin America, 2004, pp. 591–597, doi:10.1109/tdc.2004.1432446.

[32] M.M. Saha, J. Izykowski, E. Rosolowski, A fault location method for application with current differential protective relays of series-compensated transmission line, IET Conference Publications, 2010 (558 CP) (2010), doi:10.1049/cp.2010.0230.

[33] Y. Liao, Some algorithms for transmission line parameter estimation, 2009 IEEE International Symposium on Sustainable Systems and Technology, ISSST 2009, 2009, pp. 127–132, doi:10.1109/SSST.2009.4806781.

[34] S. Abdelkader, Online Thevenin's Equivalent Using Local PMU Measurements, International Conference on Renewable Energies and Power Quality (ICREPQ'11), 2011 1229–1232, doi:10.24084/repqj09.604.

[35] S.M. Abdelkader, D.J. Morrow, Online thévenin equivalent determination considering system side changes and measurement errors, IEEE Trans. Power Syst. 30 (5) (2015) 2716–2725, doi:10.1109/TPWRS.2014.2365114.

[36] S.M. Abdelkader, D.J. Morrow, Online tracking of Thévenin equivalent parameters using PMU measurements, IEEE Trans. Power Syst. 27 (2) (2012) 975–983, doi:10.1109/TPWRS.2011.2178868.

Intelligent fault diagnosis technique for distribution grid

10.1 Introduction

Efficient diagnosis of the abnormalities in the electric networks at their commencement is crucial for optimal management and utilization of the power system assets. Better diagnosis of power system faults provides precise information that helps achieve the revenue target by timely dispatching the crews to the respective sites. Such actions expedite the service restoration processes, reducing the outage duration and the customer minute losses. Fault diagnosis in electric networks consists of three parts, for example, fault detection, classification, and location. Fault detection allows the relays to isolate the faulty part of the network from the healthy part to protect the assets of the faulty part and to continue the power supply to the healthy part. Alternatively, knowledge of the fault class offers essential insights about the fault location, whereas fault location information is crucial for expediting the restoration process [1–4].

Fault diagnosis in the electric transmission networks got maturity due to extensive research and investment over the last couple of decades. However, such diagnosis schemes cannot be applied to the electric distribution networks immediately due to their inherent complexities, including various laterals (single-phase, double-phase, and three-phase), shorter line length, and nonhomogeneity. However, the researchers detected and classified distribution grid faults using conventional and advanced approaches throughout the years. The traditional methods are primarily based on search operations that are computationally burdensome as they are based on complex mathematical equations. On the other hand, the later approaches involve relatively simple calculations to achieve higher speed without compromising the accuracy required for the smart grid initiatives [5–7]. Different fault detection and classification approaches were reported in Refs. [8–15]. Likewise, the researchers employed three major categories of fault location schemes for the distribution grids, namely, the impedance-based [16–21], traveling wave-based [22–27], and knowledge-based [28–32]. The impedance-based fault location techniques are primarily based on hectic iterative processes, whereas the traveling wave-based schemes require sophisticated communication channels and applicable for long transmission lines. Considering the mentioned notes, knowledge-based intelligent fault diagnosis (IFD) schemes offer promising opportunities in dealing with the inherent complexities of distribution grids [2–4]. However, many of the reported fault diagnosis schemes did not consider uncertainty (loading condition and fault information) and the presence of measurement noises. Besides, instead of developing a universal fault diagnosis model, a few selected fault types were diagnosed in most of the reported literature.

This chapter presents an IFD scheme for a four-node test distribution feeder combining advanced signal processing techniques and machine learning tools (MLTs). It starts with modeling the mentioned test feeder and faults by varying the prefault loading conditions and fault information (resistance and inception angle). Then, it extracts useful features from the recorded current signals employing discrete wavelet transform (DWT) and ST. Finally, it fetches the extracted features into three different MLT namely, the artificial neural networks (ANNs), support vector machines (SVMs), and extreme learning machines (ELMs) for the development of fault detection, classification, location schemes. Moreover, the employed MLT control parameters are also tuned using a metaheuristic optimization algorithm, for example, constriction factor particle swarm optimization (CF-PSO), for better generalization performance. The CF-PSO is an updated version of the well-known PSO technique that produces universal solutions without getting trapped into the local optima. The obtained results confirm the efficacy of the IFD schemes and their independence in prefault loading conditions, fault information, and the presence of measurement noises.

The remaining parts of this chapter are organized as follows: Section 10.2 demonstrates the step-by-step modeling of the test distribution feeder and applied fault. It also discusses the data generation and feature extraction procedures. Section 10.3 illustrates the IFD approach. Sections 10.4 and 10.5 present the obtained fault diagnosis results with nonoptimized and optimized MLTs under balanced loading conditions, respectively. Section 10.6 presents the obtained results considering unbalanced loading conditions. Finally, an overall summary of this chapter is presented in Section 10.7.

10.2 Four-node test distribution feeder modeling

This section presents the step-by-step modeling of the four-node test distribution feeder in the MATLAB/SIMULINK platform. It also discusses the load demand and fault information (inception angle and resistance) uncertainty modeling, data generation, and data recording techniques. Finally, it presents the feature extraction processes from the recorded signals employing DWT and ST.

10.2.1 Test feeder modeling in MATLAB/SIMULINK environment

The selected test distribution feeder consists of four nodes, two distribution transformers, one lumped load, one feeder, and one distribution line, as shown in Fig. 10.1. Detailed information regarding the test distribution feeder is tabulated

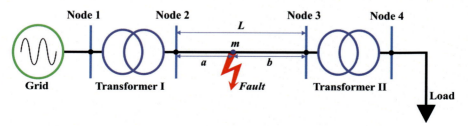

Fig. 10.1 Four-node test distribution feeder.

Intelligent fault diagnosis technique for distribution grid

Table 10.1 Details of the four-node test distribution network.

Item	Details
Transformer I	12 MVA, 120/25 kV, Yg-Δ
Transformer II	12 MVA, 25 kV/575 V, Yg-Δ
Load	5 MW, 2.5 MVAR (inductive)
Distribution line	30 km
System frequency	60 Hz

in Table 10.1 and can be found in Refs. [33–35]. MATLAB/SIMULINK (R2020b) is used to develop the mentioned test feeder. In addition, the Simscape Electrical Toolbox that provides the component libraries (programmable power and load sources, motors, transformers, renewable energy resources, overhead transmission lines, underground cables, etc.) is used for modeling and simulating the electric power system networks [36,37]. It also supports C-code generation to deploy the models on other simulation environments, including hardware-in-the-loop systems. One can master himself/herself on this modeling platform by following different tutorials available online.

This chapter picked different power system components to model the test feeder, as depicted in Fig. 10.2. Eventually, the four-node test distribution feeder model was

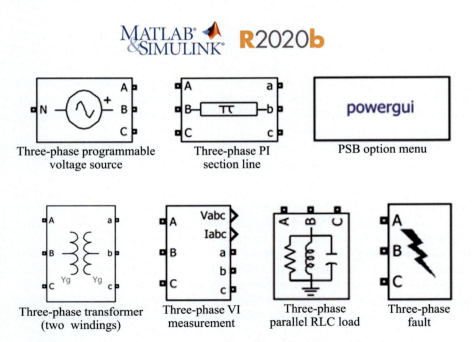

Fig. 10.2 Major components of the four-node test feeder in MATLAB/SIMULINK environment.

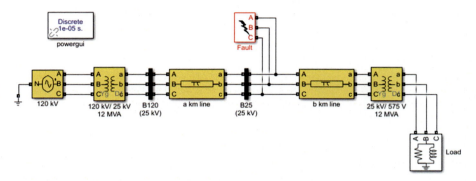

Fig. 10.3 Four-node test feeder in MATLAB/SIMULINK environment.

developed, as shown in Fig. 10.3. Among different options of the "PSB option menu" block, the "discrete" option was selected where the sample time was 0.00001 seconds (equivalent to 100 kHz sampling frequency and 1667 samples/cycle). It is worth noting that this chapter modeled the distribution line with two blocks where the distances of the first block and the second block were "a" and "b" kilometers (km), respectively. These two parameters facilitated the faults in different locations on the 30-km long distribution line.

Likewise, the parameters of the load block were also set to variable numbers to facilitate dynamic loading conditions of the selected test feeder (load demand uncertainty). Other parameters were chosen in the three-phase fault block to incorporate different fault types and fault resistance uncertainty. All mentioned parameters and the fault inception time were generated from the specified ranges in a MATLAB script file. Moreover, the developed SIMULINK file was opened and simulated from the same script file. Finally, the recorded three-phase current signals from the SIMULINK file were exported to MATLAB Workspace using the 'To Workspace' block for feature extraction purposes.

10.2.2 Fault modeling and data generation

This study selected a total of 59 locations on the distribution line of the test feeder starting from 0.50 km to 29.50 km with a step size of 0.50 km. Different types of faults for four cycles were then applied to the selected locations by varying the prefault loading conditions, fault information (resistance and inception angle), and fault types. The IFD scheme picked fault resistance from "0 Ω" to "15 Ω" randomly and varied prefault loading conditions in a range of ±10% of the rated loading conditions. Besides, it applied different types of faults, including single-line-to-ground (SLG), line-to-line-to-ground (LLG), and three-phase-to-ground (LLLG) faults. The SLG faults included phase-A-to-ground (AG), phase-B-to-ground (BG), and phase-C-to-ground (CG) faults. In contrast, the LLG faults included phase-A-to-phase-B-to-ground (ABG), phase-B-to-phase-C-to-ground (BCG), and phase-C-to-phase-A-to-ground (CAG) faults, and LLLG faults included only phase-A-to-phase-B-to-phase-C-to-ground

(ABCG) faults. Finally, the IFD scheme recorded three-phase faulty current signals (two-cycle: one cycle before and the other cycle after the fault occurrence) for feature extraction employing the DWT and the ST.

10.2.3 Feature extraction

From the recorded three-phase current signals, a total of 144 statistical measures, also known as features, were extracted through a seven-level DWT decomposition. Detailed illustrations of the DWT-based feature extraction process can be found in Chapter 4 and Refs. [33–35]. Likewise, the ST approach collected 36 features from the same signals. Detailed illustrations of the ST-based feature extraction process can be found in Chapter 4 and Refs. [38–42].

10.3 Intelligent fault diagnosis approach

The IFD scheme starts with modeling the test distribution feeder in MATLAB/SIMULINK platform, as stated in the previous section. Then, it applies different types of faults on the modeled feeder by varying the prefault loading conditions and fault information. Faults in distribution grids can be diagnosed with both current and voltage signals as they carry the characteristic signatures of the power system transients [43–46]. The IFD scheme of this chapter utilizes the three-phase current signals from the sending end (Node 2 to Node 3) as most of the distribution grid protection schemes are based on overcurrent protection and record three-phase branch currents through microprocessor-based relays in the substation during fault conditions. In addition, such current signals can also be acquired from the deployed phasor measurement units (PMUs) for other purposes, for example, state estimation and system observability. Therefore, the IFD scheme does not require the installation of further measurement devices [44].

After recording the three-phase current signals for two cycles (one cycle before and the other cycle after the fault occurrence), the IFD scheme treasured the valuable features using advanced signal processing techniques. Then, the features were fetched into the MLTs for the discovery of any fault occurrence. In the case of fault detection, the IFD scheme identified the fault classes; otherwise, it continued measuring three-phase current signals. Based on the fault classes, the IFD discovered the fault location distance from the sending end. Fig. 10.4 presents the simplified flowchart of the IFD scheme discussed in this chapter.

10.4 Fault diagnosis results

This section presents the fault diagnosis (detection, classification, and location) results of the IFD scheme for the modeled four-node test distribution feeder considering load demand and fault information uncertainty. The presence of measurement noises in the data recording processes was also considered.

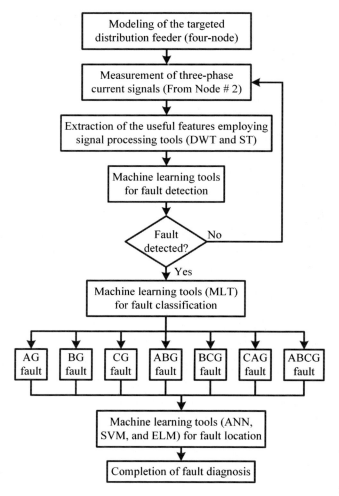

Fig. 10.4 Flowchart of the intelligent fault diagnosis (IFD) scheme.

10.4.1 Fault detection results

The IFD scheme recorded current signals from the sending end of the distribution line for 1050 faulty cases consisting of SLG (AG, BG, and CG), LLG (ABG, BCG, and CAG), and LLLG (ABCG) faults. The scheme incorporated the uncertainty in prefault loading conditions, fault information (inception angle, resistance, and location) during fault generation. In addition to the faulty data, it also recorded current signals for the same number of nonfaulty scenarios generated by incorporating load demand uncertainty only. Then, the scheme extracted features from the recorded (both faulty and nonfaulty) signals using the DWT and the ST. The extracted features were fetched

Table 10.2 Fault detection results using the DWT and ST-based approaches for noise-free and noisy data.

Item	Samples detected successfully							
	Noise free		40 dB SNR		30 dB SNR		20 dB SNR	
	DWT	ST	DWT	ST	DWT	ST	DWT	ST
Faulty cases	1050	1050	1050	1050	1050	1050	1050	1050
Nonfaulty cases	1050	1050	1050	1050	1050	1050	1050	1050
Overall accuracy (%)	100	100	100	100	100	100	100	100

DWT, discrete wavelet transform; SNR, signal-to-noise ratio; ST, stockwell transform.

into the multilayer perceptron neural network (MLP-NN) to develop a fault detection scheme. The scheme tested the MLP-NN with various neurons in the hidden layer and obtained the best performance with five hidden neurons in terms of overall accuracy and minimum mean squared error (MSE). Besides, the scheme chose the tan sigmoid as the squashing functions and resilient backpropagation ("trainrp") as the training algorithms via a systematic trial and error approach for both DWT and ST-based approaches. It should be noted that the IFD scheme utilized 70% of the available data for training and the rest of them for testing and validation purposes.

Table 10.2 presents the results obtained from the fault detection scheme. As can be seen, the scheme was able to separate the faulty cases from the healthy cases with 100% accuracy under a noise-free measurement situation. To investigate the efficacy of the fault detection scheme under a noisy environment, different levels of additive white Gaussian noise (AWGN) were added to the recorded current signals before feature extraction. Both DWT and ST-based MLP-NN were able to differentiate the faulty data from the healthy data without any deterioration of accuracy even in the presence of measurement noises. Finally, it is worth noting that the scheme required less than a cycle of the 60 Hz power system to detect a single case (on average) that suggests the real-time application potential of the intelligent fault detection scheme.

10.4.2 Fault classification results

Like the fault detection scheme, the fault classification approach acquired inputs (DWT and ST extracted features) for 700 faulty scenarios for each type by incorporating uncertainty associated with prefault loading conditions and fault information. Then, it fetched them into the MLP-NN with the various number of neurons in the hidden layer and selected the best configuration (seven neurons) in terms of overall accuracy. Besides, tan sigmoid and resilient backpropagation ("trainrp") were selected via a systematic trial and error process as the squashing functions and training algorithms, respectively, for both DWT and ST-based approaches. The classification scheme also employed 70% of the available data for training and the rest for testing and validation purposes. Tables 10.3 and 10.4 present the 7 × 7 confusion matrices obtained under noise-free measurement using DWT and ST-based approaches, respectively.

Table 10.3 Classification results of the DWT-based approach for noise-free measurement.

	AG	BG	CG	ABG	BCG	CAG	ABCG
AG	700	0	0	0	0	0	0
BG	0	700	0	0	0	0	0
CG	0	0	700	0	0	0	0
ABG	0	0	0	700	0	0	0
BCG	0	1	0	0	699	0	0
CAG	0	0	0	0	0	700	0
ABCG	0	0	0	0	0	0	700
Overall accuracy = 99.979%

DWT, discrete wavelet transform.

Table 10.4 Classification results of the ST-based approach for noise-free measurement.

	AG	BG	CG	ABG	BCG	CAG	ABCG
AG	700	0	0	0	0	0	0
BG	0	700	0	0	0	0	0
CG	0	0	700	0	0	0	0
ABG	0	0	0	700	0	0	0
BCG	0	0	0	0	700	0	0
CAG	0	0	0	0	0	700	0
ABCG	0	0	0	0	0	1	699
Overall accuracy = 99.979%

ST, stockwell transform.

The sizes of the confusion matrices were 7 × 7 as the MLP-NN classified seven types of faults. The successful and unsuccessful classifications of a particular kind of fault were denoted by the diagonal and off-diagonal elements of the confusion matrices, respectively. The scheme was able to classify faults with nearly 100% accuracy for both approaches under noise-free measurement.

To investigate the efficacy of the classification scheme under a noisy environment, different levels of AWGN were added to recorded current signals before feature extraction. Table 10.5 presents the results achieved with various signal-to-noise ratio (SNR) levels for the DWT and the ST-based MLP-NN. The DWT-based approach ended up with the overall accuracies of 99.979%, 99.979%, and 99.959% in the presence of 40 dB, 30 dB, and 20 dB SNR, respectively. In contrast, the ST-based approach ended up with the overall accuracies of 99.979%, 99.959%, and 99.959% for 40 dB, 30 dB, and 20 dB SNR, respectively. Thus, the presented results validated the efficacy of the intelligent fault classification scheme under both noisy and noise-free measurements. Finally, it is worth noting that the intelligent fault classification scheme required less than a cycle of the 60 Hz power system to classify a single case (on average) that demonstrates the real-time application potential of the discussed method.

Intelligent fault diagnosis technique for distribution grid

Table 10.5 Classification results of DWT and ST-based MLP-NN under noisy environment.

Fault type	Samples classified successfully					
	40 dB SNR		30 dB SNR		20 dB SNR	
	DWT	ST	DWT	ST	DWT	ST
AG	700	700	700	700	700	700
BG	700	700	700	700	700	700
CG	700	700	700	699	700	700
ABG	700	699	700	700	700	700
BCG	699	700	700	700	700	700
CAG	700	700	700	700	699	700
ABCG	700	700	699	699	699	698
Overall accuracy (%)	99.979	99.979	99.979	99.959	99.959	99.959

DWT, discrete wavelet transform; MLP-NN, multilayer perceptron neural network; SNR, signal-to-noise ratio; ST, stockwell transform.

10.4.3 Fault location results

The fault location scheme fetched the signal processing extracted features into three different MLTs to locate the faults on the distribution line. Among many MLTs, the MLP-NN, support vector regression (SVR), and ELM were selected in this chapter to locate various types of faults. It is worth noting that it collected 1000 faulty scenarios of each type and employed 70% of them to train the selected MLT and the rest for testing and validation purposes. Like fault detection and classification schemes, it also generated datasets considering the presence of measurement noise. However, this chapter formulated the fault location scheme as the regression problem, unlike the fault detection and classification schemes. Besides, it used MLP-NN regression toolbox ("newff") from the MATLAB library, MATLAB SVM-KM toolbox from Ref. [47], and MATLAB ELM toolbox from Ref. [48].

At the beginning of the fault location scheme of this chapter, the control parameters of the employed MLT were picked through the trial-and-error bases. For instance, it selected four nodes in the hidden layers for the DWT-based MLP-NN. Similarly, it set tan sigmoid as the squashing function and Levenberg-Marquardt backpropagation ("trainlm") as the training algorithm. The SVR control parameters such as the kernel option (K_O), tolerance of termination criterion (ε), and regularization coefficients (C and λ) were chosen as 500, 6×10^{-2}, 0.15, and 1000, respectively. In contrast, the Gaussian RBF kernel was used to create the separation surface. On the contrary, the ELM control parameters such as the regularization coefficient (C_R) and kernel parameter (K_p) values were selected as 5×10^{10} and 7×10^{12}, respectively. Gaussian RBF kernel was used to create the separation surface for the ELM. The ST-based fault location scheme also selected the control parameters of the employed MLT by following a similar approach. Table 10.6 summarizes the selected control parameters of the employed MLT for both DWT and ST-based approaches.

Table 10.6 Selected MLT control parameters for the DWT and the ST-based fault location approaches.

SPT	MLT	Values of the control parameters
DWT	MLP-NN	Number of hidden neurons = 4, Squashing function = tan sigmoid, and training algorithm = Levenberg-Marquardt backpropagation ("trainlm").
	SVR	kernel = Gaussian RBF, $C = 1000$, $\lambda = 6 \times 10^{-2}$, $\varepsilon = 0.15$, $K_O = 500$
	ELM	kernel = Gaussian RBF, $C_R = 5 \times 10^{10}$, $K_p = 7 \times 10^{12}$
ST	MLP-NN	Number of hidden neurons = 6, Squashing function = tan sigmoid, and training algorithm = Levenberg-Marquardt backpropagation ("trainlm").
	SVR	kernel = Gaussian RBF, $C = 1000$, $\lambda = 10^{-8}$, $\varepsilon = 0.1$, $K_O = 100$
	ELM	kernel = Gaussian RBF, $C_R = 10^{10}$, $K_p = 1.5 \times 10^8$

DWT, discrete wavelet transform; ELM, extreme learning machine; MAPE, mean absolute percentage error; MLP-NN, multilayer perceptron neural network; MLT, machine learning tool; SVR, support vector regression; ST, stockwell transform.

As the requirement of any regression problem, the fault location scheme evaluated several statistical performance indices (SPI), including root mean squared error (RMSE), mean absolute percentage error (MAPE), RMSE-observations standard deviation ratio (RSR), coefficient of determination (R^2), Willmott's index of agreement (WIA), Nash–Sutcliffe model efficiency coefficient (NSEC), and percent bias (PBIAS) to verify the employed signal processing based MLT efficacy. Relevant equations for the selected SPIs are appended in the appendix of this book. However, according to Lewis [49], any regression model is considered accurate as long as it has a less than 10% MAPE. In addition, lower values of RSR and RMSE show the strength of the model [50]. If the evaluated PBIAS value of any regression model is 0.0, the model can predict the targeted outputs accurately. Negative and positive values of the PBIAS refer to overestimation and underestimation of the forecasted outputs, respectively. Besides, NSEC, R^2, and WIA values vary from "0" to "1." The value "1" indicates the perfect match between the actual and the predicted outputs, while the value "0" shows the output cannot be predicted from available inputs [50–52]. Like the SPI, the scatter plot of the predicted and the targeted values provides a comprehensive summary of the bivariate data and is frequently used to obtain the possible relations between the variables. The resulting pattern illustrates the type and strength of the association between two variables. The more data points located about the identity line, for example, $y = x$, the more the data sets agree with each other. If the actual data and the model output are the same, then all data points fall on the identity line.

Fig. 10.5 shows the targeted and the DWT-based MLT predicted fault locations for forty randomly chosen observations from the test dataset of the AG faults for comparison purposes. It visually appeared that the MLP-NN and the ELM adequately predicted the desired outputs in most of the cases. However, the SVR estimated the

Intelligent fault diagnosis technique for distribution grid

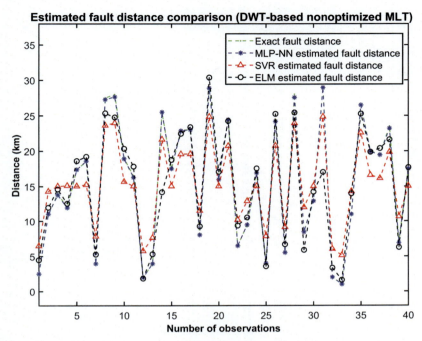

Fig. 10.5 Comparing DWT-based MLT predicted fault and actual fault distances for forty cases from the AG fault test dataset. *DWT*, discrete wavelet transform; *MLT*, machine learning tool.

fault locations with relatively lower accuracy than the other two techniques. Fig. 10.6 shows the scatter plots of the targeted versus MLT predicted fault locations of the test dataset of AG faults. As can be seen, the data points are located closer to the identity line for the MLP-NN as expected for any efficient model. The points for the ELM model are distributed about the identity line but not as close as the MLP-NN model. The SVR model estimated relatively larger values if the actual fault locations were less than 15 km and estimated somewhat smaller values in case of fault locations greater than 15 km.

However, this chapter refrained from presenting fault distance comparisons and scatter plots for other types of faults to avoid repeating similar figures and discussions. Instead, operational times and evaluated SPI values for the test datasets of DWT-based MLT for all kinds of faults are summarized in Table 10.7. As can be observed, the SVR required almost three-fold time to be trained compared to the MLP-NN, whereas the training process of the ELM was a hundred times faster than that of the MLP-NN. On the other hand, all the trained MLT required almost equal time for the testing purposes. The evaluated MAPE, RMSE, and RSR values were relatively low, whereas NSCE, R^2, and WIA values were close to unity for the MLP-NN. Besides, the PBIAS values were positive for a few fault types and negative for others

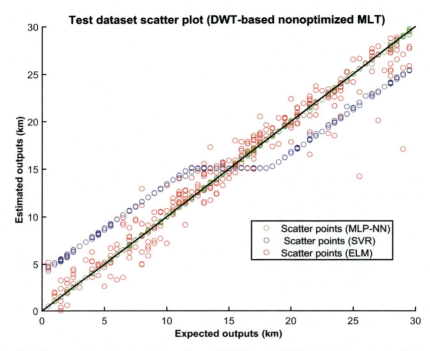

Fig. 10.6 Scatter plots of targeted (actual) and DWT-based MLT predicted outputs for the test dataset of AG fault. *DWT*, discrete wavelet transform; *MLT*, machine learning tool.

that demonstrated overestimations and underestimations of the fault distances, respectively. Nonetheless, these values were small and close to zero; hence, such small overestimations and underestimations can be neglected. Such satisfactory values of the SPI for all types of faults indicated the strength of the DWT-based MLP-NN in locating different kinds of faults. In contrast, the SPI values for the other two MLTs (SVR and ELM) were not satisfactory for the DWT-based approaches.

Like the DWT-based approach, Fig. 10.7 shows the targeted and the ST-based MLT predicted fault locations for forty randomly chosen observations from the AG fault test dataset. It visually appeared that the MLP-NN sufficiently predicted the desired outputs in most of the cases. The SVR also predicted the fault locations with little lower accuracy. On the contrary, the accuracy of the ELM was poor compared to the other two techniques. Fig. 10.8 presents the scatter plot of the targeted versus MLT predicted fault locations for the AG fault test dataset. As can be seen, the data points were about the identity line for the MLP-NN as expected for any efficient model. The points for the SVR model were distributed closer to the identity line but not as close as the data points of the MLP-NN model. However, the scatter points of the ELM model illustrated its lower efficacy compared to other models. Like DWT-based approaches, targeted and estimated fault location comparisons and scatter plots for other types of faults were not presented to avoid repetitive figures and discussions. However, Table 10.8 summarizes the operational times and the evaluated

Intelligent fault diagnosis technique for distribution grid   261

Table 10.7 Operation time and SPI values for the test datasets of DWT-based MLT.

Fault type	Technique	Time (s) Training	Time (s) Testing	RMSE (km)	MAPE	RSR	PBIAS	R²	WIA	NSCE
AG	MLP-NN	12.6875	0.0156	0.2351	0.0206	0.0287	0.0406	0.9996	0.9998	0.9992
	SVR	31.9531	0.0313	3.2540	0.4758	0.3972	1.9563	0.9794	0.9606	0.8422
	ELM	0.0781	0.0156	1.6034	0.2123	0.1957	-0.522	0.9807	0.9904	0.9617
BG	MLP-NN	10.4219	0.0313	0.2287	0.0272	0.0279	0.0150	0.9996	0.9998	0.9992
	SVR	31.1563	0.0313	3.2379	0.4716	0.3953	1.8434	0.9782	0.9609	0.8438
	ELM	0.1094	0.0313	1.8687	0.2896	0.2281	-0.799	0.9739	0.9870	0.9480
CG	MLP-NN	11.5938	0.0156	0.2116	0.0211	0.0258	-0.065	0.9997	0.9998	0.9993
	SVR	30.8594	0.0313	3.2553	0.4746	0.3974	2.0289	0.9790	0.9605	0.8421
	ELM	0.1250	0.0313	1.7990	0.2437	0.2196	0.3261	0.9756	0.9879	0.9518
ABG	MLP-NN	12.2031	0.0156	0.1332	0.0216	0.0157	0.0429	0.9999	0.9999	0.9998
	SVR	30.3438	0.0313	3.4117	0.6037	0.4010	-0.947	0.9811	0.9598	0.8392
	ELM	0.1563	0.0156	0.8912	0.1453	0.1047	-0.253	0.9946	0.9973	0.9890
BCG	MLP-NN	10.4219	0.0313	0.3300	0.0443	0.0378	-0.067	0.9993	0.9996	0.9986
	SVR	31.8438	0.0156	3.4516	0.6720	0.3951	-1.264	0.9804	0.9610	0.8439
	ELM	0.0625	0.0469	0.8878	0.1447	0.1016	0.2618	0.9948	0.9974	0.9897
CAG	MLP-NN	11.2813	0.0313	0.1206	0.0198	0.0146	-0.073	0.9999	0.9999	0.9998
	SVR	31.6875	0.0142	3.3744	0.6005	0.4092	-1.859	0.9788	0.9581	0.8325
	ELM	0.1406	0.0469	0.8467	0.1152	0.1027	-0.246	0.9947	0.9974	0.9895
ABCG	MLP-NN	9.1719	0.0469	0.2618	0.0327	0.0320	-0.024	0.9995	0.9997	0.9990
	SVR	28.6563	0.0313	3.3307	0.4872	0.4066	1.8971	0.9791	0.9587	0.8347
	ELM	0.1250	0.0313	1.5670	0.2325	0.1913	-0.243	0.9816	0.9909	0.9634

DWT, discrete wavelet transform; ELM, extreme learning machine; MAPE, mean absolute percentage error; MLP-NN, multilayer perceptron neural network; MLT, machine learning tool; PBIAS, percent bias; RMSE, root mean squared error; RSR, root mean square error-observations standard deviation ratio; SPI, statistical performance indices; SVR, support vector regression; WIA, Willmott's index of agreement.

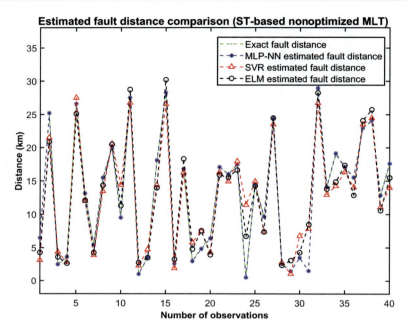

Fig. 10.7 Comparing ST-based MLT predicted fault and actual fault distances for forty cases from AG fault test dataset. *MLT*, machine learning tool; *ST*, stockwell transform.

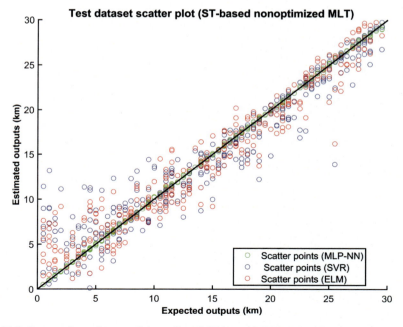

Fig. 10.8 Scatter plots of targeted (actual) and ST-based MLT predicted outputs for the test dataset of AG fault. *MLT*, machine learning tool; *ST*, stockwell transform.

Table 10.8 Operation time and SPI values for the test datasets of ST-based MLT.

Fault type	Technique	Time (s) Training	Time (s) Testing	RMSE (km)	MAPE	RSR	PBIAS	R²	WIA	NSCE
AG	MLP-NN	12.5625	0.0156	0.2228	0.0311	0.0272	−0.049	0.9996	0.9998	0.9993
	SVR	31.9375	0.0313	1.6923	0.2711	0.2066	1.2553	0.9787	0.9893	0.9573
	ELM	0.1250	0.0469	3.0928	0.4747	0.3776	0.7176	0.9261	0.9644	0.8575
BG	MLP-NN	10.2813	0.0313	0.2885	0.0482	0.0356	−0.010	0.9994	0.9997	0.9987
	SVR	27.5781	0.0156	1.4347	0.2726	0.1769	−0.463	0.9848	0.9922	0.9687
	ELM	0.0781	0.0313	3.3668	0.5685	0.4151	−0.575	0.9099	0.9569	0.8277
CG	MLP-NN	11.9063	0.0156	0.2150	0.0382	0.0252	−0.141	0.9997	0.9998	0.9994
	SVR	29.6563	0.0156	1.2212	0.2266	0.1430	−0.816	0.9901	0.9949	0.9796
	ELM	0.0625	0.0313	3.0670	0.4692	0.3590	0.4258	0.9336	0.9678	0.8711
ABG	MLP-NN	9.9844	0.0156	0.1823	0.0180	0.0210	−0.085	0.9998	0.9999	0.9996
	SVR	29.8750	0.0313	2.9545	0.5592	0.3405	1.0599	0.9759	0.9710	0.8841
	ELM	0.1094	0.0156	1.8146	0.2884	0.2091	−0.246	0.9779	0.9891	0.9563
BCG	MLP-NN	9.3438	0.0156	0.1470	0.0200	0.0177	0.0068	0.9998	0.9999	0.9997
	SVR	31.8594	0.0313	2.9251	0.5375	0.3516	0.5363	0.9789	0.9691	0.8764
	ELM	0.1094	0.0156	1.7692	0.2729	0.2127	−1.266	0.9774	0.9887	0.9548
CAG	MLP-NN	10.6094	0.0313	0.1809	0.0226	0.0217	0.0130	0.9998	0.9999	0.9995
	SVR	31.9531	0.0313	2.8767	0.5398	0.3458	0.5927	0.9770	0.9701	0.8804
	ELM	0.0625	0.0156	1.4871	0.2205	0.1788	−1.250	0.9842	0.9920	0.9680
ABCG	MLP-NN	9.7656	0.0625	0.3253	0.0647	0.0385	−0.201	0.9993	0.9996	0.9985
	SVR	30.9688	0.0313	3.4686	0.7676	0.4107	0.1279	0.9846	0.9578	0.8313
	ELM	0.1406	0.0313	2.4698	0.5183	0.2924	0.5786	0.9564	0.9786	0.9145

ELM, extreme learning machine; MAPE, mean absolute percentage error; MLP-NN, multilayer perceptron neural network; MLT, machine learning tool; PBIAS, percent bias; RMSE, root mean squared error; RSR, root mean square error-observations standard deviation ratio; SPI, statistical performance indices; ST, stockwell transform; SVR, support vector regression; WIA, Willmott's index of agreement.

SPI values of the test datasets of all seven types of faults for the ST-based MLT. As can be seen, the SVR required almost three-fold time to get trained compared to the MLP-NN, whereas the training process of the ELM was a hundred times faster than that of the MLP-NN. In contrast, the trained MLT required almost equal times for testing purposes.

As shown in Table 10.8, evaluated MAPE, RMSE, and RSR values were relatively low, whereas NSCE, R^2, and WIA values were almost close to unity for the MLP-NN in locating all types of faults. Besides, the PBIAS values were positive for a few fault types and negative for other types that demonstrated overestimations and underestimations of the fault distances, respectively. Nonetheless, these values were small and close to zero; hence, such overestimations and underestimations can be neglected. Therefore, all the evaluated SPI values indicated the strength of the ST-based MLP-NN in locating different types of faults. On the other hand, the SPI values for the other two techniques (SVR and ELM) were not satisfactory. The dissatisfactory performance of the SVR and ELM models for both DWT and ST-based approaches drove the authors to optimize their control parameters to achieve better generalization performance.

10.5 Optimized machine learning tools for fault location

This section presents the formulation of the optimization problem to tune the control parameter of the MLTs. Then, it shows the obtained fault location results employing the optimized MLTs under both noisy and noise-free measurements.

10.5.1 Machine learning tools optimization problem formulation

The MAPE of the test data sets were considered as the objective function while tuning the critical parameters of the employed MLTs, for example, SVR and ELM.

10.5.1.1 Support vector machine tuning problem formulation

In the previous section, this chapter employed the SVM-KM [47] toolbox developed in the MATLAB environment and picked the control parameters on a trial and error basis. In this section, an optimization problem was formulated to tune the SVM key parameters such as the tolerance of termination criterion (ε), kernel option (K_O), and regularization coefficients (λ and C) where the MAPE of the test data sets was considered as the objective function. Consequently, it formulated the following optimization problem to tune the SVM key parameters:

$$\text{Minimize } J$$
$$J = \text{MAPE} \qquad (10.1)$$

Subjected to:

$$\lambda_{min} \leq \lambda \leq \lambda_{max}$$
$$\varepsilon_{min} \leq \varepsilon \leq \varepsilon_{max}$$
$$C_{min} \leq C \leq C_{max}$$
$$KO_{min} \leq KO \leq KO_{max-}$$

10.5.1.2 Extreme learning machine tuning problem formulation

Likewise, this chapter utilized a MATLAB-based ELM toolbox [48] in the previous section to predict the fault locations of the four-node test distribution feeder where the control parameters were chosen through a trial and error basis. This section formulated an optimization problem to tune the ELM key parameters such as kernel parameter (K_P) and regularization coefficients (C_R). At the same time, the considered objective function was the MAPE of test datasets. Eventually, the following optimization problem was formulated:

$$\text{Minimize } J$$
$$J = \text{MAPE} \quad (10.2)$$

Subjected to:

$$C_{Rmin} \leq C_R \leq C_{Rmax}$$
$$K_{Pmin} \leq K_P \leq K_{Pmax_-}$$

After formulating the optimization problems, an efficient metaheuristic algorithm, namely the CF-PSO, was employed to determine the MLT optimal parameters. Detailed about the CF-PSO can be found in Chapter 2 and Ref. [53–57].

10.5.2 Fault location results

This chapter employed a swarm of 20 particles for 100 iterations to evaluate the optimal parameters of the SVR toolbox. Likewise, it used a swarm of 80 individuals for 100 iterations to achieve the ELM control parameters. Table 10.9 presents the achieved optimal control parameters for both SVR and ELM with DWT and ST-based approaches. It is worth noting that this chapter neither optimized the number of hidden neurons nor the biases and connecting weights of the MLP-NN because of its satisfactory performance with the systematically chosen parameters for all types of faults. Hence, the same parameters of Table 10.6 for the MLP-NN are presented in Table 10.9.

10.5.2.1 Fault location in noise-free measurement

After optimizing the SVM and ELM control parameters, this chapter employed them to locate faults in a noise-free environment. Fig. 10.9 shows the targeted and

Table 10.9 CF-PSO optimized control parameters for DWT and ST-based MLT.

SPT	MLT	Values of the control parameters
DWT	MLP-NN	Number of hidden neurons = 4, Squashing function = tan sigmoid, and training algorithm = Levenberg-Marquardt backpropagation ("trainlm").
	SVR	$C = 1000$, $\lambda = 5 \times 10\text{-}8$, $\varepsilon = 2 \times 10 - 1$, $K_O = 100$ and kernel = Gaussian RBF
	ELM	$C_R = 2.5 \times 1012$, $K_p = 2.5 \times 10^{11}$ and kernel = Gaussian RBF
ST	MLP-NN	Number of hidden neurons = 6, Squashing function = tan sigmoid, and training algorithm = Levenberg-Marquardt backpropagation ("trainlm").
	SVR	$C = 5000$, $\lambda = 3 \times 10^{-12}$, $\varepsilon = 10^{-12}$, $K_O = -80$, and kernel = Gaussian RBF
	ELM	$C_R = 2 \times 10^{12}$, $K_p = 9.5 \times 10^7$, and kernel = Gaussian RBF

CF-PSO, constriction factor particle swarm optimization; DWT, discrete wavelet transform; ELM, extreme learning machine; MLP-NN, multilayer perceptron neural network; MLT, machine learning tool; ST, stockwell transform; SVR, support vector regression.

the DWT-based optimized MLT predicted fault locations for forty randomly chosen observations from the test dataset of AG faults for comparison purposes. As can be seen, the MLT sufficiently predicted the desired outputs for almost all cases. Fig. 10.10 shows the scatter plot of the DWT-based MLT predicted versus targeted

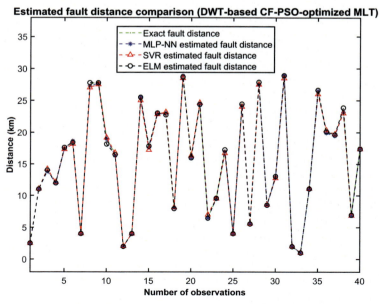

Fig. 10.9 Comparing DWT-based optimized MLT predicted fault and actual fault distances for 40 cases from AG fault test dataset. ***DWT***, discrete wavelet transform; ***MLT***, machine learning tool.

Intelligent fault diagnosis technique for distribution grid

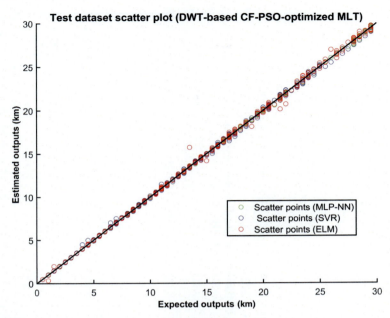

Fig. 10.10 Scatter plots of targeted (actual) and DWT-based optimized MLT predicted outputs for the test dataset of AG fault. *DWT*, discrete wavelet transform; *MLT*, machine learning tool.

fault locations for the AG fault test data. As can be observed, most of the data points were located closer to the identity line for all employed MLT as expected for the efficient regression models. Therefore, it can be concluded that metaheuristic optimized MLT showed better generalization performance than the nonoptimized MLT for AG faults.

Table 10.10 presents the operational times and the evaluated SPI values for the test datasets of the DWT-based MLT for all seven types of faults. As can be seen, the SVR required almost three-fold time to get trained compared to the MLP-NN, whereas the training process of the ELM was a hundred times faster than that of the MLP-NN. Besides, the test times for the employed MLT were almost equal. Evaluated MAPE, RMSE, and RSR values were relatively low, whereas NSCE, R^2, and the WIA were near unity for MLT. However, the PBIAS values were positive for a few scenarios and negative for the others that demonstrated overestimations and underestimations of the fault distances, respectively. Nonetheless, these values were minimal and close to zero for all MLT; such negligible overestimations and underestimations were acceptable. Therefore, the evaluated SPI values confirmed the strength and the efficacy of the optimized MLT for different fault types. Thus, the tuning of the MLT key parameters employing the metaheuristic algorithm was justified.

Fig. 10.11 shows the targeted and ST-based MLT predicted fault locations for 40 randomly chosen observations from the test dataset of AG faults. The employed MLT

Table 10.10 Operation time and SPI values for the test datasets of DWT-based optimized MLT.

Fault type	Technique	Time (s) Training	Time (s) Testing	RMSE (km)	MAPE	RSR	PBIAS	R²	WIA	NSCE
AG	MLP-NN	12.6875	0.0156	0.2351	0.0206	0.0287	0.0406	0.9996	0.9998	0.9992
	SVR	27.6875	0.0156	0.4898	0.0664	0.0598	0.2746	0.9995	0.9991	0.9964
	ELM	0.0625	0.0469	0.2022	0.0212	0.0247	−0.077	0.9997	0.9998	0.9994
BG	MLP-NN	10.3125	0.0156	0.7184	0.0904	0.0877	−0.068	0.9962	0.9981	0.9923
	SVR	32.4063	0.0313	0.1985	0.0268	0.0242	0.0658	0.9998	0.9999	0.9994
	ELM	0.0938	0.0469	0.4751	0.0439	0.0580	0.1764	0.9983	0.9992	0.9966
CG	MLP-NN	12.5938	0.0156	0.2116	0.0211	0.0258	−0.065	0.9997	0.9998	0.9993
	SVR	34.0781	0.0313	0.1987	0.0268	0.0243	0.1207	0.9999	0.9999	0.9994
	ELM	0.0625	0.0469	0.3483	0.0378	0.0425	−0.012	0.9991	0.9995	0.9982
ABG	MLP-NN	13.5469	0.0156	0.1332	0.0216	0.0157	0.0429	0.9999	0.9999	0.9998
	SVR	35.8750	0.0156	0.4948	0.0804	0.0582	−0.118	0.9996	0.9992	0.9966
	ELM	0.0625	0.0313	0.2010	0.0263	0.0236	−0.057	0.9997	0.9999	0.9994
BCG	MLP-NN	13.4219	0.0313	0.3300	0.0443	0.0378	−0.067	0.9993	0.9996	0.9986
	SVR	35.0156	0.0625	0.4950	0.0889	0.0567	−0.185	0.9996	0.9992	0.9968
	ELM	0.1250	0.0469	0.2359	0.0386	0.0270	0.0726	0.9996	0.9998	0.9993
CAG	MLP-NN	11.2813	0.0142	0.1206	0.0198	0.0146	−0.073	0.9999	0.9999	0.9998
	SVR	32.1250	0.0469	0.4937	0.0808	0.0599	−0.258	0.9995	0.9991	0.9964
	ELM	0.0625	0.0469	0.2061	0.0274	0.0250	0.0170	0.9997	0.9998	0.9994
ABCG	MLP-NN	9.1719	0.0469	0.2618	0.0327	0.0320	−0.024	0.9995	0.9997	0.9990
	SVR	33.8125	0.0156	0.2000	0.0270	0.0244	0.1007	0.9999	0.9999	0.9994
	ELM	0.0781	0.0469	0.2325	0.0174	0.0284	0.0073	0.9996	0.9998	0.9992

DWT, discrete wavelet transform; ELM, extreme learning machine; MAPE, mean absolute percentage error; MLP-NN, multilayer perceptron neural network; MLT, machine learning tool; PBIAS, percent bias; RMSE, root mean squared error; RSR, root mean square error-observations standard deviation ratio; SPI, statistical performance indices; SVR, support vector regression; WIA, Willmott's index of agreement.

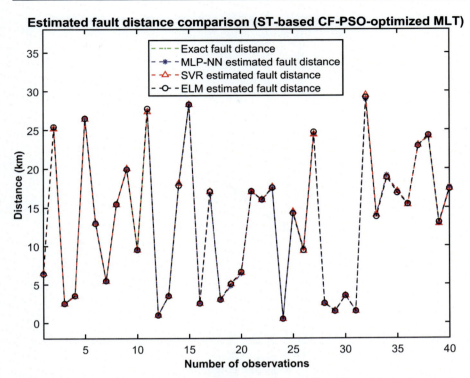

Fig. 10.11 Comparing ST-based optimized MLT predicted fault and actual fault distances for forty cases from AG fault test dataset. *MLT*, machine learning tool; *ST*, stockwell transform.

sufficiently predicted the targeted outputs for almost all cases. Besides, Fig. 10.12 shows the scatter plots of the targeted and MLT predicted fault locations of the AG fault test dataset, where most of the data points were located closer to the identity line as required for the efficient regression models. Therefore, it can also be concluded that metaheuristic optimized MLT showed better generalization performance than the trial-and-error basis optimized MLT for AG faults.

Table 10.11 summarizes the operational times and the evaluated SPI values of the test datasets for ST-based MLT schemes for seven types of faults. Similar to previous results, the SVR required almost threefold time for training compared to the MLP-NN, whereas the training process of the ELM was nearly a hundred times faster than that of the MLP-NN. Besides, the testing times for all techniques were almost equal. Moreover, the evaluated MAPE, RMSE, and RSR values were relatively low, whereas NSCE, R^2, and WIA were near unity for the MLT. However, the PBIAS values were positive for a few scenarios and negative for the others that demonstrated overestimations and underestimations of the fault distances, respectively. Nonetheless, these values were minimal and close to zero for all MLT; such negligible overestimations and underestimations were acceptable. Therefore, the evaluated SPI values confirmed the strength and the efficacy of the optimized MLT for all types

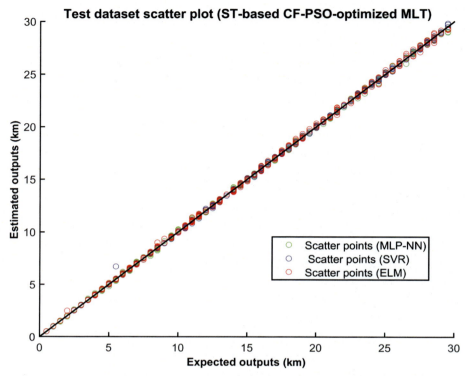

Fig. 10.12 Scatter plots of targeted (actual) and ST-based optimized MLT predicted outputs for the test dataset of AG fault. *MLT*, machine learning tool; *ST*, stockwell transform.

of faults. Thus, the tuning of the MLT key parameters employing the metaheuristic algorithm was justified for ST-based approaches as well.

10.5.2.2 Fault location considering measurement noise

This section investigates the efficacy of the intelligent fault location scheme in the presence of measurement noises. To do so, it adds different levels of AWGN with recorded three-phase current signals and extracted features employing both DWT and ST. Tables 10.12 and 10.13 show the evaluated SPI values together with the operating times of the DWT-based optimized MLT under 30 dB and 20 dB SNR, respectively. Likewise, Tables 10.14 and 10.15 presents the SPI values and the operating times of the ST-based optimized MLT under the 30 dB and 20 dB SNR, respectively. As can be seen from the mentioned tables, the training processes of the SVR and MLP-NN were computationally more expensive compared to the ELM. In contrast, testing times of all MLT were almost equivalent. The evaluated MAPE, PBIAS, RMSE, and RSR values were relatively low, whereas NSCE, R^2, and WIA values were almost unity for all types of faults. However, the DWT and ST-based MLT performances deteriorated a bit in the presence of the measurement noises in terms of the evaluated SPI values. However, still, they located different types of faults with convincing accuracy.

Intelligent fault diagnosis technique for distribution grid 271

Table 10.11 Operation time and SPI values for the test datasets of ST-based optimized MLT.

Fault type	Technique	Time (s) Training	Time (s) Testing	RMSE (km)	MAPE	RSR	PBIAS	R²	WIA	NSCE
AG	MLP-NN	12.5625	0.0156	0.2228	0.0311	0.0272	−0.049	0.9996	0.9998	0.9993
	SVR	28.1719	0.0313	0.2123	0.0066	0.0259	0.0748	0.9997	0.9998	0.9993
	ELM	0.1406	0.0313	0.1592	0.0188	0.0194	−0.057	0.9998	0.9999	0.9996
BG	MLP-NN	10.2813	0.0156	0.2885	0.0482	0.0356	−0.010	0.9994	0.9997	0.9987
	SVR	29.7969	0.0313	0.2640	0.0153	0.0326	0.0911	0.9995	0.9997	0.9989
	ELM	0.0781	0.0469	0.2634	0.0472	0.0325	0.3282	0.9995	0.9997	0.9989
CG	MLP-NN	10.9063	0.0156	0.2150	0.0382	0.0252	−0.141	0.9997	0.9998	0.9994
	SVR	27.2031	0.0156	0.5267	0.0687	0.0617	−0.122	0.9981	0.9990	0.9962
	ELM	0.0781	0.0469	0.2471	0.0352	0.0289	0.0754	0.9996	0.9998	0.9992
ABG	MLP-NN	9.9844	0.0156	0.1823	0.0180	0.0210	−0.085	0.9998	0.9999	0.9996
	SVR	29.4375	0.0469	0.4899	0.0774	0.0564	−0.044	0.9990	0.9992	0.9968
	ELM	0.0938	0.0156	0.1674	0.0163	0.0193	0.0082	0.9998	0.9999	0.9996
BCG	MLP-NN	9.3438	0.0156	0.1470	0.0200	0.0177	0.0068	0.9998	0.9999	0.9997
	SVR	31.5000	0.0156	0.4933	0.0787	0.0593	0.0950	0.9991	0.9991	0.9965
	ELM	0.0469	0.0313	0.1536	0.0193	0.0185	0.0737	0.9998	0.9999	0.9997
CAG	MLP-NN	9.6094	0.0156	0.1809	0.0226	0.0217	0.0130	0.9998	0.9999	0.9995
	SVR	28.4688	0.0313	0.4917	0.0785	0.0591	0.0491	0.9990	0.9991	0.9965
	ELM	0.0781	0.0313	0.1533	0.0151	0.0184	0.0103	0.9998	0.9999	0.9997
ABCG	MLP-NN	9.7656	0.0156	0.3253	0.0647	0.0385	−0.201	0.9993	0.9996	0.9985
	SVR	26.6719	0.0313	0.5094	0.0910	0.0603	0.0144	0.9992	0.9991	0.9964
	ELM	0.0938	0.0156	0.3241	0.0439	0.0384	0.0350	0.9993	0.9996	0.9985

ELM, extreme learning machine; MAPE, mean absolute percentage error; MLP-NN, multilayer perceptron neural network; MLT, machine learning tool; PBIAS, percent bias; RMSE, root mean squared error; RSR, root mean square error-observations standard deviation ratio; SPI, statistical performance indices; ST, stockwell transform; SVR, support vector regression; WIA, Willmott's index of agreement.

Table 10.12 Operation time and SPI values for the test datasets of DWT-based optimized MLT in the presence of 30 dB SNR.

Fault type	Technique	Time (s) Training	Time (s) Testing	Statistical performance indices RMSE (km)	MAPE	RSR	PBIAS	R^2	WIA	NSCE
AG	MLP-NN	9.1875	0.0625	0.2123	0.0206	0.0259	0.0189	0.9997	0.9998	0.9993
	SVR	29.8125	0.0469	0.4731	0.0640	0.0577	0.3514	0.9989	0.9992	0.9967
	ELM	0.0313	0.0381	0.3244	0.0272	0.0396	−0.001	0.9992	0.9996	0.9984
BG	MLP-NN	10.3125	0.0469	0.7184	0.0904	0.0877	−0.068	0.9962	0.9981	0.9923
	SVR	32.4063	0.0313	0.1985	0.0268	0.0242	0.0658	0.9998	0.9999	0.9994
	ELM	0.0938	0.0469	0.4751	0.0439	0.0580	0.1764	0.9983	0.9992	0.9966
CG	MLP-NN	12.5000	0.0156	0.1647	0.0133	0.0201	0.0478	0.9998	0.9999	0.9996
	SVR	32.1719	0.0625	0.1960	0.0264	0.0239	0.0080	0.9998	0.9999	0.9994
	ELM	0.0781	0.0469	0.4725	0.0474	0.0577	−0.076	0.9983	0.9992	0.9967
ABG	MLP-NN	11.7969	0.0381	0.2783	0.0295	0.0327	0.0646	0.9995	0.9997	0.9989
	SVR	34.6250	0.0469	0.4937	0.0803	0.0580	−0.107	0.9996	0.9992	0.9966
	ELM	0.0938	0.0313	0.4248	0.0448	0.0499	−0.034	0.9988	0.9994	0.9975
BCG	MLP-NN	12.2188	0.0625	0.3426	0.0378	0.0392	0.1757	0.9992	0.9996	0.9985
	SVR	35.2188	0.0469	0.4952	0.0889	0.0567	−0.171	0.9996	0.9992	0.9968
	ELM	0.0781	0.0381	0.4432	0.0589	0.0507	−0.001	0.9987	0.9994	0.9974
CAG	MLP-NN	9.2969	0.0469	0.1293	0.0144	0.0157	−0.046	0.9999	0.9999	0.9998
	SVR	33.1875	0.0313	0.4937	0.0808	0.0599	−0.248	0.9995	0.9991	0.9964
	ELM	0.1250	0.0156	0.4329	0.0522	0.0525	−0.243	0.9986	0.9993	0.9972
ABCG	MLP-NN	10.9688	0.0313	0.3045	0.0296	0.0372	−0.004	0.9993	0.9997	0.9986
	SVR	34.3594	0.0313	0.2968	0.0401	0.0362	0.1780	0.9998	0.9997	0.9987
	ELM	0.0781	0.0469	0.2580	0.0145	0.0315	−0.033	0.9995	0.9998	0.9990

DWT, discrete wavelet transform; ELM, extreme learning machine; MAPE, mean absolute percentage error; MLP-NN, multilayer perceptron neural network; MLT, machine learning tool; PBIAS, percent bias; RMSE, root mean squared error; RSR, root mean square error-observations standard deviation ratio; SNR, signal-to-noise ratio; SPI, statistical performance indices; ST, stockwell transform; SVR, support vector regression; WIA, Willmott's index of agreement.

Table 10.13 Operation time and SPI values for the test datasets of DWT-based optimized MLT in the presence of 20 dB SNR.

Fault type	Technique	Time (s) Training	Time (s) Testing	Statistical performance indices RMSE (km)	MAPE	RSR	PBIAS	R²	WIA	NSCE
AG	MLP-NN	9.9375	0.0156	0.3411	0.0364	0.0416	−0.064	0.9991	0.9996	0.9983
	SVR	32.5000	0.0156	0.4788	0.0648	0.0584	0.3333	0.9991	0.9991	0.9966
	ELM	0.0469	0.0469	0.5661	0.0616	0.0691	0.2388	0.9976	0.9988	0.9952
BG	MLP-NN	11.2031	0.0156	0.5035	0.0635	0.0615	−0.106	0.9981	0.9991	0.9962
	SVR	32.9063	0.0313	0.1978	0.0267	0.0241	0.1023	0.9998	0.9999	0.9994
	ELM	0.1406	0.0469	0.7093	0.0758	0.0866	1.0695	0.9965	0.9981	0.9925
CG	MLP-NN	9.0938	0.0156	0.3370	0.0318	0.0411	−0.027	0.9992	0.9996	0.9983
	SVR	34.7188	0.0156	0.1969	0.0266	0.0240	0.0453	0.9998	0.9999	0.9994
	ELM	0.1250	0.0625	0.6418	0.0662	0.0784	0.1967	0.9969	0.9985	0.9939
ABG	MLP-NN	11.9531	0.0156	0.4169	0.0544	0.0490	0.2101	0.9988	0.9994	0.9976
	SVR	32.4531	0.0156	0.4945	0.0804	0.0581	−0.121	0.9996	0.9992	0.9966
	ELM	0.0781	0.0625	0.5748	0.0606	0.0676	0.2775	0.9977	0.9989	0.9954
BCG	MLP-NN	13.6719	0.0156	0.6678	0.0906	0.0764	0.1802	0.9971	0.9985	0.9942
	SVR	34.7813	0.0469	0.4947	0.0889	0.0566	−0.182	0.9996	0.9992	0.9968
	ELM	0.0938	0.0625	0.5364	0.0673	0.0614	−0.133	0.9981	0.9991	0.9962
CAG	MLP-NN	12.0781	0.0156	0.3858	0.0484	0.0468	0.1941	0.9989	0.9995	0.9978
	SVR	35.3281	0.0313	0.4937	0.0808	0.0599	−0.249	0.9995	0.9991	0.9964
	ELM	0.0625	0.0313	0.5724	0.0640	0.0694	0.2047	0.9976	0.9988	0.9952
ABCG	MLP-NN	11.0781	0.0156	0.5178	0.0391	0.0632	0.0634	0.9980	0.9990	0.9960
	SVR	34.9375	0.0313	0.2973	0.0402	0.0363	0.1713	0.9998	0.9997	0.9987
	ELM	0.0781	0.0313	0.5011	0.0252	0.0612	−0.014	0.9981	0.9991	0.9963

DWT, discrete wavelet transform; ELM, extreme learning machine; MAPE, mean absolute percentage error; MLP-NN, multilayer perceptron neural network; MLT, machine learning tool; PBIAS, percent bias; RMSE, root mean squared error; RSR, root mean square error-observations standard deviation ratio; SNR, signal-to-noise ratio; SPI, statistical performance indices; ST, stockwell transform; SVR, support vector regression; WIA, Willmott's index of agreement.

Table 10.14 Operation time and SPI values for the test datasets of ST-based optimized MLT in the presence of 30 dB SNR.

Fault type	Technique	Time (s) Training	Time (s) Testing	Statistical performance indices RMSE (km)	MAPE	RSR	PBIAS	R²	WIA	NSCE
AG	MLP-NN	12.4688	0.0469	0.5505	0.0972	0.0672	−0.182	0.9978	0.9989	0.9955
	SVR	27.5156	0.0156	0.2689	0.0142	0.0328	0.0701	0.9995	0.9997	0.9989
	ELM	0.0938	0.0781	0.3312	0.0346	0.0404	0.0507	0.9992	0.9996	0.9984
BG	MLP-NN	12.3281	0.0313	0.4292	0.0456	0.0529	−0.169	0.9986	0.9993	0.9972
	SVR	32.0000	0.0469	0.2995	0.0375	0.0369	−0.199	0.9993	0.9997	0.9986
	ELM	0.0781	0.0313	0.3366	0.0465	0.0415	−0.168	0.9992	0.9996	0.9983
CG	MLP-NN	10.4063	0.0313	0.5706	0.0971	0.0668	−0.012	0.9978	0.9989	0.9955
	SVR	27.2500	0.0313	0.3290	0.0219	0.0385	−0.203	0.9993	0.9996	0.9985
	ELM	0.0625	0.0469	0.3860	0.0484	0.0452	0.0803	0.9990	0.9995	0.9980
ABG	MLP-NN	11.5469	0.0156	0.2121	0.0248	0.0244	−0.032	0.9997	0.9999	0.9994
	SVR	27.9375	0.0313	0.4888	0.0776	0.0563	−0.042	0.9991	0.9992	0.9968
	ELM	0.0469	0.0469	0.2563	0.0210	0.0295	−0.054	0.9996	0.9998	0.9991
BCG	MLP-NN	11.0313	0.0469	0.1711	0.0252	0.0206	0.0208	0.9998	0.9999	0.9996
	SVR	28.9375	0.0313	0.4940	0.0786	0.0594	0.1087	0.9990	0.9991	0.9965
	ELM	0.1094	0.0156	0.2225	0.0206	0.0267	0.0401	0.9996	0.9998	0.9993
CAG	MLP-NN	9.3906	0.0313	0.1686	0.0210	0.0203	−0.021	0.9998	0.9999	0.9996
	SVR	28.2188	0.0313	0.4930	0.0788	0.0593	0.0465	0.9990	0.9991	0.9965
	ELM	0.0938	0.0313	0.2571	0.0231	0.0309	−0.119	0.9995	0.9998	0.9990
ABCG	MLP-NN	11.8281	0.0469	0.2556	0.0392	0.0303	0.0351	0.9996	0.9998	0.9991
	SVR	32.9688	0.0469	0.5089	0.0909	0.0603	−0.006	0.9992	0.9991	0.9964
	ELM	0.0781	0.0313	0.4726	0.0588	0.0560	−0.121	0.9984	0.9992	0.9969

ELM, extreme learning machine; MAPE, mean absolute percentage error; MLP-NN, multilayer perceptron neural network; MLT, machine learning tool; PBIAS, percent bias; RMSE, root mean squared error; RSR, root mean square error-observations standard deviation ratio; SNR, signal-to-noise ratio; SPI, statistical performance indices; ST, stockwell transform; SVR, support vector regression; WIA, Willmott's index of agreement.

Table 10.15 Operation time and SPI values for the test datasets of ST-based optimized MLT in the presence of 20 dB SNR.

Fault type	Technique	Time (s) Training	Time (s) Testing	Statistical performance indices RMSE (km)	MAPE	RSR	PBIAS	R^2	WIA	NSCE
AG	MLP-NN	11.1719	0.0313	0.7181	0.1127	0.0877	−0.305	0.9962	0.9981	0.9923
	SVR	27.0469	0.0313	0.5145	0.0566	0.0628	−0.102	0.9980	0.9990	0.9961
	ELM	0.1094	0.0313	0.4740	0.0485	0.0579	0.1930	0.9983	0.9992	0.9967
BG	MLP-NN	10.1406	0.0156	0.7303	0.1233	0.0900	−0.314	0.9960	0.9980	0.9919
	SVR	28.0469	0.0469	0.5588	0.0300	0.0689	−0.150	0.9976	0.9988	0.9953
	ELM	0.1250	0.0313	0.4516	0.0543	0.0557	−0.212	0.9985	0.9992	0.9969
CG	MLP-NN	12.7344	0.0156	0.7144	0.0923	0.0836	0.1110	0.9965	0.9983	0.9930
	SVR	28.8594	0.0313	0.7436	0.0424	0.0870	−0.213	0.9962	0.9981	0.9924
	ELM	0.0938	0.0156	0.4566	0.0489	0.0534	0.1954	0.9986	0.9993	0.9971
ABG	MLP-NN	12.4375	0.0469	0.3055	0.0423	0.0352	−0.181	0.9994	0.9997	0.9988
	SVR	29.1719	0.0313	0.4975	0.0781	0.0573	0.0283	0.9990	0.9992	0.9967
	ELM	0.0938	0.0313	0.3723	0.0331	0.0429	−0.036	0.9991	0.9995	0.9982
BCG	MLP-NN	11.5938	0.0313	0.2135	0.0406	0.0257	−0.029	0.9997	0.9998	0.9993
	SVR	29.4531	0.0469	0.4982	0.0788	0.0599	0.1352	0.9990	0.9991	0.9964
	ELM	0.0938	0.0313	0.3804	0.0345	0.0457	0.0465	0.9990	0.9995	0.9979
CAG	MLP-NN	10.0156	0.0156	0.2474	0.0366	0.0297	−0.034	0.9996	0.9998	0.9991
	SVR	29.1719	0.0469	0.4987	0.0791	0.0599	0.0553	0.9989	0.9991	0.9964
	ELM	0.0938	0.0313	0.4055	0.0380	0.0487	0.0108	0.9988	0.9994	0.9976
ABCG	MLP-NN	11.0625	0.0156	0.4573	0.0634	0.0541	−0.022	0.9985	0.9993	0.9971
	SVR	32.3281	0.0156	0.5059	0.0906	0.0599	0.0450	0.9992	0.9991	0.9964
	ELM	0.0781	0.0156	0.7247	0.0861	0.0858	−0.237	0.9963	0.9982	0.9926

ELM, extreme learning machine; MAPE, mean absolute percentage error; MLP-NN, multilayer perceptron neural network; MLT, machine learning tool; PBIAS, percent bias; RMSE, root mean squared error; RSR, root mean square error-observations standard deviation ratio; SNR, signal-to-noise ratio; SPI, statistical performance indices; ST, stockwell transform; SVR, support vector regression; WIA, Willmott's index of agreement.

Reported testing times of Tables 10.12–10.15 were for predicting fault locations for 300 cases. Therefore, the optimized MLT required less than a cycle of the 60 Hz power system network to predict fault location for a single case (on average). Such prompt and efficient responses of the intelligent fault location schemes demonstrated their potential in real-time applications.

10.6 Fault diagnosis under unbalanced loading condition

Presented results and discussions of the IFD scheme of the previous sections considered balanced loading conditions of the modeled four-node test distribution feeder. The efficacy of the IFD scheme under unbalanced loading conditions was also investigated by changing the balanced loading of Table 10.1 with the unbalanced loading of Table 10.16. This section followed similar procedures for fault modeling, data recording, and feature extraction like the previous sections.

10.6.1 Fault detection and classification results

The intelligent fault detection scheme acquired inputs (DWT and ST extracted features) for 1050 faulty cases consisting of SLG (AG, BG, and CG), LLG (ABG, BCG, and CAG), and LLLG (ABCG) faults applied on random locations of the distribution line incorporating prefault loading condition and fault information uncertainty. In the same way, it acquired features for 1050 nonfaulty scenarios. The scheme fetched the extracted features into the MLP-NN to distinguish the faulty signals from their healthy counterparts. The detection scheme chose six and five neurons in the hidden layers of the MLP-NN in terms of overall performance (accuracy and MSE) for the DWT and ST-based approaches, respectively. Besides, tan sigmoid and resilient backpropagation ("trainrp") were employed as the squashing functions and training algorithms, respectively. Like the previous detection scheme, it utilized 70% of the available data for training and the rest for testing purposes. Table 10.17 summarizes the fault detection results for both DWT and ST-based approaches under noisy and noise-free environments. The scheme separated the faulty and healthy scenarios with 100% accuracy for both DWT and ST-based approaches. Therefore, the presented results confirmed the effectiveness of the intelligent fault detection scheme.

Like balanced systems, the intelligent fault classification scheme also acquired inputs (DWT and ST extracted features) for 700 faulty scenarios of each type incorporating prefault loading conditions and fault information uncertainty. It fetched the

Table 10.16 Unbalanced loading of the four-node test distribution feeder of Fig. 10.1.

Phase	Real power (MW)	Reactive power (MVAR)
Phase A	1.00	0.48
Phase B	2.00	0.65
Phase C	2.5	1.20

Table 10.17 Fault detection results for the noise-free and noisy data of DWT and ST-based MLP-NN under unbalanced loading conditions.

Item	Samples classified successfully							
	Noise free		40 dB SNR		30 dB SNR		20 dB SNR	
	DWT	ST	DWT	ST	DWT	ST	DWT	ST
Faulty cases	1050	1050	1050	1050	1050	1050	1050	1050
Nonfaulty cases	1050	1050	1050	1050	1050	1050	1050	1050
Overall accuracy (%)	100	100	100	100	100	100	100	100

DWT, discrete wavelet transform; MLP-NN, multilayer perceptron neural network; SNR, signal-to-noise ratio; ST, stockwell transform.

features into the MLP-NN and selected eight and seven hidden layer neurons in terms of overall performance (accuracy and MSE) for DWT and ST-based approaches, respectively. Besides, tan sigmoid and resilient backpropagation ("trainrp") were selected as the squashing function and the training algorithm through systematic trial and error processes for both DWT and ST-based classification schemes. Finally, 70% of the available data were used for training and the rest for testing purposes. Table 10.18 summarizes the obtained results of the intelligent fault classification schemes. As can be seen, both DWT and ST-based approaches classified the faults with almost 100 accuracies for noise-free data. Besides, the DWT-based method ended up with overall accuracies of 99.979%, 99.959%, and 99.959% for 40 dB, 30 dB, and 20 dB SNR, respectively. Similar accuracies were also achieved employing the ST-based approach under noisy measurements. Therefore, presented results validated the fruitfulness of the adopted intelligent fault classification scheme under noisy and noise-free measurements during unbalanced loading conditions as well.

Table 10.18 Classification results of DWT and ST-based MLP-NN under noise-free and noisy environment during unbalanced loading conditions.

Fault type	Samples classified successfully							
	No noise		40 dB noise		30 dB SNR		20 dB SNR	
	DWT	ST	DWT	ST	DWT	ST	DWT	ST
AG	700	700	700	700	700	700	700	700
BG	700	700	700	700	700	700	700	700
CG	700	700	700	700	700	700	700	700
ABG	700	700	699	700	700	700	699	700
BCG	700	700	700	700	700	699	700	700
CAG	700	700	700	700	699	700	700	699
ABCG	699	700	699	699	699	699	699	699
Overall accuracy (%)	99.98	100.0	99.96	99.98	99.96	99.96	99.96	99.96

DWT, discrete wavelet transform; MLP-NN, multilayer perceptron neural network; SNR, signal-to-noise ratio; ST, stockwell transform.

10.6.2 Fault detection and classification results comparison

This section compares the DWT and ST-based intelligent fault detection and classification schemes with the referenced techniques in terms of overall accuracy. Fault detection and classification results comparisons are presented in Tables 10.19 and 10.20, respectively. As can be observed, most of the referenced techniques detected and classified distribution grid faults with satisfactory accuracy, and only a few of them diagnosed faults with lower accuracy. However, only a few of the referenced works assessed their performance considering the presence of measurement noises. The intelligent fault detection and classification schemes of this chapter exceeded 99.00% accuracy in all cases for both DWT and ST-based approaches for the selected four-node test distribution feeder under balanced and unbalanced loading conditions. The presented intelligent schemes detected and classified distribution grid faults with better or a competitive accuracy over the referenced works. Furthermore, the schemes showed their independence on uncertainty associated with prefault loading conditions, fault information (resistance, inception angle, and location), and presence of measurement noises. Finally, it is worth noting that the presented schemes approach required less than a cycle (average for a single case on average) of the 60 Hz power system network to extract characteristics features employing the signal processing techniques and to detect and classify the faults utilizing the trained neural networks after receiving the three-phase current signals. Therefore, the presented intelligent schemes can be implemented for the detection and classification of distribution grid faults.

Table 10.19 Fault detection results comparison.

Technique name and reference	Overall accuracy (%) Noise-free environment	Noisy environment
Neural network [8]	95.00	Not Assessed
Temporal attribute QSSVM [14]	98.10	Not Assessed
Wavelet singular entropy based fuzzy logic theory [14]	100.0	Not assessed
Principal components analysis based SVM [58]	99.74	99.79 (30 dB) 99.77 (20 dB)
Hybrid DWT approach (four-node system: balanced load)	100.0	100.0 (40 dB) 100.0 (30 dB) 100.0 (20 dB)
Hybrid ST approach (four-node system: balanced load)	100.0	100.0 (40 dB) 100.0 (30 dB) 100.0 (20 dB)
Hybrid DWT approach (four-node system: balanced load)	100.0	100.0 (40 dB) 100.0 (30 dB) 100.0 (20 dB)
Hybrid ST approach (four-node system: balanced load)	100.0	100.0 (40 dB) 100.0 (30 dB) 100.0 (20 dB)

DWT, discrete wavelet transform; QSSVM, quarter-sphere support vector machine; SVM, support vector machine; ST, stockwell transform.

Intelligent fault diagnosis technique for distribution grid

Table 10.20 Fault classification results comparison.

Technique name and reference	Overall accuracy (%)	
	Noise-free environment	Noisy environment
Wavelet-based fuzzy logic algorithm [15]	89.50	Not assessed
Wavelet multi-resolution approach [59]	94.73	Not Assessed
Wavelet-based neural network [60]	98.40	Not assessed
Wavelet-based adaptive neuro-fuzzy inference system [61]	99.84	Not assessed
Attribute QSSVM [14]	99.05	Not assessed
Time-time transform-based ART neural network [12]	99.18	Not assessed
Combination of wavelet singular entropy theory and fuzzy logic [14]	100.0	Not Assessed
Principal components analysis based SVM [58]	99.93	99.77 (30 dB) 99.70 (20 dB)
Hybrid DWT Approach [(four-node system: balanced load)	99.98	99.98 (40 dB) 99.98 (30 dB) 99.96 (20 dB)
Hybrid ST approach (four-node system: balanced load)	99.98	99.98 (40 dB) 99.96 (30 dB) 99.96 (20 dB)
Hybrid DWT approach (four-node system: balanced load)	99.98	99.96 (40 dB) 99.96 (30 dB) 99.96 (20 dB)
Hybrid ST approach (four-node system: balanced load)	100.0	99.98 (40 dB) 99.96 (30 dB) 99.96 (20 dB)

DWT, discrete wavelet transform; QSSVM, quarter-sphere support vector machine; ST, stockwell transform.

10.6.3 Fault location results under noise-free measurement

This section presents fault location results of the four-node test distribution under unbalanced loading conditions. Based on the experience of the balanced system, the critical parameters of the SVR and ELM were optimized employing the CF-PSO. In contrast, the MLP-NN parameters were chosen through a systematic trial and error basis. The objective function of the optimization problems was the minimization of MAPE values of the actual and the predicted fault locations. The scheme optimized the SVR control parameters with a swarm of 20 particles for 100 iterations. Similarly, the EML control parameters were optimized, employing a swarm of 80 particles for 100 iterations. Table 10.21 presents the evaluated SVR and ELM control parameters utilizing CF-PSO and MLP-NN parameters through a systematic trial and error basis. It is worth noting that the MLP-NN parameters were not optimized using the CF-PSO as it provided satisfactory results, even with the parameters selected via the trial-and-error approach.

Actual and predicted fault location comparisons and the scatter plots are not presented in this section to avoid similar figures and discussions. However, Tables 10.22 and 10.23 present the evaluated SPI values obtained from the test datasets of different types of faults for the unbalanced system using the DWT and ST-based MLT schemes,

Table 10.21 CF-PSO optimized control parameters for DWT and ST-based MLT under unbalanced loading conditions.

SPT	MLT	Values of the control parameters
DWT	MLP-NN	Number of hidden neurons = 8, Squashing function = tan sigmoid, and training algorithm = Levenberg-Marquardt backpropagation ("trainlm").
	SVR	kernel = Gaussian RBF $C = 1000$, $K_O = 100$, $\varepsilon = 5.0 \times 10^{-1}$, $\lambda = 9.5 \times 10^{-3}$
	ELM	kernel = Gaussian RBF, $K_p = 9.5 \times 10^{10}$, $C_R = 2.5 \times 10^{12}$
ST	MLP-NN	Number of hidden neurons = 6, Squashing function = tan sigmoid, and training algorithm = Levenberg-Marquardt backpropagation ("trainlm").
	SVR	kernel = Gaussian RBF, $C = 1000$, $K_O = 100$, $\varepsilon = 4.75 \times 10^{-1}$, $\lambda = 8.5 \times 10^{-3}$
	ELM	kernel = Gaussian RBF, $K_p = 3.5 \times 10^9$, $C_R = 5 \times 10^{11}$

CF-PSO, constriction factor particle swarm optimization; DWT, discrete wavelet transform; ELM, extreme learning machine; MLP-NN, multilayer perceptron neural network; MLT, machine learning tool; ST, stockwell transform; SVR, support vector regression.

respectively. Like the previous cases, the SVR was the most computationally expensive technique followed by the MLP-NN, whereas the ELM was much faster in terms of training time. However, all of the employed MLT required almost similar times for testing purposes. Evaluated MAPE, RMSE, RSR, and PBIAS values were relatively low, whereas NSCE, R^2, and WIA were close to unity for all types of faults for the employed MLT based on both DWT and ST-based approaches. The satisfactory SPI values validated the effectiveness of the presented intelligent fault location scheme of the four-node unbalanced test distribution feeder.

10.6.4 Fault location results under the presence of measurement noise

The efficacy of the intelligent fault location schemes under unbalanced loading conditions was also investigated considering the presence of measurement noises. Like the previous section, actual and predicted fault location comparisons and the scatter plots are not presented in this section to avoid similar figures and discussions. However, Tables 10.24 and 10.25 present the evaluated SPI values and operating times for the DWT-based MLT in the presence of 30 dB and 20 dB SNR, respectively. Tables 10.26 and 10.27 present the results of ST-based MTL in the presence of 30 dB and 20 dB SNR, respectively. As can be seen, the SVR required almost three-fold time compared to the MLP-NN for the training purpose, and the ELM was much faster than the other two for both DWT and ST-based approaches. However, all three MLT required similar times for testing purposes. The evaluated MAPE, RMSE, RSR, and PBIAS were relatively low, whereas NSCE, R^2, and WIA were close to unity. Presented results confirmed the efficacy of the employed DWT and ST-based MLT in locating different

Table 10.22 Operation time and SPI values for the test datasets of DWT-based optimized MLT under unbalanced loading conditions.

Fault type	Technique	Time (s) Training	Time (s) Testing	RMSE (km)	MAPE	RSR	PBIAS	R^2	WIA	NSCE
AG	MLP-NN	11.6406	0.0156	0.4844	0.0627	0.0545	0.1099	0.9985	0.9993	0.9970
	SVR	30.1406	0.0313	0.3140	0.0463	0.0353	0.1400	0.9999	0.9997	0.9988
	ELM	0.0781	0.0469	0.2094	0.0265	0.0235	0.0492	0.9997	0.9999	0.9994
BG	MLP-NN	12.3906	0.0313	0.3497	0.0384	0.0394	0.1082	0.9992	0.9996	0.9984
	SVR	29.0938	0.0313	0.5519	0.0991	0.0622	0.1256	0.9996	0.9990	0.9961
	ELM	0.1094	0.0156	0.1665	0.0181	0.0188	−0.037	0.9998	0.9999	0.9996
CG	MLP-NN	11.5781	0.0313	0.1719	0.0187	0.0195	0.0062	0.9998	0.9999	0.9996
	SVR	29.5156	0.0156	0.5529	0.1012	0.0626	−0.101	0.9996	0.9990	0.9961
	ELM	0.1250	0.0313	0.2654	0.0374	0.0300	−0.128	0.9996	0.9998	0.9991
ABG	MLP-NN	12.4375	0.0156	0.1444	0.0173	0.0173	−0.015	0.9999	0.9999	0.9997
	SVR	28.3594	0.0313	0.4923	0.0695	0.0589	−0.025	0.9995	0.9991	0.9965
	ELM	0.0938	0.0781	0.1689	0.0172	0.0202	−0.089	0.9998	0.9999	0.9996
BCG	MLP-NN	11.2344	0.0313	0.2667	0.0277	0.0313	0.0025	0.9995	0.9998	0.9990
	SVR	28.9531	0.0313	0.4940	0.0776	0.0605	0.1043	0.9995	0.9991	0.9963
	ELM	0.1250	0.0469	0.1691	0.0194	0.0207	0.0039	0.9998	0.9999	0.9996
CAG	MLP-NN	12.1094	0.0313	0.1535	0.0191	0.0183	−0.047	0.9998	0.9999	0.9997
	SVR	26.7188	0.0469	0.4969	0.0687	0.0592	0.2240	0.9996	0.9991	0.9965
	ELM	0.1094	0.0313	0.1876	0.0203	0.0224	0.0450	0.9998	0.9999	0.9995
ABCG	MLP-NN	12.8750	0.0156	0.1493	0.0148	0.0172	0.0032	0.9999	0.9999	0.9997
	SVR	28.7813	0.0313	0.6953	0.0981	0.0800	0.1278	0.9992	0.9984	0.9936
	ELM	0.1094	0.0469	0.4999	0.0464	0.0576	0.0403	0.9984	0.9992	0.9967

DWT, discrete wavelet transform; ELM, extreme learning machine; MAPE, mean absolute percentage error; MLP-NN, multilayer perceptron neural network; MLT, machine learning tool; PBIAS, percent bias; RMSE, root mean squared error; RSR, root mean square error-observations standard deviation ratio; SPI, statistical performance indices; ST, stockwell transform; SVR, support vector regression; WIA, Willmott's index of agreement.

Table 10.23 Operation time and SPI values for the test datasets of ST-based optimized MLT under unbalanced loading conditions.

Fault type	Technique	Time (s) Training	Time (s) Testing	Statistical performance indices RMSE (km)	MAPE	RSR	PBIAS	R²	WIA	NSCE
AG	MLP-NN	11.5469	0.0156	0.3333	0.0442	0.0378	0.0642	0.9993	0.9996	0.9986
	SVR	26.2813	0.0156	0.3646	0.0094	0.0414	0.0304	0.9991	0.9996	0.9983
	ELM	0.1250	0.0469	0.2517	0.0266	0.0286	0.0388	0.9996	0.9998	0.9992
BG	MLP-NN	12.5000	0.0156	0.4419	0.0816	0.0517	−0.193	0.9987	0.9993	0.9973
	SVR	26.3281	0.0156	0.2271	0.0058	0.0265	0.1045	0.9996	0.9998	0.9993
	ELM	0.0781	0.0625	0.3172	0.0335	0.0371	−0.035	0.9993	0.9997	0.9986
CG	MLP-NN	12.2344	0.0313	0.3327	0.0589	0.0385	−0.151	0.9993	0.9996	0.9985
	SVR	29.8438	0.0156	0.2263	0.0030	0.0262	0.2070	0.9997	0.9998	0.9993
	ELM	0.0938	0.0313	0.1668	0.0221	0.0193	−0.001	0.9998	0.9999	0.9996
ABG	MLP-NN	11.0000	0.0156	0.2627	0.0337	0.0298	0.0849	0.9996	0.9998	0.9991
	SVR	28.1250	0.0156	0.4898	0.0814	0.0555	0.0985	0.9992	0.9992	0.9969
	ELM	0.0781	0.0156	0.1899	0.0209	0.0215	−0.012	0.9998	0.9999	0.9995
BCG	MLP-NN	12.3750	0.0469	1.9296	0.1843	0.2240	1.2046	0.9748	0.9875	0.9498
	SVR	30.1875	0.0313	0.4868	0.0799	0.0565	0.1896	0.9991	0.9992	0.9968
	ELM	0.0625	0.0469	0.1638	0.0201	0.0190	0.0309	0.9998	0.9999	0.9996
CAG	MLP-NN	12.1406	0.0469	0.2019	0.0261	0.0228	−0.030	0.9997	0.9999	0.9995
	SVR	26.5156	0.0313	0.4878	0.0874	0.0551	0.0111	0.9991	0.9992	0.9970
	ELM	0.1094	0.0313	0.1714	0.0188	0.0194	−0.012	0.9998	0.9999	0.9996
ABCG	MLP-NN	11.6406	0.0313	0.3496	0.0327	0.0425	−0.001	0.9991	0.9995	0.9982
	SVR	30.7813	0.0469	0.5138	0.0852	0.0625	0.3396	0.9994	0.9990	0.9961
	ELM	0.0625	0.0313	0.2942	0.0255	0.0358	0.2254	0.9994	0.9997	0.9987

ELM, extreme learning machine; MAPE, mean absolute percentage error; MLP-NN, multilayer perceptron neural network; MLT, machine learning tool; PBIAS, percent bias; RMSE, root mean squared error; RSR, root mean square error-observations standard deviation ratio; SPI, statistical performance indices; ST, stockwell transform; SVR, support vector regression; WIA, Willmott's index of agreement.

Table 10.24 Operation time and SPI values for the test datasets under unbalanced loading conditions in the presence of 30 dB SNR (DWT approach).

Fault type	Technique	Time (s) Training	Time (s) Testing	Statistical performance indices RMSE (km)	MAPE	RSR	PBIAS	R^2	WIA	NSCE
AG	MLP-NN	12.0000	0.0313	0.6216	0.0767	0.0699	−0.137	0.9976	0.9988	0.9951
	SVR	25.7188	0.0156	0.3111	0.0442	0.0350	0.3116	0.9996	0.9997	0.9988
	ELM	0.0938	0.0781	0.2557	0.0242	0.0287	0.1020	0.9996	0.9998	0.9992
BG	MLP-NN	12.3906	0.0313	0.5485	0.0498	0.0618	0.0118	0.9981	0.9990	0.9962
	SVR	29.0938	0.0313	0.5826	0.1012	0.0657	0.0150	0.9984	0.9989	0.9957
	ELM	0.1094	0.0156	0.3955	0.0466	0.0446	−0.104	0.9990	0.9995	0.9980
CG	MLP-NN	12.9531	0.0156	0.2810	0.0359	0.0318	−0.012	0.9995	0.9997	0.9990
	SVR	28.8438	0.0313	0.5363	0.0981	0.0607	0.0369	0.9988	0.9991	0.9963
	ELM	0.0781	0.0156	0.2405	0.0260	0.0272	−0.285	0.9996	0.9998	0.9993
ABG	MLP-NN	12.8906	0.0156	0.6070	0.0688	0.0727	−0.122	0.9974	0.9987	0.9947
	SVR	27.0469	0.0313	0.4920	0.0695	0.0589	−0.010	0.9995	0.9991	0.9965
	ELM	0.0938	0.0625	0.3761	0.0343	0.0450	0.0102	0.9990	0.9995	0.9980
BCG	MLP-NN	11.5469	0.0469	0.2225	0.0342	0.0272	0.0368	0.9996	0.9998	0.9993
	SVR	26.2188	0.0313	0.4932	0.0776	0.0604	0.1077	0.9995	0.9991	0.9964
	ELM	0.0938	0.0469	0.3476	0.0362	0.0425	0.1184	0.9991	0.9995	0.9982
CAG	MLP-NN	11.1563	0.0156	0.5032	0.0618	0.0600	0.1166	0.9982	0.9991	0.9964
	SVR	27.1094	0.0313	0.4961	0.0686	0.0591	0.2246	0.9996	0.9991	0.9965
	ELM	0.0938	0.0313	0.3648	0.0331	0.0435	0.0669	0.9991	0.9995	0.9981
ABCG	MLP-NN	12.9688	0.0156	0.6896	0.0772	0.0794	−0.078	0.9968	0.9984	0.9937
	SVR	31.5469	0.0313	0.6953	0.0981	0.0801	0.1333	0.9992	0.9984	0.9936
	ELM	0.0781	0.0469	0.8586	0.0697	0.0988	−0.527	0.9952	0.9976	0.9902

DWT, discrete wavelet transform; ELM, extreme learning machine; MAPE, mean absolute percentage error; MLP-NN, multilayer perceptron neural network; MLT, machine learning tool; PBIAS, percent bias; RMSE, root mean squared error; RSR, root mean square error-observations standard deviation ratio; SPI, statistical performance indices; ST, stockwell transform; SVR, support vector regression; WIA, Willmott's index of agreement.

Table 10.25 Operation time and SPI values for the test datasets under unbalanced loading conditions in the presence of 20 dB SNR (DWT approach).

Fault type	Technique	Time (s) Training	Time (s) Testing	RMSE (km)	MAPE	RSR	PBIAS	R^2	WIA	NSCE
AG	MLP-NN	11.4375	0.0156	0.6972	0.0827	0.0784	0.2221	0.9969	0.9985	0.9939
	SVR	28.1563	0.0313	0.3148	0.0461	0.0354	0.2197	0.9997	0.9997	0.9987
	ELM	0.0625	0.0625	0.2737	0.0252	0.0308	−0.056	0.9995	0.9998	0.9991
BG	MLP-NN	11.1250	0.0156	0.5440	0.0644	0.0613	0.4372	0.9982	0.9991	0.9962
	SVR	29.5469	0.0313	0.6066	0.1077	0.0684	−0.271	0.9986	0.9988	0.9953
	ELM	0.0469	0.0625	0.4866	0.0565	0.0548	0.0675	0.9985	0.9992	0.9970
CG	MLP-NN	12.3594	0.0156	0.3470	0.0445	0.0393	−0.059	0.9992	0.9996	0.9985
	SVR	29.5000	0.0313	0.5606	0.0994	0.0634	0.0327	0.9988	0.9990	0.9960
	ELM	0.0781	0.0469	0.2532	0.0312	0.0287	−0.030	0.9996	0.9998	0.9992
ABG	MLP-NN	11.6563	0.0156	0.3784	0.0411	0.0453	0.1662	0.9990	0.9995	0.9979
	SVR	27.8594	0.0625	0.4920	0.0695	0.0589	−0.019	0.9995	0.9991	0.9965
	ELM	0.1094	0.0313	0.5631	0.0532	0.0674	−0.412	0.9978	0.9989	0.9955
BCG	MLP-NN	12.2500	0.0156	0.3542	0.0388	0.0434	0.0769	0.9991	0.9995	0.9981
	SVR	28.4531	0.0313	0.4945	0.0777	0.0605	0.1193	0.9995	0.9991	0.9963
	ELM	0.1875	0.0156	0.5315	0.0516	0.0651	0.1495	0.9979	0.9989	0.9958
CAG	MLP-NN	10.0156	0.0156	0.3868	0.0401	0.0461	0.0424	0.9989	0.9995	0.9979
	SVR	26.9219	0.0313	0.4958	0.0686	0.0591	0.2261	0.9996	0.9991	0.9965
	ELM	0.0938	0.0469	0.6008	0.0681	0.0716	0.0479	0.9974	0.9987	0.9949
ABCG	MLP-NN	12.4688	0.0156	0.8533	0.0617	0.0982	0.1009	0.9952	0.9976	0.9903
	SVR	30.4531	0.0469	0.6953	0.0981	0.0800	0.1299	0.9992	0.9984	0.9936
	ELM	0.0625	0.0469	0.7656	0.0466	0.0881	0.0204	0.9961	0.9981	0.9922

DWT, discrete wavelet transform; ELM, extreme learning machine; MAPE, mean absolute percentage error; MLP-NN, multilayer perceptron neural network; MLT, machine learning tool; PBIAS, percent bias; RMSE, root mean squared error; RSR, root mean square error-observations standard deviation ratio; SNR, signal-to-noise ratio; SPI, statistical performance indices; SVR, support vector regression; WIA, Willmott's index of agreement.

Table 10.26 Operation time and SPI values for the test datasets under unbalanced loading conditions in the presence of 30 dB SNR (ST approach).

Fault type	Technique	Time (s) Training	Time (s) Testing	RMSE (km)	MAPE	RSR	PBIAS	R^2	WIA	NSCE
AG	MLP-NN	12.3906	0.0156	0.4775	0.0761	0.0542	−0.075	0.9986	0.9993	0.9971
	SVR	30.4688	0.0313	0.4179	0.0564	0.0474	−0.006	0.9989	0.9994	0.9978
	ELM	0.1250	0.0156	0.4959	0.0606	0.0563	0.1173	0.9984	0.9992	0.9968
BG	MLP-NN	12.0000	0.0156	0.5551	0.0993	0.0649	−0.443	0.9979	0.9989	0.9958
	SVR	26.6406	0.0313	0.2184	0.0042	0.0255	−0.056	0.9997	0.9998	0.9993
	ELM	0.0781	0.0156	0.3641	0.0529	0.0426	−0.243	0.9991	0.9995	0.9982
CG	MLP-NN	12.2813	0.0313	0.2729	0.0495	0.0316	−0.145	0.9995	0.9998	0.9990
	SVR	27.3438	0.0313	0.7008	0.1291	0.0811	0.5864	0.9969	0.9984	0.9934
	ELM	0.0781	0.0625	0.2815	0.0357	0.0326	0.1662	0.9995	0.9997	0.9989
ABG	MLP-NN	12.9531	0.0156	0.2220	0.0264	0.0252	0.0921	0.9997	0.9998	0.9994
	SVR	31.4063	0.0156	0.4898	0.0814	0.0555	0.0758	0.9991	0.9992	0.9969
	ELM	0.0625	0.0156	0.2551	0.0307	0.0289	0.0261	0.9996	0.9998	0.9992
BCG	MLP-NN	11.9375	0.0313	0.1421	0.0229	0.0165	−0.094	0.9999	0.9999	0.9997
	SVR	33.1094	0.0156	0.4876	0.0802	0.0566	0.1567	0.9991	0.9992	0.9968
	ELM	0.0781	0.0156	0.2271	0.0244	0.0264	0.0301	0.9997	0.9998	0.9993
CAG	MLP-NN	12.2813	0.0469	0.1599	0.0265	0.0181	0.0164	0.9998	0.9999	0.9997
	SVR	28.9844	0.0156	0.4880	0.0876	0.0551	0.0313	0.9991	0.9992	0.9970
	ELM	0.0938	0.0313	0.2265	0.0241	0.0256	0.0912	0.9997	0.9998	0.9993
ABCG	MLP-NN	10.7344	0.0156	0.1960	0.0252	0.0238	0.0435	0.9997	0.9999	0.9994
	SVR	29.3906	0.0156	0.5138	0.0852	0.0625	0.3203	0.9993	0.9990	0.9961
	ELM	0.0469	0.0469	0.4437	0.0408	0.0540	0.1201	0.9985	0.9993	0.9971

ELM, extreme learning machine; MAPE, mean absolute percentage error; MLP-NN, multilayer perceptron neural network; MLT, machine learning tool; PBIAS, percent bias; RMSE, root mean squared error; RSR, root mean square error-observations standard deviation ratio; SNR, signal-to-noise ratio; ST, stockwell transform; SPI, statistical performance indices; SVR, support vector regression; WIA, Willmott's index of agreement.

Table 10.27 Operation time and SPI values for the test datasets under unbalanced loading conditions in the presence of 20 dB SNR (ST approach).

Fault type	Technique	Time (s) Training	Time (s) Testing	RMSE (km)	MAPE	RSR	PBIAS	R^2	WIA	NSCE
AG	MLP-NN	12.7344	0.0313	0.6308	0.1032	0.0716	0.1823	0.9975	0.9987	0.9949
	SVR	26.7344	0.0313	0.4518	0.0635	0.0513	0.0368	0.9987	0.9993	0.9974
	ELM	0.1250	0.0313	0.5413	0.0684	0.0614	−0.190	0.9981	0.9991	0.9962
BG	MLP-NN	11.3438	0.0156	0.7414	0.1215	0.0867	0.0410	0.9962	0.9981	0.9925
	SVR	31.5781	0.0313	0.5687	0.0144	0.0665	0.4296	0.9978	0.9989	0.9956
	ELM	0.1094	0.0156	0.6717	0.0775	0.0785	1.1419	0.9971	0.9985	0.9938
CG	MLP-NN	10.6563	0.0313	0.4570	0.0625	0.0529	−0.129	0.9986	0.9993	0.9972
	SVR	33.8906	0.0313	0.7456	0.1352	0.0863	0.1454	0.9964	0.9981	0.9926
	ELM	0.0781	0.0313	0.3366	0.0425	0.0390	−0.118	0.9992	0.9996	0.9985
ABG	MLP-NN	11.6250	0.0313	0.3126	0.0357	0.0354	0.0672	0.9994	0.9997	0.9987
	SVR	32.5938	0.0156	0.4989	0.0823	0.0566	0.1467	0.9991	0.9992	0.9968
	ELM	0.0625	0.0156	0.3678	0.0360	0.0417	−0.032	0.9991	0.9996	0.9983
BCG	MLP-NN	11.4375	0.0313	0.4005	0.0778	0.0465	0.0445	0.9989	0.9995	0.9978
	SVR	28.4531	0.1920	0.4934	0.0801	0.0573	0.1787	0.9990	0.9992	0.9967
	ELM	0.0781	0.0313	0.3747	0.0346	0.0435	0.0331	0.9991	0.9995	0.9981
CAG	MLP-NN	11.8594	0.0156	0.2753	0.0407	0.0311	0.0033	0.9995	0.9998	0.9990
	SVR	27.5938	0.4898	0.4972	0.0890	0.0562	−0.146	0.9990	0.9992	0.9968
	ELM	0.1094	0.1920	0.3388	0.0392	0.0383	0.1323	0.9993	0.9996	0.9985
ABCG	MLP-NN	10.3281	0.0313	0.4188	0.0453	0.0509	0.1053	0.9987	0.9994	0.9974
	SVR	29.1250	0.0156	0.5161	0.0854	0.0628	0.3574	0.9994	0.9990	0.9961
	ELM	0.0938	0.4898	0.6710	0.0504	0.0816	0.2233	0.9967	0.9983	0.9933

ELM, extreme learning machine; MAPE, mean absolute percentage error; MLP-NN, multilayer perceptron neural network; MLT, machine learning tool; PBIAS, percent bias; RMSE, root mean squared error; RSR, root mean square error-observations standard deviation ratio; SNR, signal-to-noise ratio; ST, stockwell transform; SPI, statistical performance indices; SVR, support vector regression; WIA, Willmott's index of agreement.

types of faults, even in the presence of measurement noises for the unbalanced distribution feeder. It is worth noting that fault location schemes performances were deteriorated a bit in the presence of measurement noises in terms of the evaluated SPI values. However, still, they located faults with convincing accuracy. Finally, the scheme required less than a cycle (on average) of the 60 Hz power system network to estimate a single fault location that signals the real-time application potential of the presented intelligent schemes.

10.7 Summary

This chapter presented the step-by-step modeling of a three-phase four-node test distribution feeder under balanced and unbalanced loading conditions. Then, fault modeling and data recording considering the load demand and fault information (resistance and inception angle) uncertainty were illustrated. Two advanced signal processing techniques, namely the discrete wavelet and stockwell transforms (DWT and ST), were employed for feature extraction from the recorded three-phase current signals. After extracting useful features, they were fetched into three different types of MLTs: MLP-NN, SVM, and ELM, to diagnose, for example, detect, classify, and locate, faults under balanced and unbalanced loading conditions. Both DWT and ST-based machine learning approaches detected and classified faults with more than 99% accuracy even in the presence of measurement noises. The MLP-NN with systematically chosen control parameters located faults with satisfactory efficacy in terms of the evaluated SPI values. However, the fault location schemes employing the other two MLTs (SVR and ELM) did not perform well when their control parameters were chosen through a trial-and-error basis. Thus, this chapter optimized their control parameters using a metaheuristic optimization algorithm, namely the CF-PSO, to achieve better generalization performance. Presented results confirmed their enhanced performance in locating faults in the distribution grid even in the presence of measurement noises. However, the performance of the IFD schemes deteriorated a bit in the presence of measurement noises in terms of the evaluated SPI values, but still, they diagnosed faults with convincing accuracy. Therefore, it can be concluded that the presented results confirmed the effectiveness of the IFD schemes and their independence in prefault loading conditions, fault information, and measurement noise uncertainty. Finally, the IFD scheme can be implemented in real-time fault diagnosis in the smart grid systems as it required less than a few cycles (on average) of the 60 Hz power system network for feature extraction and detection, classification, and location of a single fault event.

References

[1] N. Sapountzoglou, J. Lago, B. Raison, Fault diagnosis in low voltage smart distribution grids using gradient boosting trees, Electr. Power Syst. Res. 182 (2020) 106254, doi:10.1016/j.epsr.2020.106254.
[2] M. Shafiullah, M.A. Abido, A review on distribution grid fault location techniques, Electr. Power Components Syst. 45 (8) (2017) 807–824, doi:10.1080/15325008.2017.1310772.

[3] M. Mirzaei, M.Z.A. Ab. Kadir, E. Moazami, H. Hizam, Review of fault location methods for distribution power system, Aust. J. Basic Appl. Sci. 3 (3) (2009) 2670–2676. http://psasir.upm.edu.my/16558/1/Review of fault location methods for distribution power system.pdf. (Accessed 26 November 2015).

[4] L. Awalin, H. Mokhlis, A. Bakar, Recent developments in fault location methods for distribution networks, Prz. Elektrotechniczny R88 (12a) (2012) 206–212. http://eprints.um.edu.my/7868/. (Accessed 26 November 2015).

[5] K. Jia, B. Yang, T. Bi, L. Zheng, An improved sparse-measurement-based fault location technology for distribution networks, IEEE Trans. Ind. Informatics 17 (3) (2021) 1712–1720, doi:10.1109/TII.2020.2995997.

[6] SmartGrid.gov. Distribution intelligence. https://www.smartgrid.gov/the_smart_grid/distribution_intelligence.html. (Accessed 17 June 2021).

[7] V. Giordano, F. Gangale, G. Fulli, and M.S. Jiménez, Smart grid projects in Europe: lessons learned and current developments, 2011. https://www.smartgrid.gov/document/smart_grid_projects_europe_lessons_learned_and_current_developments. (Accessed 20 June 2021).

[8] L.C. Acacio, P.A. Guaracy, T.O. Diniz, D.R.R.P. Araujo, L.R. Araujo, Evaluation of the impact of different neural network structure and data input on fault detection, IEEE PES Innovative Smart Grid Technologies Conference - Latin America (ISGT Latin America), IEEE, 2017, pp. 1–5, doi:10.1109/ISGT-LA.2017.8126699.

[9] Z. Wang, F. Wang, Earth fault detection in distribution network based on wide-area measurement information, 2011 International Conference on Electrical and Control Engineering, IEEE, 2011, pp. 5855–5859, doi:10.1109/ICECENG.2011.6057020.

[10] Q. Cui, K. El-Arroudi, G. Joos, An effective feature extraction method in pattern recognition based high impedance fault detection, 19th International Conference on Intelligent System Application to Power Systems (ISAP), IEEE, 2017, pp. 1–6, doi:10.1109/ISAP.2017.8071380.

[11] R. Li, C. Li, Y. Su, A cost-effective approach for feeder fault detection and location of radial distribution networks, IEEE Advanced Information Technology, Electronic and Automation Control Conference, IEEE, 2015, pp. 657–663, doi:10.1109/IAEAC.2015.7428636.

[12] I. Nikoofekr, M. Sarlak, S.M. Shahrtash, Detection and classification of high impedance faults in power distribution networks using ART neural networks 2013, 21st Iranian Conference on Electrical Engineering (ICEE), IEEE, 2013, pp. 1–6, doi:10.1109/IranianCEE.2013.6599760.

[13] X. Qin, P. Wang, Y. Liu, L. Guo, G. Sheng, X. Jiang, Research on distribution network fault recognition method based on time-frequency characteristics of fault waveforms, IEEE, 2017, pp. 1–1, doi:10.1109/ACCESS.2017.2728015.

[14] M. Dehghani, M.H. Khooban, T. Niknam, Fast fault detection and classification based on a combination of wavelet singular entropy theory and fuzzy logic in distribution lines in the presence of distributed generations, Int. J. Electr. Power Energy Syst. 78 (2016) 455–462, doi:10.1016/j.ijepes.2015.11.048.

[15] J. Klomjit, A. Ngaopitakkul, Selection of proper input pattern in fuzzy logic algorithm for classifying the fault type in underground distribution system, IEEE Region 10 Conference (TENCON), IEEE, 2016, pp. 2650–2655, doi:10.1109/TENCON.2016.7848519.

[16] E.C. Senger, G. Manassero, C. Goldemberg, E.L. Pellini, Automated fault location system for primary distribution networks, IEEE Trans. Power Deliv. 20 (2) (2005) 1332–1340, doi:10.1109/TPWRD.2004.834871.

[17] X. Li, Y. Lu, High frequency impedance based fast directional pilot protection scheme for distribution network with wind DGs, Asia-Pacific Power and Energy Engineering Conference, 2020, APPEEC, 2020, doi:10.1109/APPEEC48164.2020.9220388.

[18] A.A. Girgis, C.M. Fallon, D.L. Lubkeman, A fault location technique for rural distribution feeders, IEEE Trans. Ind. Appl. 29 (6) (1993) 1170–1175, doi:10.1109/28.259729.

[19] R.H. Salim, et al., Extended fault-location formulation for power distribution systems, IEEE Trans. Power Deliv. 24 (2) (2009) 508–516, doi:10.1109/TPWRD.2008.2002977.

[20] R.H. Salim, K.C.O. Salim, A.S. Bretas, Further improvements on impedance-based fault location for power distribution systems, IET Gener. Transm. Distrib. 5 (4) (2011) 467, doi:10.1049/iet-gtd.2010.0446.

[21] R. Dashti, M. Ghasemi, M. Daisy, Fault location in power distribution network with presence of distributed generation resources using impedance based method and applying Π line model, Energy 159 (2018) 344–360, doi:10.1016/j.energy.2018.06.111.

[22] R. Kumar, D. Saxena, A traveling wave based method for fault location in multi-lateral distribution network, Proceedings of 2018 IEEE International Conference on Power, Instrumentation, Control and Computing, PICC 2018, 2018, pp. 1–5, doi:10.1109/PICC.2018.8384749.

[23] J. Tang, X. Yin, Z. Zhang, Traveling-Wave-Based Fault Location in Electrical Distribution Systems With Digital Simulations, TELKOMNIKA (Telecommunication Comput. Electron. Control. 12 (2) (2014) 297, doi:10.12928/telkomnika.v12i2.67.

[24] F. Yan, W. Liu, L. Tian, Fault location for 10kV distribution line based on traveling wave-ANN theory, 2011 IEEE Power Engineering and Automation, Proceedings of Conference on, 2011, pp. 437–440, doi:10.1109/PEAM.2011.6135094.

[25] Y. Sheng-nan, Y. Yi-han, B. Hai, Study on fault location in distribution network based on C-type traveling-wave scheme, Relay 35 (10) (2007) 1–5.

[26] Y. Wang, T. Zheng, C. Yang, L. Yu, Traveling-wave based fault location for phase-to-ground fault in non-effectively earthed distribution networks, Energies 13 (19) (2020) 5028, doi:10.3390/en13195028.

[27] Y. Wei, P. Sun, Z. Song, P. Wang, Z. Zeng, X. Wang, Fault location of VSC based DC distribution network based on traveling wave differential current with Hausdorff distance and cubic spline interpolation, IEEE (2021), doi:10.1109/ACCESS.2021.3059935.

[28] O.O. Babayomi, P.O. Oluseyi, Intelligent fault diagnosis in a power distribution network, Adv. Electr. Eng. 2016 (2016) 1–10, doi:10.1155/2016/8651630.

[29] F. Dehghani, H. Nezami, A new fault location technique on radial distribution systems using artificial neural network, 22nd International Conference and Exhibition on Electricity Distribution (CIRED 2013), Proceedings of the Conference on, 2013, pp. 1–4, doi:10.1049/cp.2013.0697.

[30] P. Ray, D. Mishra, Artificial Intelligence Based Fault Location in a Distribution System, 2014 International Conference on Information Technology, Proceedings of the Conference on, 2014, pp. 18–23, doi:10.1109/ICIT.2014.10.

[31] H. Mokhlis, H.Y. Li, A.R. Khalid, The application of voltage sags pattern to locate a faulted section in distribution network, Int. Rev. Electr. Eng. 5 (2010) 173–179.

[32] S.R. Naidu, E. Guedes da Costa, G.V. Andrade, Fault location in distribution systems using the voltage sag-duration table, 11th IEEE/IAS International Conference on Industry Applications, 2014, pp. 1–7, doi:10.1109/INDUSCON.2014.7059401.

[33] Y.D. Mamuya, Y.-D. Lee, J.-W. Shen, M. Shafiullah, C.-C. Kuo, Application of machine learning for fault classification and location in a radial distribution grid, Appl. Sci. 10 (14) (2020) 4965, doi:10.3390/app10144965.

[34] M. Shafiullah, M.A. Abido, Z. Al-Hamouz, Wavelet-based extreme learning machine for distribution grid fault location, IET Gener. Transm. Distrib. 11 (17) (2017) 4256–4263, doi:10.1049/iet-gtd.2017.0656.
[35] M. Shafiullah, M. Ijaz, M.A. Abido, Z. Al-Hamouz, Optimized support vector machine & wavelet transform for distribution grid fault location 2017, 11th IEEE International Conference on Compatibility, Power Electronics and Power Engineering (CPE-POWERENG), IEEE, 2017, pp. 77–82, doi:10.1109/CPE.2017.7915148.
[36] MATLAB & Simulink, MathWorks - makers of MATLAB and Simulink, 2021. https://www.mathworks.com/. (Accessed 21 March 2021).
[37] MATLAB & Simulink, Simscape Electrical, 2021. https://www.mathworks.com/products/simscape-electrical.html. (Accessed 21 March 2021).
[38] M. Shafiullah, M.A.M. Khan, S.D. Ahmed, PQ disturbance detection and classification combining advanced signal processing and machine learning tools, in: P. Sanjeevikumar, C. Sharmeela, J.B. Holm-Nielsen, P. Sivaraman (Eds.), Power Quality in Modern Power Systems, Academic Press, Massachusetts, USA, 2021, pp. 311–335.
[39] A. Aljohani, T. Sheikhoon, A. Fataa, M. Shafiullah, and M.A. Abido, Design and implementation of an intelligent single line to ground fault locator for distribution feeders, 2019, doi:10.1109/ICCAD46983.2019.9037950.
[40] M. Shafiullah, M. Abido, T. Abdel-Fattah, Distribution grids fault location employing ST-based optimized machine learning approach, Energies 11 (9) (2018) 2328, doi:10.3390/en11092328.
[41] A. Aljohani, A. Aljurbua, M. Shafiullah, M.A. Abido, Smart fault detection and classification for distribution grid hybridizing ST and MLP-NN, 15th International Multi-Conference on Systems, Signals & Devices (SSD), IEEE, Hammamet, Tunisia, 2018, pp. 1–5.
[42] M. Shafiullah, M.A. Abido, S-Transform based FFNN approach for distribution grids fault detection and classification, IEEE Access 6 (2018) 8080–8088, doi:10.1109/ACCESS.2018.2809045.
[43] Y. Liao, Generalized fault-location methods for overhead electric distribution systems, IEEE Trans. Power Deliv. 26 (1) (2011) 53–64, doi:10.1109/TPWRD.2010.2057454.
[44] K.V Lout, Development of a fault location method based on fault induced transients in distribution networks with wind farm connections, University of Bath, 2015.
[45] A.C. Adewole, Investigation of methodologies for fault detection and diagnosis in electric power system protection, Cape Peninsula University of Technology, 2012.
[46] A. Dwivedi, X. Yu, Fault location in radial distribution lines using travelling waves and network theory, 2011 IEEE International Symposium on Industrial Electronics, 2011, pp. 1051–1056, doi:10.1109/ISIE.2011.5984305.
[47] S. Canu, Y. Grandvalet, V. Guigue, A. Rakotomamonjy, SVM and kernel methods matlab toolbox, Percept. Systmes Inf. 69 (2003) 70. https://scholar.google.com/scholar?cluster=641952285995963962&hl=en&as_sdt=2005#0. (Accessed 24 February 2016).
[48] MATLAB codes of ELM algorithm. http://www.ntu.edu.sg/home/egbhuang/elm_kernel.html. (Accessed 29 February 2016).
[49] C.D. Lewis, Industrial and Business Forecasting Methods: A Practical Guide to Exponential Smoothing and Curve Fitting, Butterworth Scientific, Oxford, UK, 1982.
[50] D. Moriasi, J. Arnold, M. Van Liew, Model evaluation guidelines for systematic quantification of accuracy in watershed simulations, Trans. (2007). http://elibrary.asabe.org/abstract.asp?aid=23153. (Accessed 13 January 2017).
[51] C.J. Willmott, S.M. Robeson, K. Matsuura, A refined index of model performance, Int. J. Climatol. 32 (13) (2012) 2088–2094, doi:10.1002/joc.2419.

[52] S. Sorooshian, Q. Duan, V.K. Gupta, Calibration of rainfall-runoff models: application of global optimization to the sacramento soil moisture accounting model, Water Resour. Res. 29 (4) (1993) 1185–1194, doi:10.1029/92WR02617.

[53] M. Clerc, The swarm and the queen: towards a deterministic and adaptive particle swarm optimization, Proceedings of the 1999 Congress on Evolutionary Computation, 3 (1999). http://ieeexplore.ieee.org/abstract/document/785513/. (Accessed 5 October 2018).

[54] R. Eberhart, Y. Shi, Comparing inertia weights and constriction factors in particle swarm optimization, Proceedings of the 2000 Congress on Evolutionary Computation, 1, 2000 84–88. http://ieeexplore.ieee.org/abstract/document/870279/. (Accessed 3 October 2018).

[55] M. Clerc, J. Kennedy, The particle swarm-explosion, stability, and convergence in a multi-dimensional complex space, IEEE Trans. Evol. (2002). http://ieeexplore.ieee.org/abstract/document/985692/. (Accessed 30 June 2018).

[56] X. Ji, M. Li, W. Li, Constriction factor particle swarm optimization algorithm with overcoming local optimum, Comput. Eng. (2011). http://en.cnki.com.cn/Article_en/CJFDTOTAL-JSJC201120073.htm. (Accessed 23 September 2018).

[57] M. Shafiullah, M.I. Hossain, M.A. Abido, T. Abdel-Fattah, A.H. Mantawy, A modified optimal PMU placement problem formulation considering channel limits under various contingencies, Meas. J. Int. Meas. Confed. 135 (2019), doi:10.1016/j.measurement.2018.12.039.

[58] N. Wang, V. Aravinthan, Y. Ding, Feeder-level fault detection and classification with multiple sensors: a smart grid scenario, 2014, IEEE Workshop on Statistical Signal Processing (SSP). IEEE, 2014, pp. 37–40, doi:10.1109/SSP.2014.6884569.

[59] U.D. Dwivedi, S.N. Singh, S.C. Srivastava, A wavelet based approach for classification and location of faults in distribution systems. In: Annual IEEE India Conference, vol. 2, IEEE, 2008, pp. 488–493, doi:10.1109/INDCON.2008.4768772.

[60] S. Jana, G. Dutta, Wavelet entropy and neural network based fault detection on a non radial power system network, IOSR J. Electr. Electron. Eng. 2 (3) (2012) 26–31.

[61] J. Zhang, Z.Y. He, S. Lin, Y.B. Zhang, Q.Q. Qian, An ANFIS-based fault classification approach in power distribution system, Int. J. Electr. Power Energy Syst. 49 (2013) 243–252, doi:10.1016/j.ijepes.2012.12.005.

Smart grid fault diagnosis under load and renewable energy uncertainty

11

11.1 Introduction

The energy-related carbon dioxide gas emission is likely to be increased from 33.6 billion metric tons in 2018 to 48.07 billion metric tons in 2050 as per the International Energy Outlook reports [1–4]. In addition to the serious environmental pollution problem, the shortage of conventional energy resources has compelled many countries to explore alternative energy resources to reduce their dependencies on traditional energy resources and ensure sustainable social and economic development [5]. Although nuclear resources are environmentally friendly but due to the vulnerabilities and insecurities of atomic power, the call for nuclear-free energy production has appeared many times in recent years [6]. The researchers and the unions highlighted two critical elements: improving the energy efficiency and augmentation of renewable energy resources (RERs) shares in the energy generation mix to accelerate the decarbonization move and produce secure energy. According to U.S. Energy Information Administration projection, almost half of the world's electricity will be delivered from the RER by 2050 [7]. The overall share of RER (modern renewables and traditional biomass) was around 18.1% of total final energy consumption globally in 2017, with an annual average increase of 0.8% since 2006 [8]. As of 2018, global RER power capacity reached 2378 GW due to billions of dollars invested in renewable power generation. Although they are mostly connected to the transmission networks, a good portion of them is also connected through the electric distribution networks. Some RER portions are added into the mini-grids and off-grid solutions to deliver cost-effective electricity access to rural people [8].

Like RERs, integration of non-RERs and energy storage systems have been increasing tremendously for several technical and economic factors, including reduction of greenhouse gases and transmission power losses, improvement of voltage profile and phase imbalance, and injecting reactive power getting integrated into in addition to RERs [9–15]. Such resources (energy storage systems, renewable, and non-RERs) are referred to as the distributed generations. Their deployment into the distribution networks disturbs their radial nature and transforms them from passive to active networks. Their operation and control schemes become more complicated than traditional passive distribution networks as their power flow changes from unidirectional to multidirectional. Furthermore, the intermittent nature of the leading RER, for example, solar and wind, affects the reliability and stability of the distribution networks. The mentioned issues affect the functionality and accuracy of the traditional protection devices and schemes (protective relays, fault diagnosis schemes, etc.) [16–21]. Therefore, it has become inevitable to upgrade the available protection schemes or design new protection schemes considering the recent and future trends of RER adoption into the smart distribution networks.

In this chapter, the applicability of the presented intelligent fault diagnosis (IFD) scheme combining the advanced signal processing and machine learning tools in Chapter 10 is extended on the IEEE 13-node test distribution feeder integrated with intermittent RER. Step-by-step modeling of the selected test feeder in the RSCAD software and simulation of the modeled feeder in the real-time digital simulator (RTDS) rack is demonstrated. This chapter also presents the load demand and renewable energy generation uncertainty modeling using the appropriate probability density function (PDF). Then, it illustrates the faulty data generation and recording processes using phasor measurement units (PMUs). Besides, it presents the obtained results (fault detection, fault classification, faulty section identification, and fault location) using the IFD scheme based on advanced signal processing and machine learning tools. Moreover, this chapter investigates the IFD scheme efficacy in the presence of measurement noise and under contingencies, for example, branch and distributed generator (DG) outages. Finally, it presents the IFD scheme validation results obtained from the RSCAD recorded and physical PMU retrieved data through RTDS Giga-Transceiver Analogue Output (GTAO) card.

The remaining parts of this chapter are organized as follows: Section 11.2 illustrates the step-by-step modeling of the test distribution feeder. Section 11.3 presents load and renewable energy generation uncertainty modeling. Section 11.4 demonstrates the fault modeling and feature extraction processes. Sections 11.5 and 11.6 present the discussions on the obtained simulated and experimental fault diagnosis results, respectively. Finally, an overall summary of this chapter is presented in Section 11.7.

11.2 IEEE 13-node test distribution feeder modeling

The distribution feeders are passive networks with unidirectional power flow from the source to the load centers. They consist of the main feeder, distribution transformers, laterals and sublaterals, spot and distributed loads, shunt capacitor banks, overhead distribution lines, and underground cables. The IEEE Power & Energy Society (IEEE-PES) published several benchmark feeders with their configurations and parameters and made them available in [22]. This chapter selects the IEEE 13-node test feeder as a single rack RTDS can simulate power system grids with a limited number of nodes. The chosen test feeder is operated at 4.16 kV and displays most of the characteristics of electrical power distribution networks. The highly loaded test distribution feeder consists of a voltage regulator at the substation, an inline transformer, shunt capacitor banks, overhead distribution lines, underground cables of various configurations, several spots, and distributed loads (balanced and unbalanced). Besides, it contains single-phase, double-phase, and

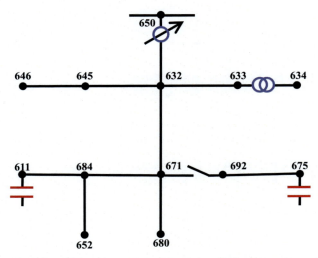

Fig. 11.1 The SLD of the IEEE 13-node benchmark distribution test feeder [22]. *SLD*, single line diagram.

three-phase laterals. A single line diagram (SLD) of the selected test feeder is shown in Fig. 11.1. The following subsections present step-by-step modeling of the IEEE 13-node test feeder in the RSCAD environment and implement the developed model in the RTDS machine. This book appends a brief discussion on RSCAD software and the RTDS machine and the technical parameters of the selected feeder in the appendix of this book.

11.2.1 Main feeder, nodes, and connecting wires

The source models often represent some portion of the electric network in a simplified way that generates three-phase alternating current (AC) power behind an internal impedance. Fig. 11.2A represents the main feeder RSCAD model of the IEEE 13-node distribution feeder that generates three-phase power for the feeder at 115 kV and 60 Hz. However, an upstream 5 MVA, 115 kV (Δ) /4.16 kV (Y-g) transformer connects the main power supply to the grid through node number 650. The connecting wires join the different components, including the main feeder, distribution lines, underground cables, transformers, loads, and capacitor banks to different nodes. Fig. 11.2B and C present the RSCAD model of the node and connecting wires, respectively.

11.2.2 Transformers

There are two transformers to step down the voltages of the selected test feeders. The 5-MVA substation transformer, placed before node 650, steps down 115 kV three-phase power supply to 4.16 kV. Conversely, the 0.5-MVA inline transformer is

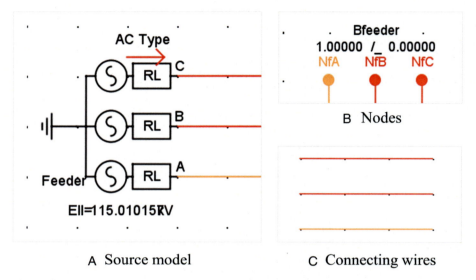

Fig. 11.2 The source model, nodes, and connecting wires in RSCAD software.

placed between nodes 633 and 634, which steps down the voltage level from 4.16 kV to 0.48 kV. This chapter models both transformers with the available "three-phase two-winding transformer" in the RSCAD library. The presented modeling divides the transformer impedances between the primary and secondary windings. Fig. 11.3 shows the RSCAD models of the used transformers.

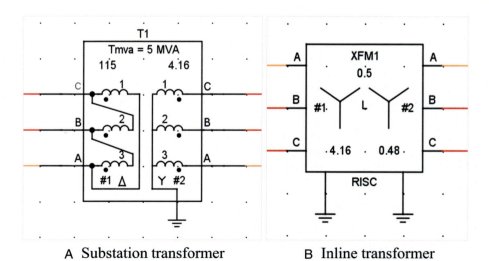

Fig. 11.3 Transformer models.

11.2.3 Voltage regulator

The voltage regulator of the selected test feeder is located between nodes 650 and 632 that is modeled with a "three-phase two winding transformer" as there is no regulator model in the RSCAD library. The primary and secondary side voltages of the transformer are set to the input and output voltages of the regulators, respectively.

11.2.4 Distribution lines and underground cables

The selected test feeder has one main distribution line (650-632-671-680), four laterals (632-633-634, 632-645-646, 671-692-675, and 671-684), and two sub-laterals (684-611 and 684-652). This chapter models the main distribution line with a 601-type overhead line configuration. Additionally, it models three laterals with other overhead line configurations and the remaining lateral with an underground cable. Also, it models one sub-lateral with an overhead line and the other one with an underground cable. The line configurations determine the types of power supply (three-phase, double-phase, or single-phase), the spacing between the conductors, overhead line or underground cable, and impedances. However, the Ref. [22] provides the impedances in a phase impedance matrix form. Thischapter models all the overhead line segments employing π-model (~pi-model) since the selected test feeder comprises short lines. The π-model of the RSCAD library uses positive and zero sequence impedances for the line segments instead of the phase impedance matrix. Consequently, this chapter uses the modified Carson's equations and Kron reduction to obtain the sequence impedances [23] from the given configuration. The following equation defines the phase impedance matrix of a specific line:

$$[Z_{abc}] = \begin{bmatrix} Z_{aa} & Z_{ab} & Z_{ac} \\ Z_{ba} & Z_{bb} & Z_{bc} \\ Z_{ca} & Z_{cb} & Z_{cc} \end{bmatrix} \Omega/\text{mile} \quad (11.1)$$

Where, Z_{aa}, Z_{bb}, and Z_{cc} are the self-impedances of phases a, b, and c, respectively, and the $Z_{ab} = Z_{ba}$, $Z_{bc} = Z_{cb}$, and $Z_{ac} = Z_{ca}$ are the mutual impedances between the phases a-b, b-c, and c-a, respectively.

The following equation converts the phase impedances into the sequence impedances:

$$[Z_{012}] = [A_s]^{-1}[Z_{abc}][A_s] = \begin{bmatrix} Z_{00} & Z_{01} & Z_{02} \\ Z_{10} & Z_{11} & Z_{12} \\ Z_{20} & Z_{21} & Z_{22} \end{bmatrix} \Omega/\text{mile} \quad (11.2)$$

Where, $[A_s] = \begin{bmatrix} 1 & 1 & 1 \\ 1 & a_s^2 & a_s \\ 1 & a_s & a_s^2 \end{bmatrix}$; $[A_s]^{-1} = \dfrac{1}{3}\begin{bmatrix} 1 & 1 & 1 \\ 1 & a_s & a_s^2 \\ 1 & a_s^2 & a_s \end{bmatrix}$; $a_s = \cos 120^0 + j \sin 120^0$;

$j = \sqrt{-1}$

The diagonal elements of the 3 × 3 matrix of the above equation are the sequence impedance of the line such that: Z_{00} refers to zero sequence impedance, Z_{11} refers to positive sequence impedance, and Z_{22} refers to negative sequence impedance. The off-diagonal elements of the equation represent the mutual coupling between the sequences, and these would be zero in an ideal situation for the transposed line. The off-diagonal elements for the distribution lines sequence matrices are not zero for most cases as they are rarely transposed. Alternatively, for a transposed line, the three diagonal elements of the phase impedance matrix can be made equal, and all the off-diagonal elements can be made equal. In the newly formed phase impedance matrix, the diagonal elements are set to the average value of the diagonal elements of the previously mentioned impedance matrix. Similarly, the off-diagonal elements are assigned to the average of the off-diagonal elements. Eventually, zero, positive and negative sequence impedances of the sequence impedance matrix are calculated [23]. The self and the mutual impedances of the newly formed impedances can be found from the following equations:

$$Z_s = \frac{Z_{aa} + Z_{bb} + Z_{cc}}{3} \; \Omega/\text{mile} \quad (11.3)$$

$$Z_m = \frac{Z_{ab} + Z_{bc} + Z_{ca}}{3} \; \Omega/\text{mile} \quad (11.4)$$

From the equation presented above, the zero, positive and negative sequence impedances are calculated employing the following equations:

$$Z_{00} = Z_s + 2\,Z_m \; \Omega/\text{mile} \quad (11.5)$$

$$Z_{11} = Z_{22} = Z_s - Z_m \; \Omega/\text{mile} \quad (11.6)$$

After calculating the sequence impedances, they are plugged into the π-model distribution lines of the RSCAD library.

11.2.5 Loads

The selected test feeder has eight-spot loads connected to different nodes, and their summation for the phases A, B, and C are 1158 kW, 973 kW, and 1135 kW, whereas the total reactive loads are 606 kVAr, 627 kVAr, and 753 kVAr, respectively. Besides, the feeder has a three-phase distributed load between nodes 632 to 671. The active loads for phases A, B, and C are 17 kW, 66 kW, and 117 kW, whereas the reactive loads are 10 kAVr, 38 kVAr, and 68 kVAr, respectively. The loads are three-phase and single-phase types and modeled as PQ, constant impedance, and constant current loads. This chapter models all of them employing the "dynamic load" model of the RSCAD library and uses sliders to incorporate their uncertainty.

11.2.6 Shunt capacitors

The IEEE 13-node test distribution feeder has one three-phase capacitor and one single-phase shunt capacitor connected at node 675 and node 611, respectively. These capacitors are placed at the extreme ends of the feeder to control the voltage.

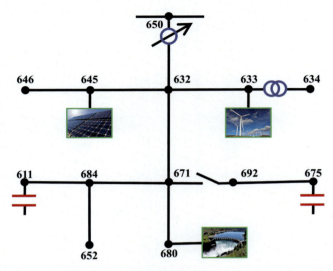

Fig. 11.4 The SLD of the IEEE 13-node test feeder after incorporation of the RER. *RER*, renewable energy resource; *SLD*, single line diagram.

11.2.7 Renewable energy resources as distributed generators

This chapter incorporates three RER as distributed generators, namely solar, wind, and hydropower generators at nodes 645, 633, and 680, respectively. This work chooses the mentioned nodes and DG type based on the experience of the literature [24]. However, the hydropower plant is considered a fixed power supplier to the test feeder. In contrast, the photovoltaic (PV) and the wind power plants supply power-based solar irradiation and wind speed using sliders from the RSCAD library. Fig. 11.4 presents the SLD of the IEEE 13-node test feeder after incorporating three RERs, and Fig. 11.5 shows the final RSCAD model of the selected test feeder.

11.3 Load and renewable energy uncertainty modeling

This section presents the background of the load demand and renewable energy generation uncertainty. Then, it illustrates the adopted uncertainty modeling approaches for the load demand, renewable energy generation, and fault information (inception angle and resistance).

11.3.1 Uncertainty background

In electricity grids, the primary operational uncertainty usually comes from the demand side. However, the integration of RER throughout the grids changes the scenario and provides uncertainties from both demand and generation sides. For

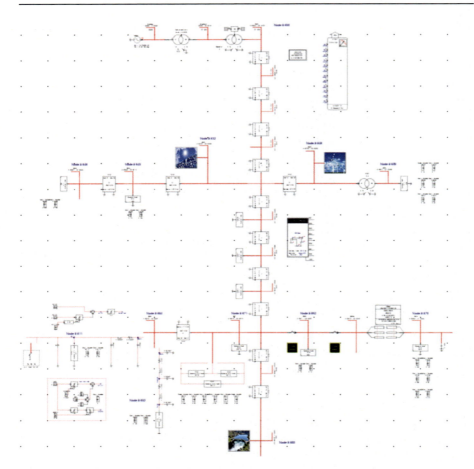

Fig. 11.5 IEEE 13-node test distribution feeder model incorporated with RER in RSCAD environment. *RER*, renewable energy resource.

instance, the performance of PV systems strongly depends on several factors, including solar irradiation, ambient temperature and humidity, dust, wind speed, PV panel orientation, PV array model, and inverter model [25–28]. Likewise, the wind speed uncertainty is usually demonstrated by the term turbulence intermittency, defined as the uneven distribution characteristics of a few statistical parameters of the wind in space and time [29]. Conversely, the load demand was mostly determined by calendar variables, as the lighting was the only end-use of electricity grids at their early commencement. However, the penetration of diversified electricity-powered appliances, including the air conditioning and refrigeration systems, to the electric networks made the role of weather (temperature and humidity) more significant in determining the load demand [30]. For instance, Germany (a winter peaking country) had an annual peak demand of 83.1 GW in

2013, whereas the lowest and the average demands of the same year were 32.47 GW and 65.75 GW, respectively [31]. Similarly, the annual peak, the lowest, and the average demands of Saudi Arabia (a summer peaking country) in 2014 were 56.5 GW, 28.25 GW, and 32.07 GW, respectively [32]. Likewise, each electricity grid experiences different peak, lowest, and average loads throughout the year. The provided information confirms that the electricity consumption for any specific grid is not constant throughout the year, even throughout a day; rather, it varies at every moment. However, the load demand and renewable energy generation uncertainty can severely affect the operation and performance of the electricity grids, sometimes may induce grid instability and lead either towards lack or waste of energy if they are not addressed in time with care [33]. Furthermore, the RER intermittencies are the biggest challenges in implementing them as reliable power supply sources [34]. Therefore, it is crucial to incorporate the intrinsic uncertainty of different planning parameters to make the electricity grids less sensitive to the variations of the grid parameters.

The researchers predicted the wind power output based on the estimated wind speed and related power curves throughout the years. However, there is no ideal strategy to forecast wind speed as it is one of the most challenging meteorological items to predict. Each strategy has its pros and cons that may be suitable in some contexts and not suitable in other cases [35]. Stochastic behavior of the wind speeds was modeled employing many approaches, including the Weibull PDF [36–39], Rayleigh PDF [40], Monte Carlo method [41], time series analysis [42], statistical method [43], and artificial intelligence [44]. Then, in the next step, the output power of the wind turbines (WTs) was calculated from the wind speed-associated power curves. The power curves are mathematical functions provided by the manufacturers to relate wind speeds with WT plants output powers precisely [45]. Similarly, the stochastic behavior of the solar irradiations was modeled using the beta PDF [26,38–40], Weibull PDF [37], probabilistic two-point estimation method [46], and artificial intelligence [47]. Likewise, the stochastic behavior of the load demand of distribution grids was modeled using the normal PDF [26,37,39,48], lognormal PDF [49], time series analysis [42], and artificial intelligence [50].

Among many researchers, Mojtahedzadeh et al. [39], considered ±10% deviation of the predicted peak values of the load demand, active power outputs of PV and WT power plants while enhancing resiliency and reliability through self-healing of a community micro-grid. Additionally, Wang et al. [51] considered a maximum of ±10% and ±15% relative errors of load demand and active power output of PV power plant, respectively, while modeling a robust voltage control for an active distribution grid. Besides, Chen et al. considered ±16% and ±20% variations of active power generation of the PV power plant and load demand, respectively, while proposing a robust restoration approach of anactive distribution grid [52]. However, the reported reactive power of a solar power plant was negative and proportional (~7%) to the active power production throughout the day, as reported in [28]. In contrast, the generated reactive power of the WT power plants was positive and approximately 45% to 65% of generated active power according to the literature [53].

11.3.2 Adopted uncertainty modeling approaches

Uncertain parameters of the modeled IEEE 13-node test distribution feeder are the power outputs of the WT and the PV power plants and the load demand. The load demand and the DG power generation uncertainties are incorporated into the test distribution feeder with probabilistic approaches, as the studies proved the lower operational cost of the probabilistic approaches over the deterministic approaches [40]. This chapter modelsthe load demand uncertainty employing the Gaussian PDF as given by the following equation:

$$f(x|\mu,\sigma) = \frac{1}{\sigma\sqrt{2\pi}} e^{-\frac{(x-\mu)^2}{2\sigma^2}}; \quad -\infty < x < \infty, \quad \mu \langle \infty, \quad \sigma \rangle 0 \quad (11.7)$$

Where, μ and σ are the mean and the standard deviation of the load demand, respectively. This study relates the variables μ, σ, and C_v using the following formula:

$$C_v = \frac{3\sigma}{\mu} \times 100\% \quad (11.8)$$

Where, C_v is the coefficient of variation. This chapter chooses μ and C_v as the rated active loads and ±15% of the rated loads, respectively. On the contrary, this chapter models the uncertainties associated with the wind speed and the solar irradiation employing the Weibull PDF as given below:

$$f(x|\alpha,\beta,\gamma) = \frac{\beta}{\alpha}\left(\frac{x-\gamma}{\alpha}\right)^{\beta-1} e^{-\left(\frac{x-\gamma}{\alpha}\right)^{\beta}}; \quad x \geq \gamma, \quad \alpha, \beta > 0 \quad (11.9)$$

Where, α, β, and γ scale, shape, and location parameters, respectively. The case where $\gamma = 0$ is known as the two parameters Weibull PDF that can be written as:

$$f(x|\alpha,\beta) = \frac{\beta}{\alpha}\left(\frac{x}{\alpha}\right)^{\beta-1} e^{-\left(\frac{x}{\alpha}\right)^{\beta}}; \quad x \geq 0, \quad \alpha, \beta > 0 \quad (11.10)$$

This chapter selects the scale (α) and the shape (β) parameters through a backward iterative process from two other variables (μ and σ) of the wind speed and the solar irradiation. Here, μ is calculated as the rated wind speed and the rated solar irradiation from the rated power outputs of the WT and the PV plants. The coefficient of variation (C_v) is assumed to be ±10% of the rated wind speed and the rated solar irradiation. Based on the Weibull PDF predicted wind speed, the output power of the WT plant could be calculated using the following equation [37,39,40]:

$$P_w(v) = \begin{cases} 0; & 0 \leq v \leq v_{ci} \quad \text{or} \quad v_{co} \leq v \\ P_{w\,rated} \times \dfrac{v - v_{ci}}{v_r - v_{ci}}; & v_{ci} \leq v \leq v_r \\ P_{w\,rated}; & v_{co} \leq v \end{cases} \quad (11.11)$$

Where, v_{ci}, v_r, v, and v_{co} are cut-in speed, rated speed, Weibull PDF predicted speed, and the cut-off speed of the WT, respectively.

Similarly, the output power of the PV plant can be calculated from the Weibull PDF predicted solar irradiation as [37]:

$$P_s(s) = \begin{cases} P_{s\,rated} \times \dfrac{G}{G_r}; & 0 \leq G \leq G_r \\ P_{s\,rated}; & G \geq G_r \end{cases} \quad (11.12)$$

Where, G_r and G are rated solar irradiation, and Weibull PDF predicted solar irradiation, respectively. However, the reactive power outputs of PV and WT plants are assumed as negative (~7%) and positive (~50%) of their generated active powers, respectively. Besides, the real and reactive powers are set to some constant values for the hydropower plant installed at node 680. Like the load demand and the DG generation uncertainties, this chapter chooses the fault information (resistance and inception angle) using the following formula:

$$R \sim U(R_{min}, R_{max}) \quad (11.13)$$

$$A_I \sim U(A_{min}, A_{max}) \quad (11.14)$$

Where, U is the uniform distribution; R, R_{min}, and R_{max} are the modeled, minimum, andmaximum values of the fault resistances, respectively; A_I, A_{min}, and A_{max} are the modeled, minimum, and maximum values of the fault inception angles, respectively.

Table 11.1 summarizes the load demand, renewable energy generation, and fault information uncertainties modeling. In this study, the load values of the selected test feeder with a variation of ±15% from the rated load using Gaussian PDF are set. Likewise, it picks wind speed and solar irradiation with ±10% variation of the rated wind speed and solar irradiation using Weibull PDF. Similarly, it employs uniform distribution to pick fault resistance values from 0 Ω to 50 Ω and fault inception angle at any point between the starting and end of the current signal cycle. However, it considers the hydropower plant as a fixed power supplier.

Table 11.1 Summary of load demand, renewable energy generation, and fault information uncertainties modeling.

Item	Variations
Load (All)	±15%
Wind power (500 kW)	±10%
Solar power (300 kW)	±10%
Hydropower (300 kW)	No variations
Fault resistance	(0 Ω to 50 Ω)
Fault inception angle	(0 to 1 cycle)

11.4 Fault modeling and feature extraction

This section briefly introduces the power flow analysis of the modeled distribution feeder that is the primary step for any further investigation of any modeled system in the RSCAD platform. Then, it illustrates the fault modeling and data recording process as required for the IFD scheme.

11.4.1 Power flow analysis

Systematic modeling of the IEEE 13-node test distribution feeder incorporated with distributed generators in the RSCAD/RTDS platform was demonstrated in Section 11.2. After successfully modeling the test network, this chapter performed the power flow analysis of the modeled feeder. The lowest node voltage (0.98817 pu) was observed at node number 634, whereas the highest value of the node voltage (1.0006 pu) was at the substation node. The acceptable values of the node voltages provided the signal to proceed with the modeled test distribution feeder. Afterward, this chapter performed the compilation to create an execution code of the modeled system for the RTDS machine by checking primary errors associated with the components and the simulation parameters. Eventually, this chapter made the modeled system for real-time simulation and disturbance recording ready from the "runtime" file.

11.4.2 Fault modeling and data recording

This chapter used script files for the "batch-mode" operation to simulate several simulation cases and to record required signals of the same fault type on the same location automatically without any manual interaction, as depicted in Fig. 11.6. It wrote the script files in a "C" like programming language that incorporated load demand, fault resistance and inception angle, and renewable energy generation uncertainties.

It is worth mentioning that this chapter applied seven different types of faults (AG, BG, CG, ABG, BCG, CAG, and ABCG) on 21 locations of the modeled test feeder as depicted in Fig. 11.7. Besides, each fault was applied for four cycles on the modeled test feeder in the RSCAD/RTDS environment. Fig. 11.8 presents the experimental setup used for test feeder modeling and data recording processes. In this study, data isrecorded using two different approaches. The first approach includes the RSCAD recorded data through "batch-mode" operation, whereas the second approach includes physical PMU recorded data. Both approaches recorded three-phase fault current data for two-cycle (one precycle and one postcycle) with a sampling frequency of 10 kHz (~167 samples/cycle).

This chapter divided RSCAD recorded data (data obtained from the first approach) into three different groups. Firstly, it employed an RSCAD built-in "PMU" to record the magnitudes and the phase angles of three-phase currents (I_A, I_B, and I_C) and associated sequence currents (I_1, I_2, and I_0) of eight selected branches (650 to 632, 632 to 671, 632 to 633, 632 to 645, 671 to 680, 671 to 684, 684 to 652, and 671 to 675) as referred to first category data. Secondly, the "runtime" file recorded two-cycle (precycle and postcycle) three-phase current signals of the main feeder and three

Smart grid fault diagnosis under load and renewable energy uncertainty 305

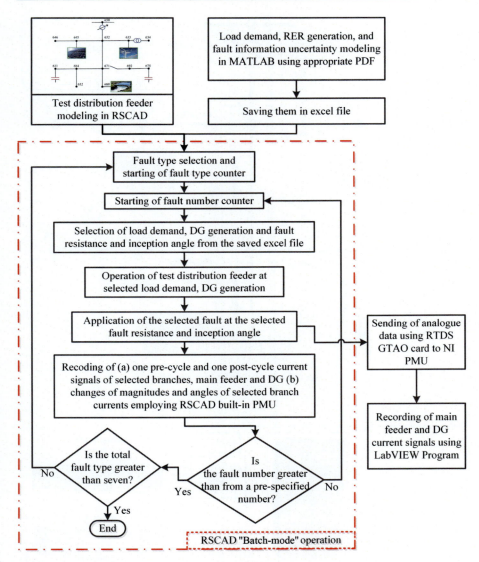

Fig. 11.6 RSCAD "batch-mode" operational flowchart to incorporate fault-related uncertainties and data recording.

DGs and referred to as second category data. Thirdly, the "*runtime*" file also recorded two-cycle (precycle and postcycle) three-phase current signals of the previously mentioned eight branches (third category data).

On the other hand, it is worth noting that the available physical PMU in the "Power System Research Lab" at King Fahd University of Petroleum and Minerals (KFUPM) is limited to measuring 12 signals only. Therefore, this chapter recorded the three-phase current signals of the main feeder and three DGs (fourth category

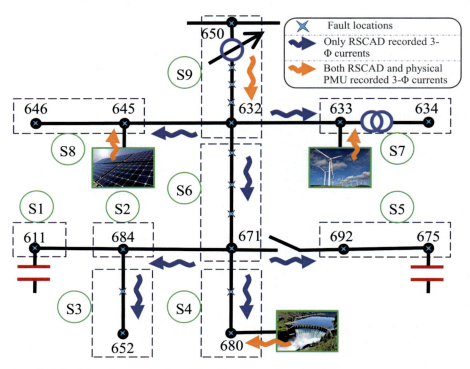

Fig. 11.7 Fault locations and selected branch currents for recording.

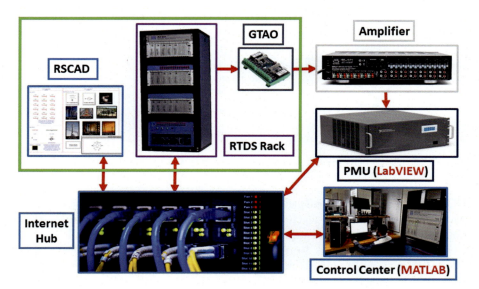

Fig. 11.8 Experimental setup used for test feeder modeling and data recording.

data) employing the physical PMU instead of measuring the branch currents. Selected current signals from the RSCAD/RTDS environment were sent through the GTAO card to the PMU and were saved using the Laboratory Virtual Instrument Engineering Workbench (LabVIEW) program. This chapter utilized the physical PMU recorded data to test the validity of the developed IFD scheme.

11.4.3 Feature extraction

As stated earlier, the RSCAD recorded first category data contained the magnitudes and the phase angles of three-phase and sequenced current signals of eight different branches. Hence, there was no need to apply any signal processing approach to collect their features; instead, this study considered the changes in those values before and after the occurrence of faults as the fault characteristics signatures. Consequently, it collected 96 (~3*2*8 for the magnitudes + 3*2*8 for the phase angles) features for each faulty case from the first data category.

Conversely, the discrete wavelet transform (DWT) approach collected 576 (~ 4 sets of three-phase current signals and 48 features from each phase current signal) and 1152 features (~ 8 sets of three-phase current signals and 48 features from each phase current signal) for each faulty scenario from the second and third categories data, respectively. Detailed illustrations of the DWT-based feature extraction process can be found in Chapter 4 and Ref. [54–58]. Likewise, the ST approach collected 144 (~4 sets of three-phase current signals and 12 features from each phase current signal) and 288 features (~8 sets of three-phase current signals and 12 features from each phase current signal) from the second and third categories of data, respectively. Detailed illustrations of the ST-based feature extraction process can be found in Chapter 4 and Ref. [58–62].

11.5 Fault diagnosis results and discussions

Previous sections of this chapter illustrated the detailed modeling of the IEEE 13-node test distribution feeder incorporated with intermittent RERs in the RSCAD/RTDS platform. Then, it demonstrated the fault modeling, data recording, and feature extraction processes using advanced signal processing techniques. This section presents the fault diagnosis results of the IFD techniques under different scenarios.

11.5.1 Fault detection

This chapter generated 1050 faulted scenarios comprising of SLG (AG, BG, and CG), LLG (ABG, BCG, and CAG), and LLLG (ABCG) faults on the RSCAD/RTDS environment as stated earlier for detection purposes. It is worth mentioning that the mentioned faults were applied on random places of the IEEE 13-node test distribution feeder, varying prefault loading conditions, DG generation, and fault resistance, inception angle, and location. After generating faulty cases, it collected the features associated with faults employing different techniques as stated

Table 11.2 Multilayer perceptron neural network parameters for fault detection approach.

Item	Specifications for the different data category
Number of hidden neurons	First: 9; second: 7 (DWT) and 6 (ST); third: 8 (DWT) and 7 (ST)
Squashing function	Tan-sigmoid
Training algorithm	"trainrp"
Objective function	Mean squared error
Training data	70%
Testing data	30%
Network name	Pattern recognition network ("patternnet")

DWT, discrete wavelet transform.

in Section 11.4.3. Likewise, this chapter collected similar features for 1050 nonfaulty cases by incorporating load demand and DG generation uncertainties. After generating faulty cases and collecting features, this chapter employed multilayer perceptron neural networks (MLP-NN) to distinguish the faulty signals from their nonfaulty counterparts. Table 11.2 summarized the technical characteristics of the MLP-NN used for fault detection. The technique tested the neural networks with many hidden neurons and chose the best configurations in terms of minimum mean squared error (MSE) and overall accuracy. Besides, it chose tan sigmoid as squashing functions and resilient backpropagation ("trainrp") as a training algorithm through a systematic trial and error process. It can be observed from the table that most of the hyperparameters of the MLP-NN were the same except for the number of neurons in the hidden layer.

11.5.1.1 Fault detection employing features collected from the first category data

Table 11.3 presents the overall confusion matrix of fault detection problem employing the features collected from the RSCAD built-in PMU recorded data (first category) into the MLP-NN with hyperparameters stated in Table 11.2. The diagonal and off-diagonal elements of the confusion matrix represent successful and unsuccessful detection of faults, respectively. It can be observed that the fault detection approach was able to differentiate the faulty cases from their nonfaulty counterparts with an

Table 11.3 Fault detection results based on the features collected from RSCAD built-in PMU recoded data (first category).

Item	Faulty cases	Nonfaulty cases
Faulty cases	923	127
Nonfaulty cases	0	1050
Overall accuracy 93.952%		

PMU, phasor measurement unit.

Table 11.4 Fault detection results based on the characteristic features of the main feeder and three DGs current signals (second category data).

| Item | Samples detected successfully ||||||||
| | Noise free || 40 dB SNR || 30 dB SNR || 20 dB SNR ||
	DWT	ST	DWT	ST	DWT	ST	DWT	ST
Faulty cases	1039	1041	1027	1035	1019	1031	992	1025
Nonfaulty cases	1050	1050	1050	1048	1042	1047	1041	1043
Overall accuracy (%)	99.47	99.57	98.90	99.19	98.14	98.95	96.80	98.47

DWT, discrete wavelet transform.

accuracy of 93.952% only. However, the presented dissatisfactory results of this section drove the authors to look for other promising techniques.

11.5.1.2 Fault detection employing features collected from the second category data

This section fetched the DWT and ST extracted characteristic features collected from the second category data (main feeder and three DGs current signals) into the MLP-NN. Table 11.2 presents the hyperparameters of the employed neural network, and Table 11.4 shows the fault detection results. Itcan be seen thatthe faulty cases from the nonfaulty cases were differentiated with almost 100% accuracy. Additionally, the technique added additive white Gaussian noise (AWGN) to the recorded pure three-phase current signals and extracted features employing DWT and ST-based approaches to test the efficacy of the trained MLP-NN under the noisy environment. Both DWT and ST-based MLP-NN approaches were able to separate faulty data from the non-faulty data with almost similar accuracy in the presence of a different level of measurement noises. However, the ST-based approach outperformed the DWT approach in terms of overall accuracy.

11.5.1.3 Fault detection employing features collected from the third category data

Likewise, this section fetched the DWT and ST extracted characteristic features collected from the third category data (eight-branch current signals) into the MLP-NN. Table 11.2 presents the hyperparameters of the employed neural network, and Table 11.5 shows the fault detection results. It can be seen that the developed approach successfully differentiated the faulty cases from their nonfaulty counterparts, even in the presence of measurement noises. Although both signal processing techniques exhibited similar accuracy in fault detection, the ST-based MLP-NN approach outperformed the DWT-based MLP-NN approach in overall accuracy.

Table 11.5 Fault detection results based on the characteristic features collected from the third category data (eight-branch current signals).

Item	Samples detected successfully							
	Noise free		40 dB SNR		30 dB SNR		20 dB SNR	
	DWT	ST	DWT	ST	DWT	ST	DWT	ST
Faulty cases	1046	1048	1038	1043	1029	1040	1023	1036
Nonfaulty cases	1050	1050	1046	1050	1043	1047	1041	1045
Overall accuracy (%)	99.81	99.91	99.24	99.67	98.67	99.38	98.29	99.09

DWT, discrete wavelet transform.

11.5.2 Fault classification

Like fault detection scheme, this chapter collected features from three different categories of data for 700 faulty cases for each type (AG, BG, CG, ABG, BCG, CAG, and ABCG). The cases were generated by applying faults on random positions of the test distribution feeder incorporating load demand, DG generation, and fault resistance and inception uncertainties. Then, it fetched the inputs to MLP-NN to classify the different types of faults. Table 11.6 summarized the technical features of the MLP-NN employed for fault classification, where they were selected in a similar way to Table 11.2.

11.5.2.1 Fault classification employing features collected from the first category data

This section fetched the useful features collected from the RSCAD built-in PMU recorded data (first category) into the classification MLP-NN with the hyperparameters stated in Table 11.6. Table 11.7 presents a 7 × 7 confusion matrix obtained from MLP-NN as it classified seven types of faults. The diagonal and off-diagonal elements of the confusion matrix represent the successful and unsuccessful classifications, respectively, for a specific fault. It can be observed that the technique was able to classify the faults with an accuracy of 90.041% based on first category data. Like the

Table 11.6 Multilayer perceptron neural network parameters for fault classification approach.

Item	Specifications for the different data category
Number of hidden neurons	First: 12; second: 9 (DWT) and 8 (ST); third: 10 (DWT) and 9 (ST)
Squashing function	Tan-sigmoid
Training algorithm	"trainrp'
Objective function	Mean squared error
Training data	70%
Testing data	30%
Network name	Pattern recognition network ("patternnet")

DWT, discrete wavelet transform.

Table 11.7 Fault classification results based on the features collected from RSCAD built-in PMU recoded data (first category).

	AG	BG	CG	ABG	BCG	CAG	ABCG
AG	631	57	1	8	2	0	1
BG	6	679	7	0	4	3	1
CG	0	54	630	0	7	8	1
ABG	20	46	4	628	0	2	0
BCG	0	65	18	0	609	5	3
CAG	6	62	13	2	5	603	9
ABCG	2	39	5	2	7	13	632
Overall accuracy = 90.041%

PMU, phasor measurement unit.

detection problem, the presented dissatisfactory results using the first category data lead the authors to look for other promising techniques.

11.5.2.2 Fault classification employing features collected from the second category data

This section employed the DWT and ST extracted features from the current signals of the main feeder and three DGs (second category data) into the classification MLP-NN with the hyperparameters stated in Table 11.6. The confusion matrices obtained from the DWT and ST-based approaches are presented in Tables 11.8 and 11.9, respectively. It can be seen that the advanced signal processing-based machine learning approaches were able to classify different types of faults with almost 100% accuracy.

This section added AWGN to the recorded three-phase current signals from the main feeder and three DGs similar to the detection problem. Then, it extracted features employing the DWT and ST to test the efficacy of the MLP-NN under a noisy environment. Table 11.10 summarizes the obtained results with the different levels of

Table 11.8 Fault classification results based on DWT extracted features from the second category data (current signals of the main feeder and three DGs).

	AG	BG	CG	ABG	BCG	CAG	ABCG
AG	699	0	0	1	0	0	0
BG	0	698	0	0	1	0	1
CG	0	0	700	0	0	0	0
ABG	1	0	0	698	0	0	1
BCG	0	0	0	0	699	0	1
CAG	0	0	0	0	0	700	0
ABCG	0	0	0	0	1	0	699
Overall accuracy = 99.857%

DWT, discrete wavelet transform.

Table 11.9 Fault classification results based on ST extracted features from the second category data (current signals of the main feeder and three DGs).

	AG	BG	CG	ABG	BCG	CAG	ABCG
AG	700	0	0	0	0	0	0
BG	0	700	0	0	0	0	0
CG	0	0	700	0	0	0	0
ABG	1	0	0	699	0	0	0
BCG	0	0	0	0	700	0	0
CAG	0	0	0	0	0	700	0
ABCG	0	0	0	0	0	0	700
Overall accuracy = 99.98%							

SNR for the DWT and the ST-based approaches. It can be seen that the DWT-based approach ended up with the overall accuracies of 99.755%, 99.653%, and 99.367% for 40 dB, 30 dB, and 20 dB SNR, respectively.

On the contrary, the ST-based approach ended up with the overall accuracies of 99.898%, 99.734%, and 99.612% for 40 dB, 30 dB, and 20 dB SNR, respectively. Consequently, the obtained results validate the efficacy of the DWT and ST-based fault classification techniques under noisy and noise-free measurement for second category data. Like fault detection problems, the ST-based approaches continued to outperform the DWT-based methods for classification problems.

11.5.2.3 Fault classification employing features collected from the third category data

The section fetched the features extracted by DWT and ST techniques from the third category data (eight-branch current signals) into the MLP-NN with the hyperparameters stated in Table 11.6. Obtained confusion matrices from the DWT and ST-based

Table 11.10 Fault classification results under noisy environment for the second category data (current signals of the main feeder and three DGs).

	Samples classified successfully					
	40 dB SNR		30 dB SNR		20 dB SNR	
Fault type	DWT	ST	DWT	ST	DWT	ST
AG	699	700	698	699	696	698
BG	697	699	698	697	695	698
CG	700	700	699	698	697	696
ABG	697	698	697	699	694	699
BCG	698	700	696	698	693	695
CAG	699	699	699	699	697	697
ABCG	698	699	696	698	697	698
Overall accuracy (%)	99.755	99.898	99.653	99.734	99.367	99.612

Table 11.11 Fault classification results based on DWT extracted features from the third category data (eight-branch current signals).

	AG	BG	CG	ABG	BCG	CAG	ABCG
AG	699	0	0	1	0	0	0
BG	0	699	0	0	0	0	1
CG	0	0	700	0	0	0	0
ABG	0	0	0	700	0	0	0
BCG	0	0	0	0	699	0	1
CAG	0	0	0	0	0	700	0
ABCG	0	0	0	0	0	0	700
Overall accuracy = 99.939%							

DWT, discrete wavelet transform.

approaches are presented in Tables 11.11 and 11.12, respectively. It can be seen that the advanced signal processing-based machine learning approaches were able to classify different types of faults with almost 100% accuracy.

Like second category data, different levels of AWGN were added to the recorded third category pure signals and extracted features employing the DWT and ST-based approaches to test the classification scheme efficacy under the noisy environment. Table 11.13 summarizes the obtained results of the DWT and the ST-based approaches with the different levels of SNR. As can be seen, the DWT-based approach ended up with the overall accuracies of 99.857%, 99.735%, and 99.571% for 40 dB, 30 dB, and 20 dB SNR, respectively. On the contrary, the ST-based approach ended up with the overall accuracies of 99.918%, 99.816%, and 99.755% for 40 dB, 30 dB, and 20 dB SNR, respectively. Consequently, the obtained results validated the efficacy of the DWT and ST-based fault classification techniques under noisy and noise-free measurement for the third category data. Finally, the presented results confirmed the superiority of the ST-based MLP-NN approach over other methods in terms of overall efficiency.

Table 11.12 Fault classification results based on ST extracted features from the third category data (eight-branch current signals).

	AG	BG	CG	ABG	BCG	CAG	ABCG
AG	700	0	0	0	0	0	0
BG	0	700	0	0	0	0	0
CG	0	0	700	0	0	0	0
ABG	0	0	0	700	0	0	0
BCG	0	0	0	0	700	0	0
CAG	0	0	0	0	0	700	0
ABCG	0	0	0	0	0	0	700
Overall accuracy = 100%							

Table 11.13 Fault classification results under noisy environment for the third category data (eight-branch current signals).

Fault type	Samples classified successfully					
	40 dB SNR		30 dB SNR		20 dB SNR	
	DWT	ST	DWT	ST	DWT	ST
AG	699	700	698	699	697	698
BG	698	699	698	698	695	699
CG	700	700	699	699	698	698
ABG	699	698	698	699	696	699
BCG	698	700	697	698	698	697
CAG	699	699	699	699	697	698
ABCG	700	700	698	699	698	699
Overall accuracy (%)	99.857	99.918	99.735	99.816	99.571	99.755

DWT, discrete wavelet transform.

11.5.3 Comparison of fault detection and classification

This section compares the DWT, and ST-based fault detection and classification approaches with the referenced techniques in terms of overall accuracy. The comparisons of the fault detection and the classification techniques are presented in Tables 11.14 and 11.15, respectively. It can be observed that most of the referenced methods detected and classified distribution grid faults with satisfactory accuracy, whereas a few of them diagnosed faults with lower accuracy. In addition, only a few of the referenced works assessed the performance of the employed techniques considering the presence of measurement noises. However, the fault detection

Table 11.14 Fault detection results comparison.

Technique name and references	Overall accuracy (%)	
	Noise-free environment	Noisy environment
Neural network [63]	95.00	Not assessed
Temporal attribute QSSVM [64]	98.10	Not assessed
Wavelet singular entropy based fuzzy logic theory [64]	100.0	Not assessed
Principal components analysis based SVM [65]	99.74	99.79 (30 dB) 99.77 (20 dB)
DWT-based MLP-NN approach (multiphase 13-node system)	99.81	99.24 (40 dB) 98.67 (30 dB) 98.29 (20 dB)
ST-based MLP-NN approach (multiphase 13-node system)	99.91	99.67 (40 dB), 99.38 (30 dB), 99.09 (20 dB)

DWT, discrete wavelet transform; MLP-NN, multilayer perceptron neural network; QSSVM, quarter-sphere support vector machine.

Table 11.15 Fault classification results comparison.

Technique name and references	Overall accuracy (%)	
	Noise-free environment	Noisy environment
Wavelet-based fuzzy logic algorithm [66]	89.50	Not assessed
Wavelet multi-resolution approach [67]	94.73	Not Assessed
Wavelet-basedneural network [68]	98.40	Not assessed
Wavelet-based adaptive neuro-fuzzy inference system [69]	99.84	Not assessed
Attribute QSSVM [64]	99.05	Not assessed
Time-time transform-based ART neural network [70]	99.18	Not assessed
Combination of wavelet singular entropy theory and fuzzy logic [64]	100.0	Not Assessed
Principal components analysis based SVM [65]	99.93	99.77 (30 dB) 99.70 (20 dB)
DWT-based MLP-NN approach (multiphase 13-node system)	99.94	99.86 (40 dB) 99.74 (30 dB) 99.57 (20 dB)
ST-based MLP-NN approach (multiphase 13-node system)	100.0	99.92 (40 dB) 99.82 (30 dB) 99.76 (20 dB)

DWT, discrete wavelet transform; MLP-NN, multilayer perceptron neural network; QSSVM, quarter-sphere support vector machine.

and classification accuracies of the presented schemes in this chapter exceeded 99.00% in all cases. Consequently, it can be concluded that the presented technique effectively diagnosed distribution grid faults with better or a competitive accuracy over the referenced works. Furthermore, the presented DWT and ST-based hybrid approaches showed their independence on prefault loading conditions, fault locations, and fault information (resistance and inception angle) uncertainty. The presented approaches also exhibited their effectiveness even in the presence of measurement noises. Finally, it is worth noting that the presented schemes required less than a cycle (on average) of 60 Hz electric networkto extract the characteristics features employing DWT and ST to detect and classify faults using the trained MLP-NN after receiving the recorded three-phase current signals from different locations of the distribution grids. Hence, the presented approaches can be implemented in real-time to detect and classify faults in electric distribution networks.

11.5.4 Faulty section identification

This chapter divided the selected IEEE 13-node test feeder into nine sections, as shown in Fig. 11.9, and made the faulty section identification problem as a classification problem. Then, it recorded three-phase currents signals (three different categories of data) for 900 faulty cases from each section by varying fault-type, resistance, inception angle, and prefault loading conditions. The detection and

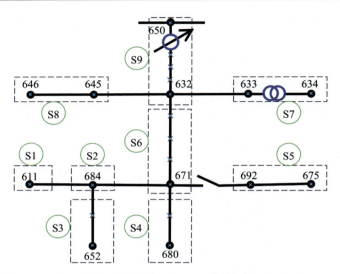

Fig. 11.9 Different sections of the IEEE 13-node test distribution feeder.

classification problems extracted useful features from the recorded data and fetched them into the MLP-NN for the faulty section. Employed MLP-NN hyperparameters for the faulty section identification schemes using three categories of data are presented in Table 11.16. The MLP-NN parameters were selected through similar ways of the detection and classification schemes (Tables 11.2 and 11.6).

11.5.4.1 Faulty section identification employing features of the first category data

After collecting useful features from the RSCAD built-in PMU recorded data (first category), the section fetched them into MLP-NN with hyperparameters of Table 11.16 to develop the faulty section identification scheme. Table 11.17 presents a 9 × 9

Table 11.16 Multilayer perceptron neural network parameters for faulty section identification approach.

Item	Specifications for the different data category
Number of hidden neurons	First: 11; second: 8 (DWT) and 7 (ST); third: 12 (DWT) and 11 (ST)
Squashing function	Tan-sigmoid
Training algorithm	"trainrp"
Objective function	Mean squared error
Training data	70%
Testing data	30%
Network name	Pattern recognition network ("patternnet")

Table 11.17 Faulty section identification results based on the features of first category data.

	S_1	S_2	S_3	S_4	S_5	S_6	S_7	S_8	S_9
S_1	822	0	0	70	0	0	0	0	8
S_2	0	784	0	69	0	0	0	0	47
S_3	0	0	828	72	0	0	0	0	0
S_4	0	0	4	891	0	0	2	1	2
S_5	0	0	0	77	821	0	0	2	0
S_6	0	0	0	69	0	826	0	5	0
S_7	1	0	0	65	0	0	834	0	0
S_8	1	1	0	64	0	0	0	834	0
S_9	155	157	0	53	1	0	0	0	534
Overall accuracy = 78.27%

confusion matrix obtained from neural networks as they classified nine sections. The diagonal and off-diagonal elements of confusion matrices represent the successful and unsuccessful classifications, respectively, for a specific faulty section. The developed scheme was able to identify the faulty sections with an accuracy of 78.27%. Like detection and classification problems, the dissatisfactory results motivated the authors to look for alternatives.

11.5.4.2 Faulty section identification employing features of the second category data

Likewise, the collected features employing DWT and ST from the second category data (three-phase current signals of the main feeder and three DGs) were fetched into the MLP-NN with hyperparameters of Table 11.16 for faulty section identification. Tables 11.18 and 11.19 present the confusion matrices obtained from the MLP-NN employing the DWT and ST-based approaches for the noise-free environment,

Table 11.18 Faulty section identification results based on DWT extracted characteristic features of the second category data.

	S_1	S_2	S_3	S_4	S_5	S_6	S_7	S_8	S_9
S_1	900	0	0	0	0	0	0	0	0
S_2	0	899	1	0	0	0	0	0	0
S_3	0	0	897	3	0	0	0	0	0
S_4	0	2	1	896	0	0	0	0	1
S_5	0	0	0	0	898	0	2	0	0
S_6	1	0	1	0	4	893	0	1	0
S_7	0	1	0	3	2	0	893	0	1
S_8	1	0	0	1	0	1	2	895	0
S_9	0	1	1	0	0	0	1	0	897
Overall accuracy = 99.605%

DWT, discrete wavelet transform.

Table 11.19 Faulty section identification results based on ST extracted characteristic features of the second category data.

	S_1	S_2	S_3	S_4	S_5	S_6	S_7	S_8	S_9
S_1	898	0	1	0	0	0	0	0	1
S_2	0	895	1	1	0	1	0	0	2
S_3	0	1	899	0	0	0	0	0	0
S_4	0	1	0	896	0	0	1	0	2
S_5	0	0	0	0	897	0	1	2	0
S_6	1	0	0	0	0	899	0	0	0
S_7	0	0	0	1	2	0	896	1	0
S_8	0	0	0	0	3	0	3	894	0
S_9	1	0	0	2	0	0	0	0	898
Overall accuracy = 99.654%									

respectively. It can be seen that the schemes were able to classify the faulty sections with almost 100% accuracy for both DWT and ST-based approaches that demonstrated the successful identification of faulty branches through the employment of the adopted approaches.

Similar to the detection and classification problems, the chapter investigated the efficacy of the DWT and ST-based faulty section identification schemes under the noisy environment for second category data. Table 11.20 summarizes the obtained results of the faulty section identification schemes in the presence of different levels of SNR. It can be seen that the DWT-based approach ended up with the overall accuracies of 99.173%, 98.198%, and 97.136% for 40 dB, 30 dB, and 20 dB SNR,

Table 11.20 Faulty section identification results based on the extracted characteristic features of the second category data under a noisy environment.

| Faulty section | Samples classified successfully ||||||
| | 40 dB SNR || 30 dB SNR || 20 dB SNR ||
	DWT	ST	DWT	ST	DWT	ST
S_1	891	896	889	893	885	890
S_2	889	897	886	885	858	865
S_3	892	896	891	897	867	870
S_4	898	898	889	889	865	881
S_5	893	896	878	889	890	891
S_6	893	897	892	899	870	878
S_7	888	893	868	889	867	869
S_8	895	886	888	884	889	886
S_9	894	895	873	892	877	879
Overall accuracy (%)	99.173	99.432	98.198	98.975	97.136	97.642

DWT, discrete wavelet transform.

Table 11.21 Faulty section identification results based on DWT extracted features from the third category data (eight-branch current signals) for noise-free measurement.

	S_1	S_2	S_3	S_4	S_5	S_6	S_7	S_8	S_9
S_1	900	0	0	0	0	0	0	0	0
S_2	0	900	0	0	0	0	0	0	0
S_3	0	0	899	0	1	0	0	0	0
S_4	0	0	0	900	0	0	0	0	0
S_5	0	0	0	0	900	0	0	0	0
S_6	0	0	0	0	0	900	0	0	0
S_7	0	0	0	0	0	0	899	1	0
S_8	0	0	0	0	0	0	0	900	0
S_9	0	0	0	0	0	0	0	0	900
Overall accuracy = 99.975%

DWT, discrete wavelet transform.

respectively. On the contrary, the ST-based approach ended up with the overall accuracies of 99.432%, 98.975%, and 97.642% for 40 dB, 30 dB, and 20 dB SNR, respectively. Consequently, the obtained results validated the efficacy of the DWT and ST-based faulty section identification techniques for both noisy and noise-free measurement. Moreover, the ST-based approach outperformed the DWT-based method in terms of overall accuracy for second category data.

11.5.4.3 Faulty section identification employing features of the third category data

This section presents the faulty section identification results employing the DWT and ST extracted features from the third category (three-phase current signals from eight branches of the test feeder) of data into the MLP-NN with hyperparameters of Table 11.16.

Obtained confusion matrices using DWT and ST extracted features are presented in Tables 11.21 and 11.22 for noise-free measurement. Besides, Table 11.23 presents the obtained results after the addition of different levels of SNR to the recorded data (three-phase current signals of eight branches). It can be seen that both approaches identified the faulty sections with almost 100% accuracy in the case of noise-free measurement. However, the DWT-based method ended up with the overall accuracies of 99.815%, 99.617%, and 97.654% for 40 dB, 30 dB, and 20 dB SNR, respectively. On the contrary, the ST-based approach ended up with the overall accuracies of 99.864%, 99.654%, and 99.370% for 40 dB, 30 dB, and 20 dB SNR, respectively.

Consequently, the obtained results validated the efficacy of the DWT and ST-based faulty section identification schemes even in the presence of measurement noises. Furthermore, obtained results of the faulty section identification

Table 11.22 Faulty section identification results based on ST extracted features from the third category data (eight-branch current signals) for noise-free measurement.

	S_1	S_2	S_3	S_4	S_5	S_6	S_7	S_8	S_9
S_1	900	0	0	0	0	0	0	0	0
S_2	0	900	0	0	0	0	0	0	0
S_3	0	0	900	0	0	0	0	0	0
S_4	0	0	0	900	0	0	0	0	0
S_5	0	0	0	0	900	0	0	0	0
S_6	0	0	0	1	0	899	0	0	0
S_7	0	0	0	0	0	0	900	0	0
S_8	0	0	0	0	0	0	0	900	0
S_9	1	0	0	0	0	0	0	0	899
Overall accuracy = 99.975%

problem confirmed the superiority of the ST-based MLP-NN technique over the DWT-based MLP-NN scheme in terms of overall efficiency for both noisy and noise-free measurements.

11.5.5 Main branch fault location

The previous sections of this chapter presented the fault detection, fault classification, and faulty section identification results based on extracted features from the recorded three-category data. This section shows the fault location results if any fault

Table 11.23 Faulty section identification results based on extracted features from the third category data (eight-branch current signals) in the presence of measurement noises.

Faulty section	40 dB SNR DWT	40 dB SNR ST	30 dB SNR DWT	30 dB SNR ST	20 dB SNR DWT	20 dB SNR ST
S_1	900	900	897	900	873	899
S_2	897	896	895	889	851	898
S_3	898	900	900	897	879	893
S_4	897	900	896	899	873	900
S_5	900	898	897	899	891	887
S_6	899	900	899	898	883	896
S_7	900	899	900	900	885	897
S_8	900	900	899	897	888	895
S_9	894	896	886	893	887	884
Overall accuracy (%)	99.815	99.864	99.617	99.654	97.654	99.370

DWT, discrete wavelet transform.

is detected on the main branch of the test distribution feeder. The total length of the main branch (650-632-671-680) of the test feeder is 5000 feet. After going through a few trials, it was observed that the main branch faults could be better located employing the features extracted from the current signals of the main feeder and three DGs (second category data) over the features extracted from the RSCAD built-in PMU recorded data (first category data) or the features extracted from the three-phase current signals of eight branches (third category data). Therefore, the main branch fault location scheme of this chapter was developed based on second category data.

However, as an attempt at the total number of inputs reduction for the main branch fault location scheme, three input sets from the second category data (three-phase current signals of the main feeder and three DGs) have been produced. The first input set (set I) contains the characteristics features extracted from the three-phase current signal of the main feeder only. The second input set (set II) includes the features extracted from the three-phase current signals of the main feeder andthe hydropower plant. Finally, the third input set (set III) contains characteristic features obtained from the three-phase current signals of the main feeder and three DGs. Table 11.24 summarized the different input sets (set I, set II, and set III) and obtained the number of features employing the DWT and the ST-based techniques as illustrated in Section 11.4.3. After selection of the features, they were fetched into the regression MLP-NN ("newff"). The hyperparameters of the regression neural networks have been selected through a systematic trial and error process based on overall accuracy and MSE values. The chosen number of neurons in the hidden layers, activation functions, and training algorithms are presented in Table 11.25 for three different types of input sets. It is worth noting that features of 1000 faults of every kind are collected by varying the prefault loading conditions, fault information (resistance and inception angle), and fault locations. It used 70% of the available samples to train the neural networks and the rest for testing purposes.

Like Chapter 10, this chapter evaluated different statistical performance indices (SPI), including mean absolute percentage error (MAPE), root mean squared error (RMSE), percent bias (PBIAS), RMSE-observations standard deviation ratio (RSR), the coefficient of determination (R^2), Willmott's index of agreement

Table 11.24 Input sets and the associated number of features for the main branch fault location schemes.

Input set	Source of current signals	Signal processing techniques	Number of features
Set I	Main feeder only	DWT	144
		ST	36
Set II	Main feeder and hydro DG	DWT	288
		ST	72
Set III	Main feeder and three DGs	DWT	576
		ST	144

DWT, discrete wavelet transform.

Table 11.25 Multilayer perceptron neural network parameters for faulty section identification approach.

Item	Specifications for different input sets
Number of hidden neurons	Input set I: 11 (DWT), 9 (ST); input set II: 8 (DWT), 7 (ST); input set III: 12 (DWT), 11 (ST)
Squashing function	Tan-sigmoid
Training algorithm	Levenberg-Marquardt backpropagation ("trainlm")
Objective function	Mean squared error
Training data	70%
Testing data	30%
Network name	Feed-forward backpropagation network ("newff")

DWT, discrete wavelet transform.

(WIA), and Nash–Sutcliffe model efficiency coefficient (NSEC) to investigate the efficacy of the signal processing based machine learning tools in locating faults on the main branch of the IEEE 13-node test feeder. Relevant equations for the selected SPIs are appended in the appendix of this book. Table 11.26 presents the operation time and the selected SPI for the DWT-based MLP-NN approach in locating faults on the main branch of the test distribution feeder for noise-free measurements employing three different input sets. It can be seen that the network training took several seconds with DWT extracted features to locate the faults. However, the trained network estimated fault locations on the main branch of the test feeder incorporated with intermittent RERs within a fraction of a second. Besides, the presented SPI of input sets II and III illustrated the effectiveness (lower values of RMSE, MAPE, RSR, and PBIAS; closer to unity values for R^2, WIA, and NSCE) of the DWT-based MLP-NN in locating different types of faults on the main branch for noise-free measurement. Input setII outperformed the other two sets for most of the fault types, and the performance of input set III was compatible with the performance of input set II. However, the performance of input set I was not satisfactory. Table 11.27 presents the SPI for the fault location scheme considering the presence of 40 dB SNR during the data measurement process. It can be seen that input set II continued to outperform the other two input sets in terms of overall accuracy. However, the obtained SPI (R^2, WIA, and NSCE are not close to unity for most cases) are not promising and sometimes provide misleading information about the fault location. Therefore, this chapter did not investigate the performance of DWT-based MLP-NN in the presence of a higher level of measurement noises, for example, 30 dB and 20 dB SNR.

Likewise, Table 11.28 presents the operation time and the SPI for the ST-based MLP-NN approaches in locating all seven types of faults on the main branch of the distribution feeder for noise-free measurements. Itcan be observed that the network training took a few seconds, whereas the trained networks estimated fault locations within a fraction of a second. Besides, the RMSE, the MAPE, and the RSR values

Smart grid fault diagnosis under load and renewable energy uncertainty 323

Table 11.26 Operation time and statistical performance indices for the test datasets of DWT-based MLP-NN approaches for noise-free measurement.

Fault types	Input feature set	Time (seconds) Training	Testing	Statistical performance measures RMSE	MAPE	RSR	PBIAS	R^2	WIA	NSCE
AG	I	22.962	0.0115	0.3979	9.6994	0.3039	1.0569	0.9571	0.9769	0.9077
	II	23.964	0.0188	0.2476	5.9360	0.1821	−0.185	0.9840	0.9917	0.9668
	III	15.397	0.0104	0.4909	13.4635	0.3495	−0.291	0.9453	0.9695	0.8778
BG	I	24.047	0.0142	0.5509	11.8252	0.4207	1.7441	0.9166	0.9558	0.8230
	II	20.956	0.0283	0.2604	6.7484	0.1979	0.5557	0.9814	0.9902	0.9608
	III	15.073	0.0213	0.2381	5.8828	0.1719	0.9817	0.9858	0.9926	0.9704
CG	I	17.063	0.0174	0.4797	9.4970	0.3663	1.7267	0.9341	0.9664	0.8658
	II	20.016	0.0147	0.2460	6.0766	0.1761	0.8100	0.9852	0.9922	0.9690
	III	24.533	0.0146	0.2774	5.1596	0.2036	0.9789	0.9797	0.9896	0.9585
ABG	I	24.614	0.0256	0.2328	7.6855	0.1778	0.3993	0.9842	0.9921	0.9684
	II	15.734	0.0120	0.1999	3.5527	0.1431	−0.590	0.9899	0.9949	0.9795
	III	24.677	0.0294	0.1599	4.3477	0.1115	−0.220	0.9939	0.9969	0.9876
BCG	I	24.529	0.0209	0.3207	9.7096	0.2449	1.9112	0.9729	0.9850	0.9400
	II	19.339	0.0177	0.1597	4.2361	0.1163	0.0514	0.9933	0.9966	0.9865
	III	22.803	0.0228	0.1958	5.3001	0.1392	−0.565	0.9908	0.9952	0.9806
CAG	I	15.561	0.0299	0.3779	9.0643	0.2886	0.0882	0.9600	0.9792	0.9167
	II	18.639	0.0195	0.1278	3.2087	0.0922	0.1464	0.9958	0.9979	0.9915
	III	24.073	0.0233	0.1559	3.9457	0.1135	0.1600	0.9935	0.9968	0.9871
ABCG	I	22.714	0.0335	0.2418	7.1471	0.1847	−0.287	0.9829	0.9915	0.9659
	II	24.554	0.0238	0.1821	4.5756	0.1304	0.5123	0.9916	0.9958	0.9830
	III	21.213	0.0247	0.1812	3.6238	0.1370	1.2689	0.9909	0.9953	0.9812

DWT, discrete wavelet transform; MAPE, mean absolute percentage error; MLP-NN, multilayer perceptron neural network; PBIAS, percent bias; RMSE, root mean squared error; RSR; RMSE-observations standard deviation ratio; WIA, Willmott's index of agreement.

Table 11.27 Operation time and statistical performance measures for the test datasets of DWT-based MLP-NN approaches in the presence of 40 dB SNR.

Fault types	Input feature set	Time (seconds) Training	Testing	RMSE	MAPE	RSR	PBIAS	R^2	WIA	NSCE
AG	I	22.421	0.0175	0.4972	19.2933	0.3797	0.6403	0.9270	0.9640	0.8559
	II	22.747	0.0158	0.5065	16.2283	0.3627	−1.433	0.9391	0.9671	0.8685
	III	16.056	0.0149	0.6010	19.5891	0.4444	0.0416	0.9135	0.9506	0.8025
BG	I	19.387	0.0143	0.9231	34.8409	0.7049	0.3255	0.7831	0.8758	0.5032
	II	18.901	0.0209	0.4975	14.7444	0.3659	2.3197	0.9406	0.9665	0.8661
	III	21.109	0.0331	0.7109	22.5145	0.5204	1.1652	0.8829	0.9323	0.7292
CG	I	21.803	0.0146	0.6906	17.4718	0.5274	1.8154	0.8649	0.9305	0.7219
	II	22.302	0.0345	0.6052	18.8653	0.4334	0.1757	0.9248	0.9530	0.8122
	III	17.036	0.0128	0.7452	22.8635	0.5633	0.7374	0.8712	0.9207	0.6827
ABG	I	21.477	0.0202	0.4522	18.5858	0.3453	0.5612	0.9412	0.9702	0.8808
	II	21.206	0.0166	0.3011	7.0871	0.2156	−0.408	0.9771	0.9884	0.9535
	III	15.789	0.0278	0.2289	7.4618	0.1709	−0.353	0.9861	0.9927	0.9708
BCG	I	15.309	0.0129	0.6960	25.1447	0.5315	−1.927	0.8821	0.9294	0.7175
	II	19.482	0.0180	0.2790	8.6978	0.2032	1.6481	0.9817	0.9897	0.9587
	III	24.557	0.0227	0.3451	10.8013	0.2460	−0.704	0.9717	0.9849	0.9395
CAG	I	17.744	0.0166	0.7029	20.3590	0.5368	−3.654	0.8828	0.9280	0.7119
	II	20.438	0.0107	0.3427	10.7695	0.2454	−1.129	0.9713	0.9850	0.9398
	III	16.462	0.0283	0.5324	14.0263	0.3876	−1.239	0.9313	0.9624	0.8497
ABCG	I	22.264	0.0245	0.6679	23.3167	0.5100	−0.529	0.8934	0.9350	0.7399
	II	16.806	0.0215	0.1978	6.0200	0.1395	0.5410	0.9904	0.9951	0.9805
	III	19.566	0.0237	0.1810	4.6495	0.1339	0.7317	0.9912	0.9955	0.9821

DWT, discrete wavelet transform; MAPE, mean absolute percentage error; MLP-NN, multilayer perceptron neural network; PBIAS, percent bias; RMSE, root mean squared error; RSR; RMSE-observations standard deviation ratio; WIA, Willmott's index of agreement.

Table 11.28 Operation time and statistical performance indices for the test datasets of ST-based MLP-NN approaches for noise-free measurement.

Fault types	Input feature set	Time Training	Time Testing	RMSE	MAPE	RSR	PBIAS	R²	WIA	NSCE
AG	I	14.752	0.0109	0.1240	1.3246	0.0947	−0.406	0.9956	0.9978	0.9910
	II	11.348	0.0153	0.1046	0.9952	0.0749	−0.347	0.9972	0.9986	0.9944
	III	17.212	0.0102	0.0562	1.4448	0.0409	0.0147	0.9992	0.9996	0.9983
BG	I	9.430	0.0125	0.0528	1.0662	0.0403	0.1742	0.9992	0.9996	0.9984
	II	15.491	0.0210	0.1084	1.1777	0.0782	−0.448	0.9970	0.9985	0.9939
	III	15.477	0.0168	0.2154	1.3634	0.1621	0.6097	0.9870	0.9934	0.9737
CG	I	14.470	0.0144	0.0479	1.0461	0.0366	0.1096	0.9993	0.9997	0.9987
	II	16.447	0.0128	0.0459	0.9343	0.0343	−0.167	0.9994	0.9997	0.9988
	III	15.868	0.0128	0.0912	1.2022	0.0663	−0.370	0.9979	0.9989	0.9956
ABG	I	12.685	0.0194	0.2790	1.7956	0.2131	0.6280	0.9774	0.9887	0.9546
	II	16.802	0.0112	0.0279	0.6035	0.0193	−0.087	0.9998	0.9999	0.9996
	III	18.294	0.0216	0.0704	1.1166	0.0503	0.0641	0.9987	0.9994	0.9975
BCG	I	13.868	0.0165	0.0881	1.7841	0.0673	−0.215	0.9978	0.9989	0.9955
	II	13.468	0.0146	0.0326	0.6612	0.0234	−0.055	0.9997	0.9999	0.9995
	III	14.085	0.0177	0.0324	0.7767	0.0231	−0.063	0.9997	0.9999	0.9995
CAG	I	17.176	0.0219	0.2775	1.6494	0.2119	0.5385	0.9775	0.9888	0.9551
	II	15.443	0.0157	0.0099	0.3212	0.0068	−0.020	1.0000	1.0000	1.0000
	III	17.116	0.0180	0.0315	0.6851	0.0225	0.0252	0.9997	0.9999	0.9995
ABCG	I	12.507	0.0241	0.1421	5.0818	0.1085	0.5419	0.9942	0.9971	0.9882
	II	17.759	0.0183	0.1839	0.8696	0.1276	0.5576	0.9919	0.9959	0.9837
	III	15.225	0.0188	0.1308	0.8687	0.0895	−0.506	0.9961	0.9980	0.9920

MAPE, mean absolute percentage error; MLP-NN, multilayer perceptron neural network; PBIAS, percent bias; RMSE, root mean squared error; RSR, RMSE-observations standard deviation ratio; WIA, Willmott's index of agreement.

are relatively low, whereas the R^2, the WIA, and the NSCE are almost unity for all three input sets that demonstrate the strength and effectiveness of the fault location approach. The PBIAS values are positive for a few cases and negative for the others that demonstrate underestimations and overestimations of the fault distances, respectively. However, these values are minimal and close to zero for all the techniques hence the smaller underestimations and overestimations are negligible. Finally, the presented results confirmed the superiority of the input set II over the other two in terms of SPI for most of the cases. Tables 11.29, 11.30 and 11.31 present SPI of the fault location scheme considering the presence of 40 dB, 30 dB, and 20 dB SNR during the data measurement process. It can be seen that the performance of the fault location scheme has been deteriorated a bit from the noise-free situation. However, the indices are still satisfactory for input Set II. In addition, it was observed that other input sets did not perform well under the presence of measurement noises. Therefore, this chapter concludes the three-phase current signals measured from two ends of the main branch (input set II: current signals of the main feeder and hydro DG) contain more characteristics features regarding the fault location on the main branch than the other two investigated input sets.

11.6 Developed intelligent fault diagnosis scheme validation

This chapter presents the developed IFD scheme results based on the different categories of data recorded in the RSCAD/RTDS environment in Section 11.5. This section validates the efficacy of the developed scheme employing RSCAD recorded and physical PMU retrieved data for several representative cases that were not seen by the schemes during their development (training and testing) phases. This section also investigates the effectiveness of the developed method under contingencies, for example, renewable energy generation and teste branch outages. It is worth noting that the representative cases were generated varying the operating conditions considering the load demand and renewable energy generation uncertainties and fault information (resistance and inception angle).

11.6.1 Intelligent fault diagnosis scheme validation using RSCAD recorded data

This section employed the detection and classification schemes developed with DWT and ST extracted features from the RSCAD recorded three-phase current signals of the eight branches (third category data) of the test distribution feeder incorporated with intermittent RERs. Besides, it employed main branch fault location schemes developed with DWT and ST extracted features from the RSCAD recorded three-phase current signals from the main feeder and the hydro DG (input

Smart grid fault diagnosis under load and renewable energy uncertainty 327

Table 11.29 Operation time and statistical performance measures for the test datasets of ST-based MLP-NN approaches in the presence of 40 dB SNR.

Fault types	Input feature set	Time Training	Time Testing	RMSE	MAPE	RSR	PBIAS	R^2	WIA	NSCE
AG	I	14.732	0.0137	0.1785	3.7628	0.1363	0.3481	0.9907	0.9954	0.9814
	II	14.378	0.0200	0.1039	3.1534	0.0777	0.1126	0.9970	0.9985	0.9940
	III	16.470	0.0163	0.0881	1.8891	0.0618	0.2785	0.9981	0.9990	0.9962
BG	I	13.042	0.0258	0.1064	3.4285	0.0813	0.1890	0.9967	0.9983	0.9934
	II	9.415	0.0242	0.1296	2.3983	0.0938	−0.497	0.9959	0.9978	0.9912
	III	19.046	0.0227	0.0991	2.3489	0.0735	−0.197	0.9973	0.9987	0.9946
CG	I	10.086	0.0142	0.1602	3.2204	0.1223	−0.445	0.9930	0.9963	0.9850
	II	12.689	0.0209	0.1654	2.9624	0.1236	−0.767	0.9926	0.9962	0.9847
	III	10.502	0.0215	0.0964	2.2012	0.0685	0.1228	0.9977	0.9988	0.9953
ABG	I	10.174	0.0205	0.1555	4.7738	0.1188	0.6400	0.9930	0.9965	0.9859
	II	14.436	0.0225	0.0341	0.8046	0.0247	0.0702	0.9997	0.9998	0.9994
	III	16.865	0.0245	0.0615	1.8526	0.0453	−0.074	0.9990	0.9995	0.9979
BCG	I	18.800	0.0153	0.1051	4.0482	0.0802	−0.221	0.9968	0.9984	0.9936
	II	16.686	0.0132	0.1447	1.2361	0.1047	0.0814	0.9945	0.9973	0.9890
	III	9.336	0.0219	0.1784	1.8218	0.1288	0.7659	0.9919	0.9959	0.9834
CAG	I	14.500	0.0177	0.1225	2.9529	0.0935	−0.283	0.9957	0.9978	0.9912
	II	18.952	0.0198	0.0508	0.8872	0.0366	−0.096	0.9993	0.9997	0.9987
	III	15.794	0.0238	0.1331	1.5395	0.0928	−0.001	0.9958	0.9978	0.9914
ABCG	I	17.860	0.0192	0.1892	7.3271	0.1445	−0.163	0.9896	0.9948	0.9791
	II	11.012	0.0138	0.0745	0.7138	0.0557	−0.048	0.9985	0.9992	0.9969
	III	18.752	0.0105	0.1821	1.8485	0.1298	−0.564	0.9917	0.9958	0.9832

MAPE, mean absolute percentage error; MLP-NN, multilayer perceptron neural network; PBIAS, percent bias; RMSE, root mean squared error; RSR, RMSE-observations standard deviation ratio; WIA, Willmott's index of agreement.

Table 11.30 Operation time and statistical performance measures for the test datasets of ST-based MLP-NN approaches in the presence of 30 dB SNR.

Fault types	Input feature set	Time Training	Time Testing	RMSE	MAPE	RSR	PBIAS	R²	WIA	NSCE
AG	I	9.929	0.0164	0.3278	7.1232	0.2503	0.5260	0.9692	0.9843	0.9374
	II	11.859	0.0228	0.1577	4.7183	0.1179	−0.333	0.9931	0.9965	0.9861
	III	13.746	0.0246	0.2443	4.2950	0.1735	1.3436	0.9853	0.9925	0.9699
BG	I	11.000	0.0142	0.3759	9.7641	0.2870	0.3926	0.9586	0.9794	0.9176
	II	10.601	0.0122	0.1207	4.6633	0.0873	−0.175	0.9963	0.9981	0.9924
	III	18.562	0.0193	0.1686	4.2072	0.1226	0.0328	0.9928	0.9962	0.9850
CG	I	15.048	0.0123	0.4849	7.9981	0.3703	2.6535	0.9340	0.9657	0.8629
	II	18.383	0.0200	0.1411	6.0667	0.1055	−0.134	0.9944	0.9972	0.9889
	III	12.860	0.0182	0.1230	3.0202	0.0911	0.0331	0.9960	0.9979	0.9917
ABG	I	13.420	0.0112	0.4672	9.1780	0.3568	1.5390	0.9364	0.9682	0.8727
	II	11.639	0.0120	0.0854	2.2436	0.0618	−0.320	0.9981	0.9990	0.9962
	III	11.023	0.0138	0.2025	2.2706	0.1412	0.2580	0.9902	0.9950	0.9801
BCG	I	13.590	0.0108	0.2059	7.5938	0.1572	0.0969	0.9877	0.9938	0.9753
	II	18.930	0.0251	0.1392	2.7148	0.1007	−0.325	0.9950	0.9975	0.9899
	III	14.400	0.0178	0.2722	3.6292	0.1912	0.5527	0.9822	0.9909	0.9635
CAG	I	12.715	0.0244	0.1925	7.1483	0.1470	0.6988	0.9893	0.9946	0.9784
	II	13.062	0.0118	0.0984	2.6369	0.0733	−0.168	0.9973	0.9987	0.9946
	III	17.583	0.0162	0.1554	2.9788	0.1084	0.4149	0.9943	0.9971	0.9883
ABCG	I	11.659	0.0165	0.3245	12.1785	0.2478	0.3109	0.9688	0.9846	0.9386
	II	10.061	0.0121	0.0611	1.6885	0.0455	−0.100	0.9990	0.9995	0.9979
	III	19.363	0.0253	0.1347	2.0088	0.0992	−0.146	0.9952	0.9975	0.9902

MAPE, mean absolute percentage error; MLP-NN, multilayer perceptron neural network; PBIAS, percent bias; RMSE, root mean squared error; RSR; RMSE-observations standard deviation ratio; WIA, Willmott's index of agreement.

Table 11.31 Operation time and statistical performance measures for the test datasets of ST-based MLP-NN approaches in the presence of 20 dB SNR.

Fault types	Input feature set	Time Training	Time Testing	RMSE	MAPE	RSR	PBIAS	R^2	WIA	NSCE
AG	I	17.421	0.0227	0.3278	7.1232	0.2503	0.5260	0.9692	0.9843	0.9374
	II	11.056	0.0178	0.1577	4.7183	0.1179	−0.333	0.9931	0.9965	0.9861
	III	13.901	0.0203	0.2443	4.2950	0.1735	1.3436	0.9853	0.9925	0.9699
BG	I	16.803	0.0221	0.5321	21.6519	0.4063	0.9798	0.9145	0.9587	0.8349
	II	12.036	0.0209	0.2965	7.0936	0.2208	−1.059	0.9769	0.9878	0.9512
	III	16.206	0.0126	0.3471	8.5563	0.2438	−1.350	0.9702	0.9851	0.9406
CG	I	10.309	0.0180	0.5212	18.0644	0.3980	2.5453	0.9210	0.9604	0.8416
	II	19.557	0.0154	0.2049	9.3568	0.1526	0.2137	0.9883	0.9942	0.9767
	III	15.438	0.0136	0.3903	8.2872	0.2890	1.9443	0.9613	0.9791	0.9165
ABG	I	17.264	0.0141	0.5347	16.4423	0.4083	0.4148	0.9187	0.9583	0.8333
	II	14.566	0.0212	0.1756	6.0718	0.1270	−0.319	0.9919	0.9960	0.9839
	III	18.800	0.0253	0.2928	6.6198	0.2244	0.4261	0.9748	0.9874	0.9497
BCG	I	15.019	0.0122	0.6462	22.8636	0.4935	−1.987	0.8791	0.9391	0.7565
	II	10.642	0.0141	0.1876	6.1606	0.1397	0.1515	0.9902	0.9951	0.9805
	III	18.248	0.0141	0.3368	6.8466	0.2581	−1.378	0.9679	0.9833	0.9334
CAG	I	17.957	0.0139	0.6753	16.8636	0.5157	0.8294	0.8728	0.9335	0.7341
	II	19.222	0.0156	0.2703	5.7115	0.1956	0.5826	0.9813	0.9904	0.9618
	III	11.163	0.0140	0.3208	7.2481	0.2253	0.0322	0.9756	0.9873	0.9492
ABCG	I	15.776	0.0176	0.5626	19.9163	0.4296	−2.084	0.9085	0.9539	0.8154
	II	12.868	0.0233	0.2711	4.1546	0.2019	−1.519	0.9813	0.9898	0.9592
	III	15.438	0.0188	0.3498	7.8855	0.2525	−0.231	0.9685	0.9841	0.9362

MAPE, mean absolute percentage error; MLP-NN, multilayer perceptron neural network; PBIAS, percent bias; RMSE, root mean squared error; RSR; RMSE-observations standard deviation ratio; WIA, Willmott's index of agreement.

set II). Tables 11.32, 11.33, 11.34, 11.35, 11.36, 11.37, 11.38, 11.39, 11.40, 11.41 and 11.42 present the IFD schemes estimated results along with the actual fault information for different operating conditions and contingencies. It can be observed from the presented results that both DWT and ST-based MLT approaches diagnosed the applied faults effectively in most cases that validated the efficacy of the developed IFD schemes.

Table 11.32 Fault diagnosis results comparison considering base loading condition.

Fault number	Item	Applied fault information	Estimated results DWT-based MLT	Estimated results ST-based MLT
1	Node	611	-	-
	Type	CG	CG	CG
	Section	1	1	1
	Main branch	No	No	No
2	Node	632a	-	-
	Type	BCG	BCG	BCG
	Section	7	7	7
	Main branch	Yes	Yes	Yes
	Location (feet)	2500	2500.3422	2513.2179
	Error (%)	-	0.0068	0.2643
3	Node	633	-	-
	Type	BG	BG	BG
	Section	6	6	6
	Main branch	No	No	No
4	Node	650b	-	-
	Type	ABG	ABG	ABG
	Section	8	8	8
	Main branch	Yes	Yes	Yes
	Location (feet)	1000	941.6677	1012.8889
	Error (%)	-	1.17	0.258
5	Node	680	-	-
	Type	CAG	CAG	CAG
	Section	3	3	3
	Main branch	Yes	Yes	Yes
	Location (feet)	5000	4999.9391	4999.1013
	Error (%)	-	0.0012	0.0179
6	Node	652	-	-
	Type	AG	AG	AG
	Section	2	2	2
	Main branch	No	No	No
7	Node	675	-	-
	Type	ABCG	ABCG	ABCG
	Section	4	6 (wrong section)	4
	Main branch	No	No	No

DWT, discrete wavelet transform.

Table 11.33 Fault diagnosis results comparison considering ±5% load uncertainty.

Fault number	Item	Applied fault information	Estimated results DWT-based MLT	Estimated results ST-based MLT
1	Node	645	-	-
	Type	CG	CG	CG
	Section	5	5	5
	Main branch	No	No	No
2	Node	632c	-	-
	Type	BG	BG	BG
	Section	7	7	7
	Main branch	Yes	Yes	Yes
	Location (feet)	3500	3384.479	3522.7847
	Error (%)	-	2.31	0.4556
3	Node	633	-	-
	Type	CAG	CAG	CAG
	Section	6	6	6
	Main branch	No	No	No
4	Node	650c	-	-
	Type	AG	AG	AG
	Section	8	8	8
	Main branch	Yes	Yes	Yes
	Location (feet)	1500	1131.9708	1502.0966
	Error (%)	-	7.36	0.0419
5	Node	680	-	-
	Type	ABCG	ABCG	ABCG
	Section	3	3	3
	Main branch	Yes	Yes	Yes
	Location (feet)	5000	4999.9621	4999.9897
	Error (%)	-	0.000758	0.000206
6	Node	675	-	-
	Type	ABG	ABG	ABG
	Section	4	6 (wrong section)	4
	Main branch	No	No	No
7	Node	-	-	-
	Type	No fault	No fault	No fault
	Section	-	-	-
	Main branch	-	-	-

DWT, discrete wavelet transform.

The ST-based approach detected and classified the applied faults and identified the faulty sections for the presented cases accurately under contingencies and uncertainties. Conversely, the DWT-based method detected the faults accurately. Still, it provided wrong decisions on the fault class and faulty sections for a few cases, for example, it identified section 11.6 as the faulty section instead of section 11.4 in Table 11.32 (for the fault number 7). It classified a fault as CG instead of BCG in

Table 11.34 Fault diagnosis results comparison considering ±10% load uncertainty.

Fault number	Item	Applied fault information	Estimated results DWT-based MLT	Estimated results ST-based MLT
1	Node	633	-	-
	Type	BG	BG	BG
	Section	6	6	6
	Main branch	No	No	No
2	Node	632b	-	-
	Type	ABG	ABG	ABG
	Section	7	7	7
	Main branch	Yes	Yes	Yes
	Location (feet)	3000	3102.9737	3015.9339
	Error (%)	-	2.05	0.3187
3	Node	646	-	-
	Type	BCG	BCG	BCG
	Section	5	5	5
	Main branch	No	No	No
4	Node	650a	-	-
	Type	CAG	CAG	CAG
	Section	8	8	8
	Main branch	Yes	Yes	Yes
	Location (feet)	500	500.0477	500.0001
	Error (%)	-	0.000954	0.000002
5	Node	680	-	-
	Type	ABCG	ABCG	ABCG
	Section	3	3	3
	Main branch	Yes	Yes	Yes
	Location (feet)	5000	4999.952	4999.8228
	Error (%)	-	0.00096	0.00354
6	Node	684	-	-
	Type	CG	CG	CG
	Section	9	9	9
	Main branch	No	No	No
7	Node	692	-	-
	Type	AG	AG	AG
	Section	4	4	4
	Main branch	No	No	No

DWT, discrete wavelet transform.

Table 11.37 (for the fault number 6). The ST-based approach located main branch faults with less than 2% error for almost all cases. On the other hand, estimated fault locations for the DWT-based methods were less accurate than the ST-based approaches. Therefore, it can be concluded that the presented results confirmed the superiority of the ST-based approach over the DWT-based approach in diagnosing faults on the investigated test distribution feeder incorporated with intermittent RERs.

Table 11.35 Fault diagnosis results comparison considering ±15% load uncertainty.

Fault number	Item	Applied fault information	Estimated results DWT-based MLT	Estimated results ST-based MLT
1	Node	611	-	-
	Type	CG	CG	CG
	Section	1	1	1
	Main branch	No	No	No
2	Node	632	-	-
	Type	AG	AG	AG
	Section	8	8	8
	Main branch	Yes	Yes	Yes
	Location (feet)	2000	2240.0672	2015.9613
	Error (%)	-	4.80	0.3192
3	Node	633	-	-
	Type	BG	BG	BG
	Section	6	6	6
	Main branch	No	No	No
4	Node	650a	-	-
	Type	CAG	CAG	CAG
	Section	8	8	8
	Main branch	Yes	Yes	Yes
	Location (feet)	500	500.2637	500.0634
	Error (%)	-	0.00527	0.00127
5	Node	680	-	-
	Type	ABCG	ABCG	ABCG
	Section	3	3	3
	Main branch	Yes	Yes	Yes
	Location (feet)	5000	4999.9763	4999.9941
	Error (%)	-	0.0012	0.0179
6	Node	646	-	-
	Type	BCG	BCG	BCG
	Section	5	5	5
	Main branch	No	No	No
7	Node	675	-	-
	Type	ABG	ABG	ABG
	Section	4	6 (wrong section)	4
	Main branch	No	No	No

DWT, discrete wavelet transform.

Table 11.36 Fault diagnosis results comparison considering ±20% load uncertainty.

Fault number	Item	Applied fault information	Estimated results DWT-based MLT	Estimated results ST-based MLT
1	Node	611	-	-
	Type	CG	CG	CG
	Section	1	1	1
	Main branch	No	No	No
2	Node	632b	-	-
	Type	ABG	ABG	ABG
	Section	7	7	7
	Main branch	Yes	Yes	Yes
	Location (feet)	3000	3158.2709	3001.7308
	Error (%)	-	3.17	0.0346
3	Node	633	-	-
	Type	CAG	CAG	CAG
	Section	6	6	6
	Main branch	No	No	No
4	Node	650c	-	-
	Type	BCG	BCG	BCG
	Section	8	8	8
	Main branch	Yes	Yes	Yes
	Location (feet)	1500	1492.0765	1501.0522
	Error (%)	-	0.1584	0.021044
5	Node	671	-	-
	Type	BG	BG	BG
	Section	7	7	7
	Main branch	Yes	Yes	Yes
	Location (feet)	4000	4007.1471	4317.377
	Error (%)	-	0.1429	6.38
6	Node	652	-	-
	Type	AG	AG	AG
	Section	2	2	2
	Main branch	No	No	No
7	Node	692	-	-
	Type	ABCG	ABCG	ABCG
	Section	4	4	4
	Main branch	No	No	No

DWT, discrete wavelet transform.

11.6.2 Intelligent fault diagnosis scheme validation using NI-PMU retrieved data

In the previous section, the efficacy of the developed IFD scheme using RSCAD recorded data and observed superior performance of the ST-based MLT approaches over the DWT-based approaches has been validated. Therefore, the effectiveness of

Table 11.37 Fault diagnosis results comparison considering ±5% renewable energy generation and ±15% load uncertainties.

Fault number	Item	Applied fault information	Estimated results DWT-based MLT	Estimated results ST-based MLT
1	Node	611	-	-
	Type	CG	CG	CG
	Section	1	1	1
	Main branch	No	No	No
2	Node	632b	-	-
	Type	ABG	ABG	ABG
	Section	7	7	7
	Main branch	Yes	Yes	Yes
	Location (feet)	3000	3117.4389	2991.6478
	Error (%)	-	2.35	0.16704
3	Node	633	-	-
	Type	CAG	CAG	CAG
	Section	6	6	6
	Main branch	No	No	No
4	Node	650a	-	-
	Type	BG	BG	BG
	Section	8	8	8
	Main branch	Yes	Yes	Yes
	Location (feet)	500	500.0011	500.0054
	Error (%)	-	0.000022	0.00011
5	Node	680	-	-
	Type	ABCG	ABCG	ABCG
	Section	3	3	3
	Main branch	Yes	Yes	Yes
	Location (feet)	5000	4999.9889	4999.9967
	Error (%)	-	0.00022	0.000066
6	Node	646	-	-
	Type	BCG	CG (wrong class)	BCG
	Section	5	5	5
	Main branch	No	No	No
7	Node	692	-	-
	Type	AG	AG	AG
	Section	4	4	4
	Main branch	No	No	No

DWT, discrete wavelet transform.

the ST-based strategies only was investigated with the physical PMU retrieveddata (fourth category data).This section used the developed intelligent fault detection and classification schemes employing ST extracted features from the RSCAD recorded second category data (three-phase current signals of the main feeder and three DGs) to validate their efficacy with the physical PMU retrieved data. Besides, it employed main branch fault location schemes developed with ST extracted features from the

Table 11.38 Fault diagnosis results comparison considering ±10% renewable energy generation and ±15% load uncertainties.

Fault number	Item	Applied fault information	Estimated results DWT-based MLT	Estimated results ST-based MLT
1	Node	646	-	-
	Type	BCG	BCG	BCG
	Section	5	5	5
	Main branch	No	No	No
2	Node	632c	-	-
	Type	CAG	CAG	CAG
	Section	7	7	7
	Main branch	Yes	Yes	Yes
	Location (feet)	3500	3594.368	3472.3532
	Error (%)	-	1.88	0.5529
3	Node	633	-	-
	Type	BG	BG	BG
	Section	6	6	6
	Main branch	No	No	No
4	Node	650b	-	-
	Type	CG	CG	CG
	Section	8	8	8
	Main branch	Yes	Yes	Yes
	Location (feet)	1000	1255.881	941.3108
	Error (%)	-	5.12	1.17
5	Node	680a	-	-
	Type	AG	AG	AG
	Section	3	3	3
	Main branch	Yes	Yes	Yes
	Location (feet)	4500	4892.5164	4438.6561
	Error (%)	-	7.85	1.22
6	Node	692	-	-
	Type	ABCG	ABCG	ABCG
	Section	4	4	4
	Main branch	No	No	No
7	Node	-	-	-
	Type	No fault	No fault	No fault
	Section	-	-	-
	Main branch	-	-	-

DWT, discrete wavelet transform.

RSCAD recorded three-phase current signals from the main feeder and the hydro DG (input set II) to validate their efficacy with the physical PMU retrieved data (experimental data). It is worth mentioning that experimental data were recorded using the physical PMU through the GTAO card of the RTDS machine.

For experimental validation of the IFD scheme, this chapter applied seven different faults on the different locations of the IEEE 13-node test distribution feeder by

Table 11.39 Fault diagnosis results comparison considering ±15% renewable energy generation and ±15% load uncertainties.

Fault number	Item	Applied fault information	Estimated results DWT-based MLT	ST-based MLT
1	Node	645	-	-
	Type	CG	CG	CG
	Section	5	5	5
	Main branch	No	No	No
2	Node	632a	-	-
	Type	ABCG	ABCG	ABCG
	Section	7	7	7
	Main branch	Yes	Yes	Yes
	Location (feet)	2500	2327.5525	2500.8153
	Error (%)	-	3.44	0.01631
3	Node	633	-	-
	Type	CAG	CAG	CAG
	Section	6	6	6
	Main branch	No	No	No
4	Node	650b	-	-
	Type	ABG	ABG	ABG
	Section	8	8	8
	Main branch	Yes	Yes	Yes
	Location (feet)	1000	1195.9309	993.1186
	Error (%)	-	3.92	0.1376
5	Node	680	-	-
	Type	BG	BG	BG
	Section	3	3	3
	Main Branch	Yes	Yes	Yes
	Location (feet)	5000	4999.9967	4999.9998
	Error (%)	-	0.000066	0.000004
6	Node	652a	-	-
	Type	AG	AG	AG
	Section	2	2	2
	Main branch	No	No	No
7	Node	692	-	-
	Type	BCG	BCG	BCG
	Section	4	4	4
	Main branch	No	No	No

DWT, discrete wavelet transform.

changing the operating conditions (i.e., ±15% variation in load demand and ±10% variation in renewable energy generation) and fault information (resistance and inception angle). Then, it recorded data (input set II: three-phase current signals of the main feeder and three DGs) in both the RSCAD environment and using physical PMU in the LabVIEW environment. Useful features from the recorded data were extracted

Table 11.40 Fault diagnosis results comparison considering the wind power plant outage with ±10% renewable energy generation and ±15% load uncertainties.

Fault number	Item	Applied fault information	Estimated results DWT-based MLT	Estimated results ST-based MLT
1	Node	611	-	-
	Type	CG	CG	CG
	Section	1	1	1
	Main branch	No	No	No
2	Node	632a	-	-
	Type	ABG	ABG	ABG
	Section	7	7	7
	Main branch	Yes	Yes	Yes
	Location (feet)	2500	2833.4187	2598.2448
	Error (%)	-	6.67	1.96
3	Node	646	-	-
	Type	BCG	BCG	BCG
	Section	5	5	5
	Main branch	No	No	No
4	Node	650a	-	-
	Type	CAG	CAG	CAG
	Section	8	8	8
	Main branch	Yes	Yes	Yes
	Location (feet)	500	516.3356	519.4524
	Error (%)	-	0.3267	0.3890
5	Node	680	-	-
	Type	BG	BG	BG
	Section	3	3	3
	Main branch	Yes	Yes	Yes
	Location (feet)	5000	4999.9952	5000
	Error (%)	-	0.000096	0
6	Node	652	-	-
	Type	AG	AG	AG
	Section	2	2	2
	Main branch	No	No	No
7	Node	675	-	-
	Type	ABCG	ABCG	ABCG
	Section	4	6 (wrong section)	4
	Main branch	No	No	No

DWT, discrete wavelet transform.

employing the ST and were fetched into the trained and tested MLP-NN of Section 11.5 to diagnose the applied faults. Table 11.43 presents the estimated results obtained from the developed IFD schemes. It can be seen that all faulty events were detected and classified accurately using extracted features from both types (RSCAD software and physical PMU recorded) of recorded data. In addition, the approach also identified the faulty section effectively. The presented results also confirmed the effectiveness

Table 11.41 Fault diagnosis results comparison considering the PV power plant outage with ±10% renewable energy generation and ±15% load uncertainties.

Fault number	Item	Applied fault information	Estimated results DWT-based MLT	Estimated results ST-based MLT
1	Node	633	-	-
	Type	ABCG	ABCG	ABCG
	Section	6	6	6
	Main branch	No	No	No
2	Node	632c	-	-
	Type	BCG	BCG	BCG
	Section	7	7	7
	Main branch	Yes	Yes	Yes
	Location (feet)	3500	2949.5343	3553.7978
	Error (%)	-	11.01	1.08
3	Node	646	-	-
	Type	CG	CG	CG
	Section	5	5	5
	Main branch	No	No	No
4	Node	650b	-	-
	Type	CAG	CAG	CAG
	Section	8	8	8
	Main branch	Yes	Yes	Yes
	Location (feet)	1000	837.2781	1084.8194
	Error (%)	-	3.25	1.70
5	Node	680	-	-
	Type	BG	BG	BG
	Section	3	3	3
	Main branch	Yes	Yes	Yes
	Location (feet)	5000	4999.7754	5000
	Error (%)	-	0.0045	0
6	Node	684	-	-
	Type	AG	AG	AG
	Section	9	9	9
	Main branch	No	No	No
7	Node	692	-	-
	Type	ABG	ABG	ABG
	Section	4	4	4
	Main branch	No	No	No

DWT, discrete wavelet transform; PV, photovoltaic.

of the scheme in locating faults on the main branch. Though the scheme was more accurate in locating faults with the RSCAD recorded data, the accuracy for physical PMU recorded data was also satisfactory. Thus, it can be concluded that the estimated fault diagnosis results validated the efficacy of the developed laboratory prototype IFD scheme.

Table 11.42 Fault diagnosis results comparison considering a branch outage (nodes 633 and 634) with ±10% renewable energy generation and ±15% load uncertainties.

Fault number	Item	Applied fault information	Estimated results DWT-based MLT	Estimated results ST-based MLT
1	Node	611	-	-
	Type	CG	CG	CG
	Section	1	1	1
	Main branch	No	No	No
2	Node	632b	-	-
	Type	BCG	BCG	BCG
	Section	7	7	7
	Main branch	Yes	Yes	Yes
	Location (feet)	3000	3070.102	2979.6298
	Error (%)	-	1.40	0.407
3	Node	645	-	-
	Type	BG	BG	BG
	Section	5	5	5
	Main branch	No	No	No
4	Node	650c	-	-
	Type	ABCG	ABCG	ABCG
	Section	8	8	8
	Main branch	Yes	Yes	Yes
	Location (feet)	1500	1755.4814	1596.8261
	Error (%)	-	5.11	1.93
5	Node	680	-	-
	Type	ABG	ABG	ABG
	Section	3	3	3
	Main branch	Yes	Yes	Yes
	Location (feet)	5000	4999.943	5000
	Error (%)	-	0.00114	0
6	Node	652a	-	-
	Type	AG	AG	AG
	Section	2	2	2
	Main branch	No	No	No
7	Node	684	-	-
	Type	CAG	CAG	CAG
	Section	9	9	9
	Main branch	No	No	No

DWT, discrete wavelet transform.

Table 11.43 Fault diagnosis results comparison employing the developed intelligent fault diagnosis scheme.

Fault number	Item	Applied fault information	Estimated results Based on RSCAD Recorded Data	Estimated results Based on physical PMU recorded data
1	Node	632	-	-
	Type	ABCG	ABCG	ABCG
	Section	8	8	8
	Main branch	Yes	Yes	Yes
	Location (feet)	2000	1988.02	2236.56
	Error (%)	-	0.2397	4.73
2	Node	632c	-	-
	Type	CAG	CAG	CAG
	Section	7	7	7
	Main branch	Yes	Yes	Yes
	Location (feet)	3500	3514.26	3696.3865
	Error (%)	-	0.2852	3.93
3	Node	633	-	-
	Type	ABG	ABG	ABG
	Section	6	6	6
	Main branch	No	No	No
4	Node	650a	-	-
	Type	AG	AG	AG
	Section	8	8	8
	Main branch	Yes	Yes	Yes
	Location (feet)	500	500	500
	Error (%)	-	0	0
5	Node	680	-	-
	Type	BCG	BCG	BCG
	Section	3	3	3
	Main branch	Yes	Yes	Yes
	Location (feet)	5000	4999.99	5000
	Error (%)	-	0.000084	0
6	Node	684	-	-
	Type	CG	CG	CG
	Section	9	9	9
	Main branch	No	No	No
7	Node	692	-	-
	Type	CAG	CAG	CAG
	Section	4	4	4
	Main branch	No	No	No

PMU, phasor measurement unit.

11.7 Summary

This chapter implemented the developed IFD scheme of the previous chapters in diagnosing faults on the IEEE 13-node test distribution feeder incorporated with intermittent RERs. It generated various fault events varying prefault loading conditions, renewable energy generation, fault location, type, and fault information (resistance and inception angle). Then, it recorded three-phase faulty current signals to extract the characteristics features of the fault events and trained different MLP-NN to detect and classify the fault events. This chapter also identified the faulty sections and located faults that occurred on the main branch of the test feeder. Besides, it tested the efficacy of the developed scheme in the presence of measurement noise. Finally, this chapter built a prototype IFD to diagnose any fault on the test distribution feeder based on current signals simulated in the RSCAD/RTDS environment and recorded using the physical PMU through the GTAO card. Presented results confirmed the robustness, scalability, and accuracy of the developed IFD scheme in terms of overall accuracy.

References

[1] U.S. Energy Information Administration (EIA), Outlook for future emissions, 2021. https://www.eia.gov/energyexplained/energy-and-the-environment/outlook-for-future-emissions.php. (Accessed 23 February 2021).

[2] U.S. Energy Information Administration (EIA), EIA projects global energy-related CO2 emissions will increase through 2050 - today in energy, 2019. https://www.eia.gov/todayinenergy/detail.php?id=41493. (Accessed 23 February 2021).

[3] U.S. Energy Information Administration, Energy-related CO2 emission. In: International Energy Outlook, Washington, DC, USA, 2016, 2016.

[4] U.S. Energy Information Administration, Data & statistics, 2018. https://www.iea.org/data-and-statistics/?country=WORLD&fuel=CO2emissions&indicator=TotCO2. (Accessed 23 February 2021).

[5] S.D. Ahmed, F.S.M. Al-Ismail, M. Shafiullah, F.A. Al-Sulaiman, I.M. El-Amin, Grid integration challenges of wind energy: a review, IEEE Access 8 (2020) 10857–10878, doi:10.1109/ACCESS.2020.2964896.

[6] Y. Zhang, W. Chen, W. Gao, A survey on the development status and challenges of smart grids in main driver countries, Renew. Sustain. Energy Rev. 79 (2017) 137–147, doi:10.1016/J.RSER.2017.05.032.

[7] M. Bowman, Today in Energy, 2019. https://www.eia.gov/todayinenergy/detail.php?id=41533. (accessed 10 January 2021).

[8] REN21, Renewables 2019 global status report, Paris, France, 2019. https://www.ren21.net/wp-content/uploads/2019/05/gsr_2019_full_report_en.pdf (accessed 17 January 2021).

[9] M. Ahmadi, O.B. Adewuyi, M.S.S. Danish, P. Mandal, A. Yona, T. Senjyu, Optimum coordination of centralized and distributed renewable power generation incorporating battery storage system into the electric distribution network, Int. J. Electr. Power Energy Syst. 125 (2021) 106458, doi:10.1016/j.ijepes.2020.106458.

[10] A. Valencia, R.A. Hincapie, R.A. Gallego, Optimal location, selection, and operation of battery energy storage systems and renewable distributed generation in medium–low voltage distribution networks, J. Energy Storage 34 (2021) 102158, doi:10.1016/j.est.2020.102158.

[11] A.S. Hassan, E.S.A. Othman, F.M. Bendary, M.A. Ebrahim, Optimal integration of distributed generation resources in active distribution networks for techno-economic benefits, Energy Rep. 6 (2020) 3462–3471, doi:10.1016/j.egyr.2020.12.004.

[12] A. Ehsan, Q. Yang, Optimal integration and planning of renewable distributed generation in the power distribution networks: a review of analytical techniques, Appl. Energy 210 (2018) 44–59, doi:10.1016/J.APENERGY.2017.10.106.

[13] S. Kakran, S. Chanana, Smart operations of smart grids integrated with distributed generation: a review, Renew. Sustain. Energy Rev. 81 (2018) 524–535, doi:10.1016/J.RSER.2017.07.045.

[14] A. Rastgou, J. Moshtagh, S. Bahramara, Improved harmony search algorithm for electrical distribution network expansion planning in the presence of distributed generators, Energy 151 (2018) 178–202, doi:10.1016/J.ENERGY.2018.03.030.

[15] M.R. Dorostkar-Ghamsari, M. Fotuhi-Firuzabad, M. Lehtonen, A. Safdarian, Value of distribution network reconfiguration in presence of renewable energy resources, IEEE Trans. Power Syst. (99) (2015) 1–10, doi:10.1109/TPWRS.2015.2457954.

[16] H. Yang, X. Liu, Y. Guo, P. Zhang, Fault location of active distribution networks based on the golden section method, Math. Probl. Eng. 2020 (2020), doi:10.1155/2020/6937319.

[17] P. Chaudhary, M. Rizwan, Voltage regulation mitigation techniques in distribution system with high PV penetration: a review, Renew. Sustain. Energy Rev. 82 (2018) 3279–3287, doi:10.1016/J.RSER.2017.10.017.

[18] R. Liang, G. Fu, X. Zhu, X. Xue, Fault location based on single terminal travelling wave analysis in radial distribution network, Int. J. Electr. Power Energy Syst. 66 (2015) 160–165, doi:10.1016/j.ijepes.2014.10.026.

[19] D. Sonoda, A.C.Z. de Souza, P.M. da Silveira, Fault identification based on artificial immunological systems, Electr. Power Syst. Res. 156 (2018) 24–34, doi:10.1016/J.EPSR.2017.11.012.

[20] J.O. Petinrin, M. Shaaban, Impact of renewable generation on voltage control in distribution systems, Renew. Sustain. Energy Rev. 65 (2016) 770–783, doi:10.1016/J.RSER.2016.06.073.

[21] M. Shafiullah, M.A. Abido, A review on distribution grid fault location techniques, Electr. Power Components Syst. 45 (8) (2017) 807–824, doi:10.1080/15325008.2017.1310772.

[22] Distribution test feeders - distribution test feeder working group - IEEE PES distribution system analysis subcommittee. https://ewh.ieee.org/soc/pes/dsacom/testfeeders/. (Accessed 6 May 2017).

[23] W.H. Kersting, Distribution System Modeling and Analysis, Taylor & Francis, Oxfordshire, UK, 2012.

[24] A.M. Abd-rabou, A.M. Soliman, A.S. Mokhtar, Impact of DG different types on the grid performance, J. Electr. Syst. Inf. Technol. 2 (2) (2015) 149–160, doi:10.1016/J.JESIT.2015.04.001.

[25] D. Atsu, I. Seres, I. Farkas, The state of solar PV and performance analysis of different PV technologies grid-connected installations in Hungary, Renew. Sustain. Energy Rev. 141 (2021) 110808, doi:10.1016/j.rser.2021.110808.

[26] M.B. Jannat, A.S. Savić, Optimal capacitor placement in distribution networks regarding uncertainty in active power load and distributed generation units production, IET Gener. Transm. Distrib. 10 (12) (2016) 3060–3067, doi:10.1049/iet-gtd.2016.0192.

[27] M.A.M. Ramli, E. Prasetyono, R.W. Wicaksana, N.A. Windarko, K. Sedraoui, Y.A. Al-Turki, On the investigation of photovoltaic output power reduction due to dust accumulation and weather conditions, Renew. Energy 99 (2016) 836–844, doi:10.1016/j.renene.2016.07.063.

[28] L. Monteiro, et al., One-year monitoring PV power plant installed on rooftop of Mineirão Fifa World Cup/Olympics Football Stadium, Energies 10 (2) (2017) 225, doi:10.3390/en10020225.

[29] G. Ren, J. Liu, J. Wan, Y. Guo, D. Yu, J. Liu, Measurement and statistical analysis of wind speed intermittency, Energy 118 (2017) 632–643, doi:10.1016/j.energy.2016.10.096.

[30] P. Wang, B. Liu, T. Hong, Electric load forecasting with recency effect: a big data approach, Int. J. Forecast. 32 (3) (2016) 585–597, doi:10.1016/j.ijforecast.2015.09.006.

[31] E. Bayer, M. Steigenberger, and M.M. Kleiner, Report on the German power system, Berlin, Germany, 2015. https://www.agora-energiewende.de/fileadmin/downloads/publikationen/CountryProfiles/Agora_CP_Germany_web.pdf. (Accessed 16 April 2017).

[32] S. Nachet and M.-C. Aoun, The Saudi electricity sector: pressing issues and challenges, 2015. https://www.ifri.org/sites/default/files/atoms/files/note_arabie_saoudite_vf.pdf. (Accessed 16 April 2017).

[33] K.P. Kumar, B. Saravanan, Recent techniques to model uncertainties in power generation from renewable energy sources and loads in microgrids – a review, Renew. Sustain. Energy Rev. 71 (2017) 348–358, doi:10.1016/j.rser.2016.12.063.

[34] S. Teleke, M.E. Baran, S. Bhattacharya, A.Q. Huang, Optimal control of battery energy storage for wind farm dispatching, IEEE Trans. Energy Convers. 25 (3) (2010) 787–794, doi:10.1109/TEC.2010.2041550.

[35] A. Lahouar, J. Ben Hadj Slama, Hour-ahead wind power forecast based on random forests, Renew. Energy 109 (2017) 529–541, doi:10.1016/j.renene.2017.03.064.

[36] S. Deep, A. Sarkar, M. Ghawat, M.K. Rajak, Estimation of the wind energy potential for coastal locations in India using the Weibull model, Renew. Energy 161 (2020) 319–339, doi:10.1016/j.renene.2020.07.054.

[37] Z. Liu, F. Wen, G. Ledwich, Optimal siting and sizing of distributed generators in distribution systems considering uncertainties, IEEE Trans. Power Deliv. 26 (4) (2011) 2541–2551, doi:10.1109/TPWRD.2011.2165972.

[38] Y.M. Atwa, E.F. El-Saadany, M.M.A. Salama, R. Seethapathy, Optimal renewable resources mix for distribution system energy loss minimization, IEEE Trans. Power Syst. 25 (1) (2010) 360–370, doi:10.1109/TPWRS.2009.2030276.

[39] S. Mojtahedzadeh, S.N. Ravadanegh, M.-R. Haghifam, Optimal multiple microgrids based forming of greenfield distribution network under uncertainty, IET Renew. Power Gener. 11 (7) (2017) 1059–1068, doi:10.1049/iet-rpg.2016.0934.

[40] A. Zakariazadeh, S. Jadid, P. Siano, Smart microgrid energy and reserve scheduling with demand response using stochastic optimization, Int. J. Electr. Power Energy Syst. 63 (2014) 523–533, doi:10.1016/j.ijepes.2014.06.037.

[41] L. Cheng, J. Lin, Y.-Z. Sun, C. Singh, W.-Z. Gao, X.-M. Qin, A model for assessing the power variation of a wind farm considering the outages of wind turbines, IEEE Trans. Sustain. Energy 3 (3) (2012) 432–444, doi:10.1109/TSTE.2012.2189251.

[42] P. Siano, G. Mokryani, Evaluating the benefits of optimal allocation of wind turbines for distribution network operators, IEEE Syst. J. 9 (2) (2015) 629–638, doi:10.1109/JSYST.2013.2279733.

[43] L. Xie, Y. Gu, X. Zhu, M.G. Genton, Short-term spatio-temporal wind power forecast in robust look-ahead power system dispatch, IEEE Trans. Smart Grid 5 (1) (2014) 511–520, doi:10.1109/TSG.2013.2282300.

[44] X. Liu, Z. Cao, Z. Zhang, Short-term predictions of multiple wind turbine power outputs based on deep neural networks with transfer learning, Energy 217 (2021) 119356, doi:10.1016/j.energy.2020.119356.

[45] S. Zolfaghari, G.H. Riahy, M. Abedi, A new method to adequate assessment of wind farms' power output, Energy Convers. Manag. 103 (2015) 585–604, doi:10.1016/j.enconman.2015.07.001.

[46] A. Baziar, A. Kavousi-Fard, Considering uncertainty in the optimal energy management of renewable micro-grids including storage devices, Renew. Energy 59 (2013) 158–166, doi:10.1016/j.renene.2013.03.026.

[47] H. Jiang, Y. Dong, Global horizontal radiation forecast using forward regression on a quadratic kernel support vector machine: case study of the Tibet autonomous region in China, Energy 133 (2017) 270–283, doi:10.1016/j.energy.2017.05.124.

[48] G. Mokryani, A. Majumdar, B.C. Pal, Probabilistic method for the operation of three-phase unbalanced active distribution networks, IET Renew. Power Gener. 10 (7) (2016) 944–954, doi:10.1049/iet-rpg.2015.0334.

[49] A. Soroudi, M. Ehsan, R. Caire, N. Hadjsaid, Hybrid immune-genetic algorithm method for benefit maximisation of distribution network operators and distributed generation owners in a deregulated environment, IET Gener. Transm. Distrib. 5 (9) (2011) 961, doi:10.1049/iet-gtd.2010.0721.

[50] X. Zhang, J. Wang, K. Zhang, Short-term electric load forecasting based on singular spectrum analysis and support vector machine optimized by Cuckoo search algorithm, Electr. Power Syst. Res. 146 (2017) 270–285, doi:10.1016/j.epsr.2017.01.035.

[51] Y. Wang, W. Wu, B. Zhang, Z. Li, W. Zheng, Robust voltage control model for active distribution network considering PVs and loads uncertainties, 2015 IEEE Power & Energy Society General Meeting, 2015, pp. 1–5, doi:10.1109/PESGM.2015.7286317.

[52] X. Chen, W. Wu, B. Zhang, X. Shi, A robust approach for active distribution network restoration based on scenario techniques considering load and DG uncertainties, IEEE Power and Energy Society General Meeting (PESGM), 2016, pp. 1–5, doi:10.1109/PESGM.2016.7741591.

[53] C. Cecati, C. Citro, P. Siano, Combined operations of renewable energy systems and responsive demand in a smart grid, IEEE Trans. Sustain. Energy 2 (4) (2011) 468–476, doi:10.1109/TSTE.2011.2161624.

[54] Y.D. Mamuya, Y.-D. Lee, J.-W. Shen, M. Shafiullah, C.-C. Kuo, Application of machine learning for fault classification and location in a radial distribution grid, Appl. Sci. 10 (14) (2020) 4965, doi:10.3390/app10144965.

[55] M. Shafiullah, M.A. Abido, Z. Al-Hamouz, Wavelet-based extreme learning machine for distribution grid fault location, IET Gener. Transm. Distrib. 11 (17) (2017) 4256–4263, doi:10.1049/iet-gtd.2017.0656.

[56] M. Shafiullah, M. Ijaz, M.A. Abido, Z. Al-Hamouz, Optimized support vector machine & wavelet transform for distribution grid fault location, 11th IEEE International Conference on Compatibility, Power Electronics and Power Engineering (CPE-POWERENG), 2017, IEEE, 2017, pp. 77–82, doi:10.1109/CPE.2017.7915148.

[57] M. Ijaz, M. Shafiullah, M.A. Abido, Classification of power quality disturbances using Wavelet Transform and Optimized ANN, 18th International Conference on Intelligent System Application to Power Systems (ISAP), Proceedings of the Conference on, 2015, pp. 1–6, doi:10.1109/ISAP.2015.7325522.

[58] M. Shafiullah, M.A.M. Khan, S.D. Ahmed, PQ disturbance detection and classification combining advanced signal processing and machine learning tools, in: P. Sanjeevikumar, C. Sharmeela, J.B. Holm-Nielsen, P. Sivaraman (Eds.), Power Quality in Modern Power Systems, Academic Press, Massachusetts, USA, 2021, pp. 311–335.

[59] A. Aljohani, T. Sheikhoon, A. Fataa, M. Shafiullah, and M.A. Abido, Design and implementation of an intelligent single line to ground fault locator for distribution feeders, 2019, doi: 10.1109/ICCAD46983.2019.9037950.

[60] M. Shafiullah, M. Abido, T. Abdel-Fattah, Distribution grids fault location employing ST based optimized machine learning approach, Energies 11 (9) (2018) 2328, doi:10.3390/en11092328.

[61] A. Aljohani, A. Aljurbua, M. Shafiullah, M.A. Abido, Smart fault detection and classification for distribution grid hybridizing ST and MLP-NN, 15th International Multi-Conference on Systems, Signals & Devices (SSD), IEEE, Hammamet, Tunisia, 2018, pp. 1–5 2018.

[62] M. Shafiullah, M.A. Abido, S-Transform based FFNN approach for distribution grids fault detection and classification, IEEE Access 6 (2018) 8080–8088, doi:10.1109/ACCESS.2018.2809045.

[63] L.C. Acacio, P.A. Guaracy, T.O. Diniz, D.R.R.P. Araujo, L.R. Araujo, Evaluation of the impact of different neural network structure and data input on fault detection, 2017, IEEE PES Innovative Smart Grid Technologies Conference - Latin America (ISGT Latin America), IEEE, 2017, pp. 1–5, doi:10.1109/ISGT-LA.2017.8126699.

[64] M. Dehghani, M.H. Khooban, T. Niknam, Fast fault detection and classification based on a combination of wavelet singular entropy theory and fuzzy logic in distribution lines in the presence of distributed generations, Int. J. Electr. Power Energy Syst. 78 (2016) 455–462, doi:10.1016/j.ijepes.2015.11.048.

[65] N. Wang, V. Aravinthan, Y. Ding, Feeder-level fault detection and classification with multiple sensors: a smart grid scenario, IEEE Workshop on Statistical Signal Processing (SSP), IEEE, 2014, pp. 37–40, doi:10.1109/SSP.2014.6884569.

[66] J. Klomjit, A. Ngaopitakkul, Selection of proper input pattern in fuzzy logic algorithm for classifying the fault type in underground distribution system, IEEE Region 10 Conference (TENCON), IEEE, 2016, pp. 2650–2655, doi:10.1109/TENCON.2016.7848519.

[67] U.D. Dwivedi, S.N. Singh, S.C. Srivastava, A wavelet based approach for classification and location of faults in distribution systems, Annual IEEE India Conference, IEEE, (2) (2008) pp. 488–493, doi:10.1109/INDCON.2008.4768772.

[68] S. Jana, G. Dutta, Wavelet entropy and neural network based fault detection on a non radial power system network, IOSR J. Electr. Electron. Eng. 2 (3) (2012) 26–31.

[69] J. Zhang, Z.Y. He, S. Lin, Y.B. Zhang, Q.Q. Qian, An ANFIS-based fault classification approach in power distribution system, Int. J. Electr. Power Energy Syst. 49 (2013) 243–252, doi:10.1016/j.ijepes.2012.12.005.

[70] I. Nikoofekr, M. Sarlak, S.M. Shahrtash, Detection and classification of high impedance faults in power distribution networks using ART neural networks, 21st Iranian Conference on Electrical Engineering (ICEE), IEEE, 2013, pp. 1–6, doi:10.1109/IranianCEE.2013.6599760.

Utility practices on fault location

12.1 Introduction

Electrical transmission networks play critical roles in the successful operation of the electric power system (EPS) networks by transferring energy from the generators to the load centers. However, recent deregulation of the electricity market and the economic and environmental requirements pushed the electrical utilities to operate their transmission networks close to their maximum limits. In addition, the electrical distribution systems are currently undergoing a significant evolution with the incorporation of distributed generation from renewable energy resources. The mentioned phenomena introduce protection and operational challenges for the system operators. However, despite the challenges mentioned above, the smooth operation of electric power transmission and distribution systems is essential for delivering power supply to consumers with minimum interruptions [1–5].

Both transmission and distribution lines experience both temporary and permanent faults. At the same time, the former is primarily self-cleared, and the latter can be detected and mitigated with the help of different commercially available fault location solutions. From the solution perspective, fault location solutions can be classified into local/device solutions and system solutions. The term "local" refers to nonrequirement for topology, and advanced network modeling considered the critical factors in the system solution. Restoration of power supply after permanent faults can be done only after repairing the damage caused by the fault. For this purpose, the fault position must be known to avoid inspecting the whole line to find the damage origin. This task becomes even more tedious in high voltage overhead transmission lines since they run up to hundreds of kilometers. In addition, underground cables must be uncovered in the event of a fault requiring more workforce and machines. Moreover, in populated areas, roads, and passageways must be blocked and dug. Therefore, accurate fault location allows saving of both time and money for the inspection and repair work and aids toward better service by utilities through faster power supply restoration [5–9].

Most of the EPS faults occur on overhead transmission lines due to their inherent characteristic of being exposed to atmospheric conditions. Around 75% of the faults occur in transmission networks, revealing the importance of fault analysis for transmission networks. Due to advances in digital metering and modern communication technology, it is now possible to pinpoint the location of a fault in a networked transmission system with very high accuracy. Nowadays, voltage and current waveforms recorded by intelligent electronic devices (IED) such as digital relays, sequence event recorders, and digital fault recorders (DFRs) can be used along with suitable fault location algorithms to estimate the location of a fault with substantial accuracy. These algorithms are incorporated in most microprocessor-based numerical relays.

Power System Fault Diagnosis. DOI: https://doi.org/10.1016/B978-0-323-88429-7.00011-4
Copyright © 2022 Elsevier Inc. All rights reserved.

In case of a fault, the IED device presents the system operator with a detailed report about the fault event and possible locations. At the distribution level, the technology to implement utility-wide fault-location systems exists today. Installed at bus-level or feeder-level, these systems can locate faults precisely across a broad spectrum of distribution systems. Feeder-level monitors are the most accurate, but good fault location can be achieved with bus-level monitors. Protective relays and other monitoring devices with low sample rates (even four samples per cycle) can still give useful fault locations, but of course, more samples are better. The PQ recorders having 128 or more samples per cycle are the best. A relay-level sampling of 16 samples per cycle is good, but it is still insufficient to allow accurate arc-voltage modeling. Metering, communications, and data aggregation continue to improve. Nowadays, fault location systems typically use substation monitors (relays, power quality [PQ] monitors, or advanced revenue-type meters). Future advances are expected to expand this to include monitoring information from line equipment like re-closers, capacitor banks, and customer meters [1,10–13].

This chapter highlights various utility practices for fault location in transmission networks, distribution systems, and underground cables. It discusses different fault location methods along with their deployment in various commercially available fault location solutions. In addition to traditional fault location methods, some advanced fault location approaches in distribution systems are also discussed. Moreover, the chapter includes examples of different utility implementations on fault finding. Furthermore, a list of commonly used and less commonly used fault location methods for underground cables is presented in this chapter. Finally, the chapter ends with a brief description of the prominent commercially available cable fault location solutions.

The remaining parts of this chapter are organized as follows: Section 12.2 presents different fault location methods adopted by the market leaders. Section 12.3 provides a discussion on distance relays, fault locators, and fault recorders. Section 12.4 illustrates different commercially available fault location solutions. Sections 12.5 and 12.6 discuss the fault locations in tapper transmission lines and distribution systems, respectively. Distribution management-based fault locations are presented in Section 12.7, whereas advanced approaches for distribution system fault location are demonstrated in Section 12.8. Different utility implementations are provided in Section 12.9, and the application of artificial intelligence (AI) for EPS fault location is presented in Section 12.10. Various fault location techniques and commercially available solutions for underground fault locations are presented in Section 12.11. Finally, an overall summary of this chapter is presented in Section 12.12.

12.2 Fault location methods

The various approaches commonly used for fault location in EPS networks can be broadly classified into four main categories: impedance-based, traveling wave (TW)-based, knowledge-based, and high frequency-based methods. Other types include the pulse methods, use of fault indicators, and monitoring the induced radiation from the EPS arcing faults using the very low frequency (VLF) and very high frequency (VHF) reception. Detailed illustrations on various EPS fault location approaches can be found in Chapter 1 and Ref. [14–17]. This section briefly discusses the prominent fault diagnosis methods used in different commercially available fault location systems.

Commercially available one-ended impedance-based methods simple reactance method, Takagi method, modified Takagi method, Eriksson method, and Novosel et al. method [1]. The **simple reactance method** avoids any arc impedance added by the fault via calculating reactance to a fault. This method is simple and requires minimum data for fault location, but its accuracy is affected by load and faults current contribution from a remote source. It was used in earlier models of the Asea Brown Boveri (ABB) distance relays. **Takagi method** shows improvement over the simple reactance method as it accounts for system load conditions by subtracting the prefault current from the total fault current recorded by the relay. It is popularly used in Schweitzer Engineering Laboratories (SEL) and general electric (GE) distance relays. **The modified Takagi method** uses source impedance parameters to calculate a correction factor accounting for current infeed from a remote source. **Eriksson method** has a superior performance compared to all other one-ended impedance-based methods as it uses prefault current and source impedance parameters to factor out load and remote source infeed. It is used in the more recent ABB distance relays. **Novosel et al. method** is a modified version of the Eriksson method and can be used to locate faults on a radial transmission line. In this method, prefault voltages and currents are utilized to calculate the impedance of the load, which is used together with the impedance of the local source to negate the effect of load current on fault location accuracy [1,5,13].

The Bonneville Power Administration (BPA) has been using the TW-based technology for fault location in their extra-high voltage networks since the 1950s. In the following two decades, the company installed an automated fault location scheme using microwave communications to send the TW arrival information to the remote terminal to locate the fault. In 1987, the BPA started to measure the TW arrival time using global positioning system (GPS). Practical fault location experience revealed that adding TW-based fault location to line protective relays improves their fault locating capability compared to relays that use only impedance-based fault locating methods [18]. Moreover, fault indicators-based fault location methods are also popular in locating faults in EPS networks as they provide valuable information on fault location. They are either installed in the substation or on towers along the transmission or distribution lines. Data from the fault indicators can be collected and used even during adverse weather conditions through radio links [13]. Fig. 12.1 shows a set of commercially available fault indicators [19–22].

12.3 Local/device fault location solutions

The fault location function can be implemented in microprocessor-based relays, stand-alone fault locators, DFRs, and postfault analysis programs. The inclusion of the fault location schemes in microprocessor-based relays is a common practice as an

A Power Delivery Products Inc.
(41-2001-302)

B Willfar Power
(JYW-60)

C ADC Energy
(SG203)

D Four Faith
(JYZ-HW V2.0)

Fig. 12.1 A set of commercially available fault indicators [19–22].

additional function. Modern relays with high computational capability and advanced communication systems with remote sites can facilitate such schemes incorporation at almost no extra cost. In addition, DFRs enable easy incorporation of the fault location function. Stand-alone fault locators are applied when sophisticated fault location algorithms are used where a higher cost of the implementation is accepted. Finally, post-fault analysis programs are used primarily to verify the operation of protective relays [13].

12.3.1 Distance relays

Distance relays are commonly used for fault location purposes amongst the different types of EPS relays. They are designed for fast and reliable indications of faults occurring in protective zones, making them the most common form of protection of transmission lines. The operation of the distance relays depends on the predetermined impedance value (voltage to current ratio). They are activated if the ratios become less than that of the predefined values. Since the transmission lines' impedances are directly proportional to their lengths thus, the distance relays can only operate if any fault occurs on the lines. However, the distance relays are most useful within reasonable line lengths (up to 20 kilometers) as their operating characteristics are based on the line parameters. Suppose any fault occurs within the predefined protective zone. In that case, a trip signal to the corresponding circuit breaker is sent immediately for quick isolation of the fault to minimize its impact on the EPS network. Distance relays have multiple protection zones for providing backup capability. Usually, the relay detecting the fault in the first zone is designed to trip first. A pair of distance relays are employed to form a pilot relaying scheme through coordination and communication to protect two-terminal lines. Both relays could trip within the first zone based on their exchanged information [5,13]. Fig. 12.2 shows a set of commercially available distance relays [23–26].

12.3.2 Fault locators

The fault locators are employed for network fault location and can also be used for fault type identification and fault impedance calculation. They are usually highly accurate and augment the protection equipment. They can be implemented as stand-alone devices, digital relays, or offline fault analysis programs. The analysis and computation segments of the fault locators form the basis of the fault location and detection techniques. Fault location algorithms are employed in the fault locators using various techniques to locate EPS faults from a reference point with ease and satisfactory accuracy. Fault locators usually use the line data obtained from the standard recording equipment such as microprocessor-based relays and IED when used with conventional protection schemes [5]. Fig. 12.3 shows a set of commercially available fault locators [27–29].

12.3.3 Fault locators versus relays

EPS protective relays and fault locators are closely related; however, they have a few distinguishing features that make them different from each other, as presented below [5,13]:

- Accuracy: Protective relays are employed to identify the faulted areas or regions, whereas fault locators can pinpoint the fault position.
- Operation speed: Protective relays require high-speed operations to mitigate the spreading of faulted current to other parts of the EPS networks using the circuit breakers and high-speed communication devices, sometimes sacrificing the relay system security and selectivity. On the other hand, the calculation in the fault locators is performed offline.
- Data transmission speed from the remote sites: Fault locators use low-speed data communication devices, whereas the protective relays require high-speed devices.

Fig. 12.2 A set of commercially available distance protection relays [23–26].

- Data window: Protective relays use a fault interval between the fault inception and clearance by a breaker around three cycles of the fundamental frequency resulting in a wide data window. On the other hand, fault locators select the best and most compatible data windows to minimize the scope of errors in calculations.
- The complexity of calculations: The high-speed operation of protective relays renders the associated calculations simple and not time-consuming. In contrast, fault locators do not possess any limitations of complexity.

12.3.4 Fault recorders

The TW schemes are highly accurate in locating faults on the transmission lines or span length of conductors. Such accuracy within a few hundred meters of the actual fault point cannot be achieved with traditional impedance-based fault location methods. Since DFRs are primarily employed to achieve high fault location accuracy, they usually use TW fault location methods. In some commercially available DFRs, the highly reliable fault location system uses an innovative and patented combination of TW and fault detection algorithms. The accuracy of the DFR is not affected by the load, line and fault impedance, mutual coupling effects in parallel lines, or reactive

Utility practices on fault location

A Qualitrol Corporation (TWFL-1)

B KEHUI International Ltd (XC-2100E)

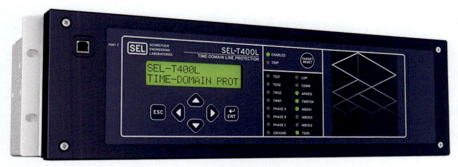

C Schweitzer Engineering Laboratories (SEL-T400L)

Fig. 12.3 A set of commercially available fault locators [27–29].

power compensation elements such as capacitor banks. The DFR Manager software automatically downloads the TW records. It computes the fault location to make it available on its Human Machine Interface (HMI) and via Modbus communication protocol for supervisory system integration. Thus, maintenance crews are dispatched to the fault site [30]. A set of commercially available DFRs is shown in Fig. 12.4 [31–34].

A SIEMENS Energy
(SIPROTEC 7KE85)

B GE Grid Solutions
(Reason DR60)

C AMETEK Power Instruments Inc.
(TR-3000)

D SATEC Power Quality Solutions
(BFM-II)

Fig. 12.4 A set of commercially available digital fault recorders [31–34].

12.4 Commercially available fault location solutions

This section illustrates different commercially available fault location solutions such as distance relays and short-circuits analysis software packages.

12.4.1 Commercially available distance relays

Most commercially available distance relays are built with one-ended impedance-based fault location algorithms as an integral feature. While SEL and GE distance relays use the Takagi method, ABB distance relays used the simple reactance method in their earlier versions and the Eriksson method in their latest models to estimate the location of the fault. Eriksson's approach compensates for fault resistance, load current, and remote in-feed in multiterminal transmission lines. If fault location is enabled in the relay firmware, transmission line length, positive-sequence line impedance, and zero-sequence line impedance must be entered as relay settings. Relays using the Eriksson method also need source impedance data. Fault location in distance relays is quick and fully automated. First, the fault condition triggers the relay to record an event report which consists of prefault and fault voltages and currents. Accuracy of fault location estimates is sufficiently maintained since prefault and fault voltages and currents are typically sampled at a rate of 16 or 32 samples per cycle.

The relay firmware uses this data to extract the required voltage and current phasors. Cosine or FFT filter is commonly used to ensure that direct current (DC) offset and harmonics do not interfere with phasor calculation. Next, the type of fault is determined depending on which relay elements operated during the fault condition. The fault locator can, alternatively, use a separate algorithm to determine the fault type. Finally, the appropriate one-ended fault location algorithms are applied to compute the distance to fault using the user-defined settings [1].

12.4.2 Commercially available short-circuit software

Commercially available short-circuit programs such as computer-aided protection engineering (CAPE), ASPEN OneLiner, SEL-Profile, and other programs can be used to conduct fault location analysis for a given network configuration. These programs are easy to use and can determine the fault location quickly. In case of challenging fault scenarios such as simultaneous faults, evolving faults, and faults with a variable fault resistance, these programs reprocess the event reports and improve the initial fault location estimates given by distance relays. Unlike distance relays which provide only the distance to fault, the distance estimate given by the short-circuit programs can be translated into a transmission line tower location, possible faulted EPS equipment, or specific geographical coordinates since these programs have the complete model of the system feeder. This helps the maintenance crew track down the faulted point quickly [1].

SEL-PROFILE is a DOS-based program allowing the user to run fault location analysis on event reports recorded by SEL relays. First, data in the event report is processed to determine phasor quantities of voltage and current at the line terminal. Then, the program determines the fault type and applies the one-ended impedance-based methods, including simple reactance and the Takagi method. If event reports are available from both ends of the line, the program can use the two-terminal negative-sequence method. Line constant values for fault location calculations are directly read from the event report, and the user can also modify these line constant values [1].

CAPE is a powerful tool that can be used to determine the location of faults in a transmission network. CAPE fault locator window is shown in Fig. 12.5. The fault locator can be accessed through the short-circuit module and performs the following functions [1]:

- Fault location using data from one terminal of the line: If data from one end of the line is available, simple reactance and Takagi methods are used. While the simple reactance method is accurate for bolted faults, the Takagi method is preferred when faults have significant arc impedance. To minimize the effect of fault resistance, zero-sequence current, or superposition current can be used. Users need to input the terminal voltage and current phasors, fault type, and line on which the fault has occurred. The program then searches the database to determine the positive- and zero-sequence impedance of the specified line. Fault location estimates using both simple reactance and Takagi methods are presented in the output window.
- Fault location using data from two terminals of the line: If data from both ends of the line is available, the CAPE program determines the location and arc impedance of the fault using the two-terminal negative-sequence method. User is prompted for fault current magnitudes at both terminals of the line, fault type, and the line on which the fault occurred.

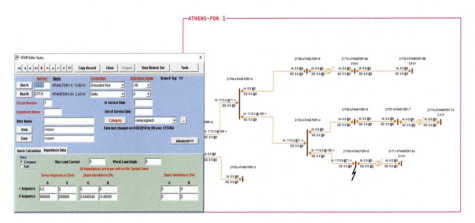

Fig. 12.5 CAPE fault locator window. *CAPE*, computer-aided protection engineering.

- Calculate fault resistance: If fault location is known and current measurements from one end of the line are available, the CAPE program can then calculate the fault resistance, which is useful for determining the root cause of faults. The principle is to use the circuit model and simulate a fault at the known location. Then, the value of fault resistance is iterated until the simulated current matches with the measured phase current.
- Fault location from event recordings: The CAPE program directly reads event reports from IED devices in the field and calculates the fault voltage and current phasors. Users shall enter, as inputs to the program, the fault type and faulted line. Fault location is determined using either one-ended or two-ended fault location algorithms depending on the data available. Then, the same fault is reproduced in the circuit model at the estimated location, and results are compared with measured values of voltages and currents. Therefore, this program can be used to determine fault location and validate the accuracy of the circuit model.
- User-defined macros: The CAPE program allows users to create user-defined macros specific to their system. These macros are written using high-level computer language commands and can be invoked to perform various tasks. These include validating the zero-sequence impedance of the transmission line using IED data, automating the fault location analysis, evaluating the performance of relays, developing short-circuit fault current profiles using other impedance-based fault locations algorithms like the Eriksson method, and more [1].

Fault locator is an advanced analysis tool in **ASPEN OneLiner** that can be used to find the exact location of a fault in a transmission line. The system model is used for performing fault location analysis. Faults are populated in the faulted line with varying fault resistance values. The location at which the simulated fault matches best with the recorded voltages and currents is reported as the location estimate. ASPEN fault locator window is shown in Fig. 12.6. Users need to enter measured values of voltages and currents during fault. Monitoring location can be either at the terminal of the faulted line or in any of the neighboring branches. ASPEN OneLiner is flexible and allows users to input voltage or current in only one phase and omit the other phases. The program's accuracy improves when more measurements are available. The program then

Utility practices on fault location

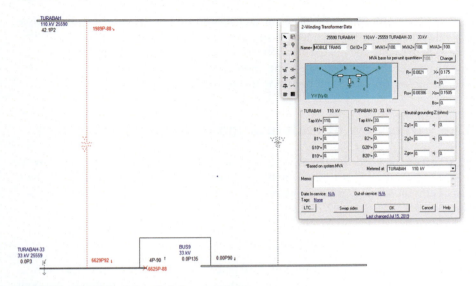

Fig. 12.6 ASPEN OneLiner fault locator window.

asks the user to indicate the type of fault and the line on which the fault has occurred. The user shall also specify the range of fault resistance and the incremental step size to be tried by the fault locator. The program then simulates several faults on the faulted line with the user-defined specifications. Cases are saved if the simulated results match comparatively well with measured values of voltages and currents. Users can select how many best-matched cases should be reported [1].

12.5 Fault location detection on tapped transmission lines

Two traditional approaches are commonly employed to locate faults on the multiple tapped transmission lines. The first approach uses the distance relays with complex algorithms to pinpoint the fault location, and the second one uses multiple reclose operations on lines closing into the faults. These solutions work well in the ideal cases; however, their complexity increases exponentially with the increase of the taps on the circuits. Conventional distance relays have issues with the complexity of this problem as relay under reach and over reach combined with multiple possible fault points that challenge the accuracy and reliability of this approach as it requires visual confirmation of the fault location before isolation. Besides, the complex solutions with multiple reclose operations and precise timing used in conjunction with switches to pick up circuits require multiple closing and reopening to pinpoint the faults through the preplanned trial and error schemes. Thus, such solutions shorten the life of capital equipment and become problematic to the users. Many utilities already reported the inefficient performance of such solutions to locate faults on the multiple tap lines that result in extended outage durations. In response, a SMART TAP™ approach to reduce the outage durations by locating faults on the tapped transmission lines was presented in Ref. [35]. The approach uses lightweight noncontact current measuring sensors or current transformers with fault directional intelligence.

12.6 Overview of fault location in distribution systems

This part is intended to give an overview of fault location in distribution systems where some light is first to shed on the used fault location techniques, followed by a discussion on some distribution system software being used by the electric utilities worldwide.

12.6.1 Fault location techniques

At distribution levels, the commonly used practices in the electric utilities, such as line investigation by maintenance crews and fault analysis-based methods, are both time-consuming and less accurate. For instance, short-circuit faults due to insulator breakdown are challenging to spot visually. Therefore, many efforts have been recently made in developing and improving fault location techniques for distribution systems. Fault location methods used to locate faults in distribution networks are mainly classified into three categories: methods based on TWs, methods using high-frequency components of voltages and currents, and impedance-based methods. The traveling-wave or high-frequency transient-based methods require a high sampling rate for the current signal. Therefore, the method's accuracy can be severely affected by the reflections of the signal waveform traveling along highly branched distribution networks. Impedance-based algorithms use the fundamental frequency voltages and currents measured during a fault at the terminals of a line to estimate the apparent impedance to a fault. Most of the distribution fault location schemes are based on impedance calculation, where many of them are outgrowths of the single-ended transmission-line fault location algorithms. The advantages of using these methods are that they are easy to implement, reasonably accurate, require data from only one end of the distribution feeder, and do not require any communication channel to connect the remote end(s) of the line. Two-terminal impedance-based methods are known to be more accurate compared to their single-ended counterpart. In these methods, voltage and current measurements from both ends of the line are used for fault location calculation. Therefore, communication channels are required to transfer data from one relay to the other. Since distribution systems have dispersed loads, the calculations are initiated in the first line section and then sequentially carried out in the network to determine the steady-state fault conditions for all the sections. The analysis can either be done with symmetrical components or in the phase domain. However, phase domain is recommended since unbalanced loads are common in distribution systems [6,12,36–39].

Digital relays have built-in fault location capability and employ an impedance-based method for fault location. They provide fault location with reasonable accuracy in the case of well-represented utility circuits. In terms of the varying network topologies, however, the method suffers from multiple estimated locations. Another approach commonly used to locate faults in distribution networks is the short-circuit fault current profile method. This method uses the short-circuit model of the distribution system to calculate the distance to the fault. It can also be implemented with the data recorded by digital relays. The circuit model of the distribution system must be available in one of the power system software like CAPE, ASPEN, CymDist, and others. In distribution systems, using measurements from field devices such as voltage sensors can improve the fault location performance in complex system conditions by comparing calculated and sensed voltages. However, these methods require a minimum number of sensing devices and are limited to the device location. Also, using fault indicators in distribution networks can help narrow the faulted area; however, the estimated area range is dictated by the deployed devices. Furthermore, under bidirectional current flow conditions, fault indicators may fail to trace the fault location. The works all suffer from shortcomings concerning the three major issues for locating a fault in emerging distribution systems: multiple location results, high impedance faults, and bidirectional current flow [6,36–39].

12.6.2 Distribution system software

The most frequently used programs by the distribution operation center (DOC) for management of the distribution network are supervisory control and data acquisition (SCADA), basic operation support system environmental (BOSSE), power delivery information system (PoDIS), and advanced distribution management system (ADMS). The SCADA is an IT system for the control and supervision of the power network. Data is recorded in primary and secondary substations throughout the network. Remote terminal units (RTUs) are deployed at all primary and a limited number of secondary substations to communicate with the SCADA system. Signals can be sent both ways through a master terminal unit (MTU), which coordinates the incoming and outgoing signals and connects the information to the SCADA system. Information about the operational conditions of the network and measurements in the primary substations is by these means made available in real-time for supervision by the DOC *via* SCADA [36].

The power grid (PG) is a network information system (NIS) program used as an administrative and documentation tool for the PG structure. The program is used as a master for the whole local grid and contains geographic and electric data. It provides a visual representation of the components and their placements in the network. The grid layout is made available in a map-depiction geographic information system (GIS) environment. All changes in the power networks are edited in PG and then continuously updated to ADMS. Moreover, calculations on the grid can be performed in PG [36]. The ADMS used by some utilities is an administrative system, including a wide range of available information and functions. It combines the real-time data capabilities of SCADA with a visual geographical representation imported from PG to create a powerful distribution management system. ADMS uses colors to show different operational conditions in the grid. If the grid is meshed or fed from another source than usual, then it is shown by different colors. For example, an electricity meter at a customer becomes red when it loses the power supply. A white power line indicates a de-energized part of the grid. A green color indicates an energized part that diverts from the typical operating structure. Progress in active errands and work orders is performed by the operator in ADMS, which then sends updates to BOSSE. The system also has several built-in applications, including fault location, among others [36].

The BOSSE is another administrative system used to log the active incidents and work orders for the local and regional grid. First, incidents in the network are sent to BOSSE and ADMS. Then, a work order is created in BOSSE, which is continuously updated with every step of the process performed by the operator to determine the duration of each step and that it is kept within an allowable predefined value. When the service is restored, the work order regarding fault location and amendment is terminated in ADMS and BOSSE. If the modification is temporary, a permanent solution is performed later to which a new work order is created. This work order is linked to the original errand but results in multiple work orders connected to the same errand. The only way to include information regarding the line restoration process is to link that work order to the previous regarding the fault clearance. The system also consists of a historical database regarding faults [36].

IBM's asset management system (MAXIMO) is another system used by the regional grid operators and is a holding register like PG for the regional grid. In addition, MAXIMO is used to send work orders to contractors, a functionality not available in PG. Persistent faults in the regional grid are also logged in MAXIMO, handled by BOSSE for the local grid. Finally, the PoDIS is a complimentary program giving a visual representation of the latest reported status from electricity meters at the customer level. This feature will be included in ADMS. Different options for the visual representation of the grid are available, and data regarding the number of customers and customer information can be easily accessed [36].

12.7 Distribution management system-based fault location

Computer-based systems used for continuous real-time monitoring and control of the energized power equipment, such as the distribution SCADA system, are commonly referred to as operations technology (OT) systems. OT systems have been operated and maintained by control room operating personnel and support staff. Traditionally, OT systems have been kept physically separate from other corporate computing systems. Corporate computer-based systems such as the customer information system (CIS), GIS, and enterprise resource planning (ERP), primarily used to maintain customer and billing information; records of network assets and static network models; accounting and human resources, have been referred to as information technology (IT) systems. The practice of keeping OT systems separate from IT systems has changed considerably in recent years. There are now more data interfaces between OT and IT systems. A distribution Management System (DMS) is a concept that integrates mostly independent facilities so that the distribution system can be operated in

Utility practices on fault location

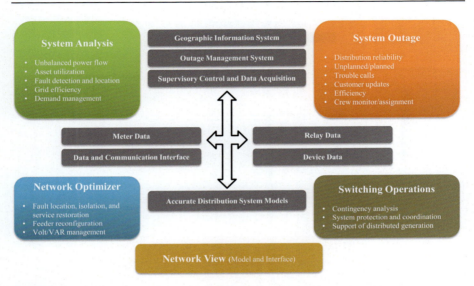

Fig. 12.7 Key DMS functions [39]. *DMS*, distribution management system.

a highly efficient and well-coordinated manner. DMS often includes predictive fault location algorithms that can determine fault location with higher accuracy than is possible with outage management system (OMS) prediction engines, faulted circuit indicators, "last gasp" messages from AMI meters, and protective relay distance-to-fault calculations. Various DMS applications, including fault location, require regular, daily, or even near real-time updates to the electric distribution network model with as-engineered and as-built data from GIS. The key DMS functions are depicted in Fig. 12.7 [39]. While there is considerable variation in the DMS functionality from utility to utility (using DMS solutions from different vendors), fault location is one of the critical applications that is being implemented today. Fig. 12.8 depicts the DMS integrated system concept for fault location application [39].

Several utilities have been using the DMS-based fault location schemes for many years due to their effectiveness. The DMS uses the existing device and data with no requirement for accurate network modeling and special devices deployment for fault location application. The network information system database includes the required data to analyze the fault current. This network modeling and fault current calculation are the DMS parts that provide the basis for fault location. The measured fault current can be acquired from the microprocessor-based relays. The algorithm finds one or more possible faulted line sections based on the available features and collected information. Since the impedance-based fault location results are not explicit, the DMS provides an excellent environment for further processing where fuzzy logic-based approaches are applied. The data from the possible fault detectors and the terrain and weather conditions can also be taken into consideration. DMS user interface is one of the essential components for fault location that provides network geographic view and fault location results. Besides, the DMS provides an entire concept of

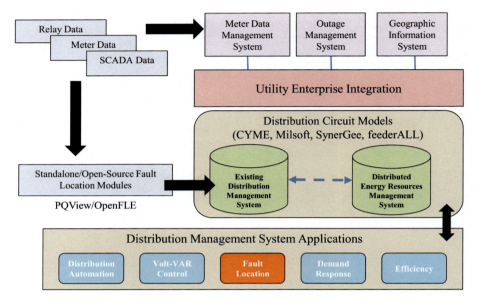

Fig. 12.8 DMS integrated system concept for fault location application [39]. *DMS*, distribution management system.

fault management, including restoration and fault reporting. Using the DMS-based schemes, customer means outage duration can be reduced significantly due to their higher accuracy. Thus, they become part of commercial DMS products for specific applications [6].

12.8 Advanced fault location approaches in distribution systems

This section presents some advanced fault location approaches deployed by different utilities worldwide for fault finding in distribution systems.

12.8.1 Fault location, isolation, and supply restoration

Fault location, isolation, and supply restoration (FLISR) technologies and systems involve automated feeder switches and reclosers, communication networks, line monitors, distribution and OMS, grid analytics, SCADA systems, and data processing tools. These technologies are employed to automate the power restoration process to reduce the interruptions durations and their associated impacts. The FLISR applications can automatically reduce the number of interrupted customers by isolating the faulty part and restoring service to remaining customers via transferring them to adjacent circuits. The FLISR operations require feeder configurations containing multiple paths to the other substations that create redundancy in power supply for the customers located downstream or upstream of the faulted or downed lines due to other disturbances. In addition, the fault isolation feature of the FLISR technology assists the maintenance crews in locating the trouble spots efficiently, thus, expedites the restoration processes and reduces the outage durations [40].

FLISR is an application that can be used within the ADMS program. The application can be run either automatically or manually where every step needs to be initiated. Different FLISR functionalities can also be run separately, provided that the previous chronological step has been completed. FLISR uses two different information sources to determine the fault location in the grid, namely fault currents and fault indicators. Typically, the fault current is retrieved from relays in the gird such as switches and breakers, which for most modern equipment, communication is available. An impedance-based method is then used within FLISR to calculate the distance to the fault, which also puts demand on accurate and updated cable and line data. When fault current is used without fault indicators, only a distance to the fault is obtained, and therefore multiple locations are provided. On the other hand, fault indicators need to be placed at strategic locations in the network, either directly at overhead lines/underground cables or in secondary substations at outgoing lines. If fault indicators are deployed but fault current is not available, then the highest accuracy is attained on the faulted line. No information can be given regarding where on the line the fault has occurred. The isolation and service restoration functions of FLISR can only be fully implemented if remote-controlled switches exist in the grid. If such equipment is not present, the steps can still be performed. A suggested action plan can be presented, which can then be mediated to the operators in the field and a manual reconfiguration executed. In some cases, FLISR is not used as an active part of the fault location process primarily due to the lack of necessary equipment in fault indicators and data about fault current levels.

Siemens Distribution Feeder Automation System (SDFA) is an example of a commercially available solution that automates the FLISR tasks. The SDFA automatically restores service to viable sections of the line, thus, minimizes outage durations, and dispatch expenses [41].

12.8.2 PQView

One of the most significant development areas for PQ monitoring is implementing intelligent systems for automatic evaluation of the disturbances and conditions. The Electrotek Concepts and Electric Power Research Institute (EPRI) developed software for such purposes, namely the PQView. The software is used for constructing, organizing, and analyzing the PQ databases. It integrates data from various sources such as protective relays, DFRs, smart meters, PQ monitors, and SCADA systems and converts them into an open relational database. The architecture of PQView is shown Fig. 12.9 [42].

Measurements in PQView are stored in Microsoft Access or Microsoft SQL Server format. Advanced users can write their queries, reports, web pages, and custom applications. PQView supports measurements from thousands of meters. PQView can store up to two billion events and hundreds of billions of steady-state measurements in

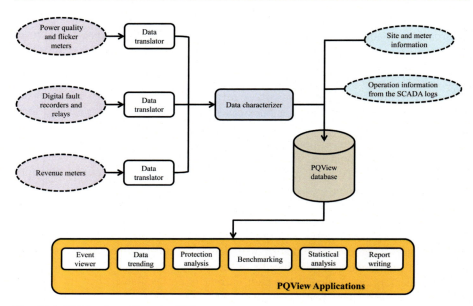

Fig. 12.9 PQView's architecture [42].

one database. With Microsoft SQL Server 2005, there is no practical limitation with database size. Querying events from databases with even millions of PQ events take few seconds to run, and the system does not bog down as the measurements database grows. The system will support 100 or more concurrent users using PQWeb with measurements in a SQL Server database. As part of this system, PQView includes a fault analysis capability that automatically combines the recorded data and other relevant information to provide an estimated fault location. A complete PQView system comprises three main applications: PQ Data Manager (PQDM), PQ Data Analyzer (PQDA), and PQWeb. The primary function of the PQDM is to build PQ databases automatically from data sources. PQDM also imports and characterizes measurements from data sources. In addition, it performs some automated functions after data is polled, such as email notifications, waveform identification, and analysis for fault location. The PQDA queries measurement databases via workstations on Ethernet and create reports. It produces trends, histograms, and statistical summary tables of more than 125 steady-state characteristics defined within the IEEE PQ data interchange format (PQDIF) standard. It also offers event lists, tables, and indices, and scores of charts for voltage sag, swell, and interruption analysis. The PQDA interfaces with Microsoft Word to produce summary documents and allow the user to filter invalid measurements from final analysis automatically. The PQWeb queries measurement databases via a World Wide Web browser on the intranet or internet. PQView features a reactance-to-fault (XTF) module that can be used to determine the distance to a fault from a substation monitor. Estimates of positive-sequence impedance between substation and feeder structures can be compared with electrical circuit model data from circuits modeling databases. PQView can then provide maps pinpointing the fault location. The estimations for fault location typically are available within minutes after the fault has occurred [10,11,39,42,43].

12.8.3 Voltage drop-based fault location

In 2003, Hydro-Québec implemented the voltage drop-based fault location (VDFL) scheme, which forms the basis of developing an application known as the MILE. The main steps of the VDFL fault-location technique include remote measurement of the voltage drop, the automated grouping of independent measurements, and contextual modeling of the distribution feeder, and determination of the fault location using an original algorithm. The VDFL technique is based on the measurement of voltage drop phenomena along the feeder caused by fault current. Fig. 12.10 presents a diagram of the overall VDFL fault location process [43].

The VDFL technique requires at least three power quality monitors (PQMs) installed, usually on the customer side at selected sites on the distribution feeder. In case of a fault, the resulting voltage drop is recorded by the PQM, and the measurements are transmitted to MILE for analysis. The technique uses symmetrical components of the voltage phasors and the balanced line model for analyzing every type of fault. In addition, the method provides the possible lateral taps and their corresponding fault currents, fault locations, and fault voltage amplitude (arcing voltage). The overall results are analyzed to determine the most accurate fault locations to be presented on the user interface. This set is then sent to the operation's control center or used by maintenance staff to analyze system performance. As of now, seven distribution feeders have been monitored with this technique, and the automated specialized software package MILE was developed because of the experiments. The MILE locates transient faults, measures PQ, and collects other information required for system monitoring and maintenance. The implemented fault location scheme is based on triangulating voltage variations at several selected measurement points along the power lines. The study showed that the deployment of four monitoring devices per distribution feeder usually suffices to accurately locate faults on most of Hydro-Québec's overhead distribution feeders. This technique's accuracy is with 1%

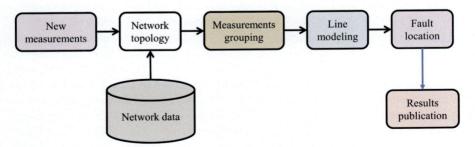

Fig. 12.10 Overall VDFL fault location process [43]. *VDFL*, voltage drop-based fault location.

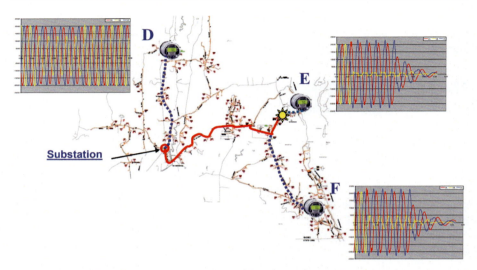

Fig. 12.11 Example of an event recorded on Hydro Quebec's monitoring system [42].

precision, and it is little affected by topology and line impedance errors. Fig. 12.11, as taken from Ref. [42], shows an example of voltage sags captured due to a line to ground fault. The dashed blue line marks the three-phase mainline. While site D records the voltage sag measured at the substation, site E records the lowest voltage as shown by the voltage waveforms given in yellow. The fault location is then determined by interpolating between the voltage at site E and the voltage at site F. The algorithm uses an iterative solution method using a model of the circuit where several synchronizations, approximations, and error corrections are performed [39,42–45].

12.8.4 Travelling wave approach

The TW approaches are primarily based on the measurement and analysis of propagation time associated with fault effects. They are less sensitive to modeling errors since they do not rely on network data such as line impedance and load levels. However, there are many challenges associated with the use of traveling-wave-based fault location methods in distribution systems. First, the significant presence of laterals in some complex distribution system topologies gives rise to many wave reflections. It thus creates the need for high-frequency sampling rates and analysis tools like Wavelet-transform. In some cases, the combined use of advanced feature extraction techniques and classification techniques is required. An example of an advanced feature extraction technique includes the use of discrete wavelet-transform combined with multiresolution analysis. Classification techniques are usually based on computational intelligence such as artificial neural networks (ANNs), fuzzy logic (FL), adaptive-network-based fuzzy inference system (ANFIS), or support vector machines (SVMs). Also, when a fault occurs close to the terminal, the time difference between a fault and its reflection will result in a short detection time. The difficulty of detecting the waves caused by discontinuities could be managed by implementing filters and high-frequency current or voltage transformers, resulting in a higher investment cost. Furthermore, the grounding resistance in the distribution system affects the TW and decreases the return. This problem can be handled by implementing adaptive filters, which can make up for signal attenuation [36,38].

Another unique approach for fault location in distribution systems is based on triangulating the radiated arc signal from a fault. This method is considered as an extension of lightning detection technologies operating on similar principles. For an arc on an overhead system, the breakdown across a gap will produce a sharp discharge radiating in all directions as the air breaks down. A sensor on the distribution circuit near the substation is used to pick up the transient as it propagates along with the distribution system. Multiple antenna stations distributed geographically are installed to pick up the radiated arc signal. These measurements can be accurately time-tagged using GPS technology. Each antenna receiving station can determine the direction of the discharge. Fault location is estimated by triangulating from multiple stations. This approach has not been implemented so far, although some tests have been done on distribution circuits with receiver stations [42].

12.8.5 Neutral/ground current ratio

A trial of a fault locating approach that used neutral current measurements has been reported. This demonstration project with Cannon Technologies, Inc. and Xcel Energy, Inc. used IED distributed along the mainline with feeder neutral current and pole down-ground current measurements recorded at each site. Data was collected in a central server, and 900-MHz spread-spectrum radios were used for communications. Six sites were selected in the joint trial demonstration along a circuit with a mainline having no branches. In a staged fault test, the fault could be located approximately as being just downstream of the IED with the most neutral current and between the IEDs with the most down-ground current. This approach is very cost-effective as only low-voltage currents are monitored. However, it only works for faults involving the ground. Moreover, some circuits may have different ground current flows during faults, and therefore more experience and possibly tuning are needed. This approach could be improved by using system models to adjust neutral currents, decreasing the number of sites required. Improvements could also be achieved by knowing the phase angle difference between the current in the neutral and the down ground [42].

12.8.6 Other approaches

Alabama Power uses switched capacitor banks as distributed fault-locating nodes. In this approach, all three phases of capacitor bank voltages and currents are monitored. The RTU can have a current level threshold to detect faults. This system works like a fault indicator where no distance estimate is attempted, and no waveforms are recorded. Further research may be possible to derive a fault distance estimate even without waveforms. RMS voltage sag and fault current information may help pinpoint the fault locations. Other advanced fault location approaches include the following [42]:

- Wireless fault-circuit sensors: In this approach, fault indicators with communications help narrow down fault locations, primarily on branched circuits. The primary obstruction in implementing this approach is the cost and communications infrastructure required. However, when putting more "smart grid" technologies in place, fault locating should take advantage of that. In addition, fault location accuracy could be improved if RMS currents are captured and used along with system models.
- Magnetic field sensors: One way to reduce sensor costs is to detect magnetic fields created by fault currents. This could be used in a line to ground faults, but cases involving line to line or three-phase faults are more difficult to detect. In addition, communications infrastructure is needed in this approach.
- Other traveling-wave approaches: Traveling-wave methods have been used successfully on transmission lines, and they have been proposed for distribution lines as well. However, distribution lines are much shorter, and their associated sensors and technology are more expensive.

12.9 Examples of utility implementations

In this section, we shed some light on four examples of various implementations that have been put in practice by different utilities worldwide for fault location.

12.9.1 Progress Carolina monitoring setup for fault location

Progress Carolina (PC) uses an advanced monitoring system for fault locating by recording the steady-state trend data and fault events using the RTU at a sampling rate of 16 samples per cycle. The PC fault-location system started as a spreadsheet assuming a constant conductor size for a given circuit. The spreadsheet would estimate the fault distance once the fault current and fault type are entered. Success with that led to the development of several more automated systems to locate faults, including [10,11]:

- Fault distance tool: A web tool available to all who provide distance to the fault considering each feeder has the same wire size.
- CYME software: It offers several possible fault locations using actual wire sizes. However, it needs more manual processing.
- OMS: It offers locations to operators using actual wire sizes and can locate feeder lockout automatically.

The reported fault location accuracy of the PC was excellent as it located faults within 0.5 miles of the actual location for almost 75% of cases for most of the remaining cases within 2 miles of the exact location.

12.9.2 Con Edison reactance-to-fault application

Con Edison has implemented a fault location system in the New York City area to maintain network reliability, reduce fault-locating time and cos, and dispatching crews efficiently. The fault location scheme system uses PQ monitoring voltages and currents on a substation transformer at a sampling rate of 128 samples per cycle. It calculates the reactive part of the impedance to the fault and compares it with the network reactance for fault location. The method uses the residual current to pick out faults involving paths to the ground. This is particularly effective for Con Edison since its load is mainly secondary network load connected through line-to-ground transformers. It does not include zero-sequence impedance; therefore, adjustment factors (k-factors) tuned for each site are used to adjust for the differences between the loop impedance to line-to-ground faults and the positive-sequence impedance. It uses PQView to download data from the recorders and estimate the reactance to the fault for fault events. An internal web application provides an operator interface. Con Edison has been happy with its performance to date that in an examination of 27 events at one location, 70% were within three manholes [10,11].

12.9.3 Open source-based fault location using OpenFLE/OpenXDA

The openFLE was developed by the Grid Protection Alliance to meet TVA's needs for automatic processing of event records to determine the distance to a fault. The openFLE development was part of an EPRI demonstration project. The goal of this demonstration project was to position an event record for analysis and make system parameters available to the process, then process the record and provide results automatically. In this project, system parameters were prepositioned for the set of lines to be studied, and parsers were developed for both COMTRADE and PQDIF event records. The openXDA is an extensible platform for processing records from PQ monitors and disturbance monitoring equipment such as DFRs. In addition to event records, openXDA is also suitable for including other types of EPS data such as SCADA and PMU data. The openXDA provides the platform to implement any appropriate algorithm for analyzing disturbance records either in the time domain or in the frequency domain. In one implementation of openXDA, the openFLE service provides the analytics to determine the fault type and calculates the distance to the fault. Initial testing of the system indicates that fault distance information can be delivered to system operators in less than two minutes [39].

12.9.4 Stockholm Fault Indicators

In Stockholm, a pilot project has been initiated in a selected area within the active grid to evaluate the FLISR system. The substations are equipped with circuit breakers, residual current breakers, and short circuit protection at both MV and LV-level to meet the fault diversion criteria set for the area. At the LV-side, the relay is triggered at around 10–25% overload of the transformer at the MV-side; the permitted overload usually is about twice the ampacity. Several fault indicators have been installed within the area and equipped with remote communications to facilitate the connection to the DOC. The deployment of fault indicators is imperative to enable the use of impedance-based fault location methods. Fault indicators react to and display the presence of fault current. From the perspective of a fault, all upstream indicators will be triggered while downstream indicators will not. This way, the location of a fault can be narrowed down. Since the inception of this pilot project, there has only been a limited number of faults in the area, which has affected the possibility of using the FLISR system. The rare fault occurrence and the whole day shift work at DOC have hindered the use of FLISR to amend an ongoing outage. Instead, the method has been used in a later purely analytic step when the actual fault location is known. The analysis was performed in a test environment of the ADMS system displaying a replica of the active system where a historical fault can be reproduced artificially. In a time-consuming activity, the fault current levels are extracted manually by calling the relays due to the unavailability of communications with the relays in the area. By inserting a fault at the actual point and with retrieved fault current levels, tests indicate a high level of accuracy within a few meters with the FLISR system when using both fault currents and fault indicators. The application FLISR can help determine the distance to a fault or just use the indicators to show the area or affected section [36].

12.10 Artificial intelligence deployment for fault location application

Employing AI tools for protection purposes is extremely limited. Some applications, such as relay initialization and coordination, were created for offline use. A recent intelligent application was developed to detect the fault that used both pattern recognition and an expert system. The proposed scheme can detect downed conductors, temporary arcing faults, and persistent arcing faults for distribution networks. However, for various reasons, neither FL-based nor ANN-based schemes have been used commercially for protection purposes. To begin with, the majority of these AI-based relaying schemes were developed to replace traditional relaying strategies that were entirely reliant on full AI-based protection functions. In addition, the majority of these applications used AI tools in a novel way, with no or minor technical improvements. Besides, the dependability of the schemes has not been emphasized. The economic perspective also plays a fundamental role. As a result, industrial acceptance of AI-based modern technologies to replace traditional technologies is still limited. These AI tools, however, have superior capabilities and performance when compared to other conventional methods. The optimum deployment of the AI tools can be realized when dedicating them to those tasks with either inefficient conventional solutions or no solutions. Therefore, it is preferable to use AI to enhance the performance of existing traditional methods rather than to replace them entirely. This can be accomplished using either the FL-based or the ANN-based approaches in fault location in the EPS networks and traditional techniques [46].

12.11 Underground cable fault location

Cable faults often arise because of insulation failure with moisture penetrating the cable. Mechanical stress due to cable laying process, manufacturing defect, excavations, the surrounding soil or water is also a common cause of cable faults. Treeing is a phenomenon that occurs in cables where a local increase in electrical stress propagates through the insulation until it fails. As in overhead lines, open circuit faults in cables are caused by a broken or discontinuity in the conductor core. The fault type can be determined by measuring the resistance between two conductors after the cable has been taken out of service and grounded. While infinite resistance indicates a fault, a resistance close to zero indicates a functional line. A zero-resistance measurement indicates short circuit faults and ground faults in cables. Short circuit faults occur only in multicored cables when the individual insulation fails, and the conductors come in contact. Ground faults can arise due to failing insulation which results in a connection to the ground. Cables often deteriorate due to chemical reactions with the soil or due to vibrations or mechanical crystallization [36].

This section identifies many fault location methods commonly used by utility fault location crews, both in terms of frequency of application by crews and available equipment on fault location trucks. The methods are classified into two main categories namely, terminal methods and tracer methods. Terminal methods rely on measurements taken at one (or both) terminals of the cable. On the other hand, tracer methods rely on measurements taken along the cable route. Moreover, this section identifies various terminal and tracer methods that are less commonly used but represent alternative fault location methods under special circumstances where the fault locating crew has the available equipment needed to perform the technique. For each technique, applications, limitations, and test equipment required are adequately addressed. Reference [47] can be referred to for a detailed description of each method.

12.11.1 Commonly used methods

This part presents five terminal methods and one tracer method commonly used by utility fault location crews to find faults in underground cables.

12.11.1.1 Voltage drop ratio method (terminal method)

This method is sometimes referred to as the Fall of Potential method. Test equipment required with this method is relatively simple. A stable ripple-free DC current source is needed to circulate the current between the faulted and un-faulted conductors. Also, suitable jumpers are required at the far end of the circuit. Finally, DC voltmeters are needed for measuring the voltages at the near end and the far end. These voltmeters must have high internal resistance relative to the resistance of the fault for the location equations to be applicable, and the two meters should have the same internal resistance. The method locates both single phase-to-ground and double phase-to-ground faults on either three-conductor shielded cable or single-conductor shielded cables. It also locates phase-to-phase short circuits on three-conductor nonshielded cables. To use the method, there must be an unfaulted phase conductor through which the DC current is circulated. Moreover, the faulted conductor must have a uniform resistance along its entire length. If the cable circuit is made from cables with different size conductors, it is challenging to use this method. One utility reported that this method successfully located the fault on a 230 kV, seven-mile-long, pipe-type cable where other fault location methods were not. In addition, the technique has also proven effective in locating damage to the polyethylene jacket on three-conductor self-contained fluid-filled (SCFF) cables and locating an oil leak on an SCFF three-conductor submarine cable. This method, however, will not locate three phase-to-ground shunt faults or open conductor faults. If the fault has burned away a significant portion of the conductor at the point of fault, the accuracy of the method can be significantly affected. Stray currents existing in the system or galvanic potentials at the point of fault can also affect the accuracy of the method. Furthermore, if the entire cable's insulation resistance is not considerable compared to the fault resistance, there can be significant error [47].

12.11.1.2 Capacitance ratio method (terminal method)

The capacitance ratio method is intended for detecting open conductors on shielded cables only. This method requires a capacitance bridge, a direct reading capacitance meter, and connects to the cable. This method locates open conductor faults on three-conductor shielded cables and single-conductor shielded cables. If there is an unfaulted conductor in parallel and of identical construction to the faulted conductor, then the measurements need only be made from one end. Performing measurements from both ends are, however, recommended for cross-checking the distance to the fault. An accuracy of 3% or better is possible with this method, which is relatively fast and inexpensive for locating open conductor faults. The technique will not work well if the fault resistances are low concerning the capacitive reactance at the frequency of the measuring device. The capacitance readings with the capacitance measuring equipment will be less accurate if the capacitance meter, capacitance bridge, or other measuring devices cannot compensate for the resistances at the open conductor. If a direct reading capacitance meter is used, the errors can be significant when the fault resistance is less than about one megaohm. When a bridge circuit is used for making the capacitance measurements, the method may be effective for fault resistances in the range from 200 to 300 ohm. However, at-fault resistances below 1000 ohm, the fault itself may have appreciable capacitance, and the results will be misleading. Interpretation of the measured capacitances can be difficult if the cable circuit is spliced with cables of different capacitances per unit length. The sections length and total circuit length are unknown [47].

12.11.1.3 Murray loop method (terminal method)

The Murray loop method is the most commonly applied bridge method. It requires the equipment for the bridge circuit, including a galvanometer or null detector. Low-voltage bridges are available from different manufacturers, and some utilities have manufactured their bridges. Most bridges operate at DC voltages of 10 kV or below,

Utility practices on fault location

but bridges have been made operating up to 25 kV DC using electronic null detectors. With the 25 kV DC bridges, faults with resistances up to 8 megaohms have been successfully located. A fault burner is needed to lower the fault resistance if the fault resistance is too high for the bridge sensitivity. The method can be used to locate single phase-to-ground faults on both single-conductor, and three-conductor shielded cables. Location of two phase-to-ground faults on three-phase circuits and phase-to-phase faults on belted cables is also possible. This method can be very effective when either fault resistance is low enough, or the bridge DC voltage is high enough to obtain the cable length sensitivity. Accuracies of better than 1% are generally possible with this method when the effects of the jumper and test leads are taken into consideration. The technique will provide suitable circuits with accurate circuit length data, such as transmission lines, submarine, and duct/manhole installations. It has been used on circuits with taps for fault prelocation by jumping the end of selected taps one at a time. However, the method cannot be used on circuits with three phase-to-ground faults unless an un-faulted conductor of known length is available from a nearby circuit. Also, it is not applicable to open conductor faults. With a high-resistance fault, an elevated DC voltage may be required in the bridge, or else the resistance of the fault must be lowered by burning such that a low-voltage bridge can be used. The fault resistance must be significantly less than the insulation resistance of the faulted cable and its terminations. The accuracy of the method can be substantially affected if there is a high resistance connection in the instrumentation loop or if the faulted conductor's resistance per unit length is not relatively constant. Stray DC currents in the ground can also affect the accuracy of this method. The method cannot be used when the far end of the circuit cannot be accessed to connect the jumper [47].

12.11.1.4 Time-domain reflectometer/cable radar (terminal method)

This method is also known as the low-voltage pulse-echo method, the low-voltage pulse reflection time method, and simply radar. Test equipment needed with this method is a low voltage radar set which is available from various manufacturers. These manufacturers also supply the necessary coaxial cables and equipment for connecting the radar set to the cable under test. A short-duration current pulse is injected into the cable by a radar set. At points of discontinuity, a portion of the pulse will reflect the set that allows estimating the distance to the fault, knowing the wave propagation velocity along the cable. Radar pulses can range from 5 ns to 5 µs wide, depending on the test set and settings. Narrower pulses give higher resolution to differentiate better the faults, reflections off splices, and other discontinuities. Cable radar or time-domain reflectometer (TDR) methods can be used on both single-conductor and three-conductor shielded cables to detect open conductor faults, low impedance shunt faults, and short circuits. If there is an unfaulted phase conductor of known length with the same characteristics as the faulted phase, this method can determine the velocity of propagation for the faulted phase. The method will work if the shunt fault resistance is less than about five times the cable surge impedance. For series faults, better known as opens, the resistances at the fault point must be higher than about ten times the characteristic impedance of the cable. The low-voltage radar will not detect a shunt fault on cables with extruded insulation since the fault resistance can be very high. The same is true for most shunt faults on paper-insulated cables. Most open conductor faults can be detected with radar methods. If there is an open in the concentric neutral of a primary distribution cable, return pulses will be produced, complicating the interpretation of the radar trace and the fault location process. Cables submerged in water-filled conduits may also attenuate the signal, making interpreting radar results more challenging. In transmission circuits with single-conductor cables and cross-bonding of sheaths, the surge impedance changes at each bonding location. These changes can affect the fault prelocation techniques, and they produce reflections that can complicate interpreting the display. Cross bonding connections may need to be temporarily reconfigured at each link box so that the cable under test uses the same conductor and concentric shield/sheath for the entire test length. Alternatively, the cable radar may be applied to each section separately, but this can add to restoration time [42,47].

12.11.1.5 Time-domain reflectometer/cable radar and thumper method (terminal method)

This method is also known as the radar with high-voltage pulse method, the arc radar method, and the arc reflection method. After a fuse or other circuit interrupter clears a cable fault, the area around the fault point recovers some insulation strength. An ohmmeter test on the cable would reveal an open circuit. Similarly, the radar pulse passes right by the fault, preventing the radar set alone to detect the fault. This challenge can be overcome using radar with a thumper. A thumper pulse breaks down the gap, and the radar superimposes a pulse that reflects off the fault arc. The rise time of the thumper waveform is on the order of a few microseconds, and the radar pulse total width may be less than 0.05 microseconds. Another approach is to continuously burn the cable with a thumper until the fault resistance is low enough to read on a radar set. This approach, however, is less attractive as it subjects the cable to many more thumps, primarily if maintenance crews use high voltages. The three significant equipment required with this method include a low-voltage cable radar set or the TDR set, an arc reflection filter/coupler, and an impulse generator or thumper. Numerous manufacturers supply both complete equipment packages and individual components for use with this method. This method locates shunt faults (shorts, high-impedance/ nonlinear) on both single-conductor shielded cables, and three-conductor shielded cables. It also locates open-conductor faults by operating only the low-voltage cable radar unit subject to knowledge of the circuit length by the operator. With this method, it may be challenging to interpret the results seen on the screen if different cable sizes are spliced together or taps and bifurcations in the circuit. This, however, can be partially overcome if the unit can compare traces with and without the thumper operating. Also, the cable radar signal on paper lead cables is attenuated more than on XLPE cables due to the higher losses. In most network systems, the only place where the radar can be connected is at the substation due to lead-sheathed cables. In addition, while the radar can be used to determine the distance to the fault, it will not indicate the direction to the fault at a tap location. The method will not work on higher-voltage transmission cables if the impulse generator output voltage is not high enough to generate a breakdown at fault [42,47].

12.11.1.6 Thumper/acoustic method (tracer method)

The thumper/acoustic method is one of the most widely used fault-pinpointing strategies. It is also known as the thumper method, the capacitor discharge method, the capacitor impulse method, the time differential method, the impulsing method, the sound directional method, and the lightning and thunder method. This method applies a pulsed DC voltage to the cable. At the fault point, the thumper discharges sound like a thumping noise. The maintenance crews can determine the fault location by listening mentioned thumping noise. Acoustic enhancement devices can help maintenance crews detect weak thumping noises. Also, antennas picking up the radio-frequency interference from the arc discharge can help find the fault location. Thumpers help to find the exact fault location so that maintenance crews can start digging and repairing the fault. On a 15 kV distribution, utilities typically thump with voltages ranging from 10 to 15 kV, but they sometimes use voltages up to 25 kV. The equipment needed for this method is the transmitter, thumper unit, and detector. The thumper unit consists of a power supply, capacitor, and a fixed or adjustable sphere gap. The detector consists of the acoustic sensor, electromagnetic detector, and associated electronics.

This method is used on secondary and primary distribution cable systems as well as underground transmission cable systems. It applies to buried cables, cables in conduits, and pipe-type cables. If a cable is located in a conduit, the thumping noise is generally louder at the conduit openings than at fault itself. Therefore, the two ends of the conduit which contain the fault will be located using the acoustic method. Thumping can be applied to various cable insulation systems, including extruded dielectric, oilpaper, and paper-insulated-lead-covered (PILC) cable insulation. Thumping can be performed on very long cable lengths up to 23 km, provided the capacitance in the thumper is large compared to the cable capacitance. The types of faults pinpointed with this method are high impedance nonlinear faults, which open when the far end of the circuit is jumpered to the ground, and low impedance faults when a compassionate detection scheme is used. Soil conditions and factors such as moisture, soil type, rock content, density, etc., will affect the effectiveness of this method. The depth of burial of the cable under test also affects the efficacy of this method since, in any soil, the traveling acoustic wave will be more strongly attenuated than if it were traveling in the air. If the cable shield/sheath or concentric neutral serves as a return path for the impulse current with breaks in the shield, arcing may occur at both the break in the shield/sheath and at the fault location, complicates pinpointing the fault. A continuous current return path should be established for this method to yield successful results in such a case. The accuracy of this method is independent of the cable length as long as fault breakdown is achieved and is in the order of a fraction of a meter. On shielded power cables, this method may be the most popular one with a higher success rate than other techniques.

The thumper/acoustic method should not be used before some prelocation work has been performed. The limited distance from which the acoustic wave will travel precludes using this method on longer cable runs without prior prelocation. This method is not appropriate to locate an actual short. It will not work if the magnitude of the discharge voltage is lower than the magnitude of the voltage it takes to arc over the gap at the fault location. Thumping should be limited on pipe-type cables as it can contaminate the pipe filling fluid because the high temperature of the arc generated at fault will break down the filling fluid. Thumping may also burn paper tapes [42,47].

12.11.2 Less commonly used terminal methods

With terminal fault location methods, measurements are made from the terminals of the cable circuit. The purpose of these methods is to prelocate the fault as accurately as possible from either one or both ends of the cable circuit. The terminal techniques can approximate the fault location but will not locate the fault with sufficient accuracy to allow digging, at least for direct-buried cables. In other words, the terminal methods will not pinpoint the fault location, but they will localize or prelocate the fault. Following fault prelocation, a tracer fault location method is generally employed to pinpoint the fault location. However, if the cable system is installed in conduits with intermediate manholes/vaults, terminal methods may provide an adequate location to identify a failed section between adjacent manholes/vaults or between the terminations and the first manhole outside of each substation. In these cases, the failed section will probably be replaced by pulling out the failed section and rebuilding the joints on either end. This part discusses different terminal methods less commonly used by utility fault location crews to find faults in underground cables. These methods represent alternative terminal fault location methods that can be used under particular circumstances when the equipment needed for these methods is available with the fault location crews.

12.11.2.1 Halfway approach method

The halfway approach method is also known as the trial-and-error method, the sectionalizing method, the cut-and-try method, and the divide-and-conquer method. On a radial tap where the fuse has blown, maintenance crews narrow down the faulted section by opening the cable at locations. The crews begin by opening the cable near the center and then replacing the fuse. The fault occurs upstream if the fuse blows; if it does not, the fault occurs downstream. The crews then open the cable near the center of the remaining section and continue bisecting the circuit at the appropriate sectionalizing points. Pad-mounted transformers are usually selected as the sectionalizing point. Whenever a cable fault occurs, more damage is encountered at the fault location, and the rest of the system is stressed by carrying the fault currents. Current-limiting fuses can reduce the fault current stress, but the additional cost is incurred in this case. The test equipment required with this method is relatively simple and available to most fault-locating personnel. A fault on a given section can be detected with an insulation resistance tester, a megohmmeter, an AC or DC voltage test set, or a dielectric test set. In those situations where the fault resistance is relatively low, a common ohmmeter can be used. This method will locate all types of shunt faults. Also, the method will identify open conductor faults if the far end of the cable section being tested is grounded. Furthermore, the method can be used on circuits with single-conductor or three-conductor cables. It is not affected, as in other methods, if the circuit has sections with different conductor sizes or dielectric materials. Because of the time required to splice the un-faulted cable sections and the availability of several terminal fault location methods, this method is generally a last resort approach with both high-voltage and medium-voltage cables that do not have some sectionalizing device or separable connectors within the cable circuit. The method cannot be effectively used on direct-buried cables, especially those running under pavement, sidewalks, or any other surface that cannot be easily excavated. In addition, the method cannot be used if the cable cannot be adequately respliced [42,47].

12.11.2.2 Charging current method

The current charging method is sometimes referred to as the AC current method. Although not widely used today, this method is intended to detect open-conductor faults. The test equipment required by this method is a variable and controllable AC high-voltage supply and a milliammeter. If the angle between the voltage applied to the cable and the current into the cable is to be measured, a suitable voltage transformer, current transformer, and phase angle meter will be required. The method locates open conductor faults on either single-conductor or three-conductor shielded cables. Since this method requires that the circuit have a uniform shunt capacitive reactance per unit of length, it will not be effective on circuits having different size phase conductors or cables with varying characteristics of insulation that are spliced together. The accuracy of the method is affected if the resistance in the fault path, either across the open conductor or from the open conductor to the ground, is of the same order of magnitude as the capacitive reactance of the cable. If all three conductors of the circuit are open, the measurements must be made from both ends of the circuit with the same voltage applied at both ends. The method is not practically applicable to belted cables where there is both phase-to-phase and phase-to-ground capacitance since the capacitances in the resulting network complicate the interpretation of the results. Also, the capacitance of terminations can contribute to location errors if it is significant compared to the cable capacitances [47].

12.11.2.3 Insulation resistance ratio method

This method requires test equipment to measure the high values of insulation resistance. The insulation resistance is usually measured using equipment designed specifically for this purpose, such as megohmmeters, wheatstone bridges, and insulation resistance testers. The insulation resistance ratio method locates open conductor faults on either three-conductor shielded cables or on three-phase circuits made with single-conductor shielded cables. The technique can be used on circuits with all conductors open by making the measurements from both ends. Moreover, this method applies to branch circuits if the lengths of all branches are known. The technique can be used on cables with oil-impregnated paper insulation. However, it cannot be used on cables with extruded insulation due to their inherently high insulation resistance. In addition, the method is not applicable if the shunt fault resistance, the fault resistance across the open conductor, or the termination resistances are low compared to the resistance of the insulation. In addition, this method cannot be used on three-conductor belted cables. Finally, using this method, inaccuracies will result if the insulation resistance is not constant along the cable length, such as may be caused by moisture [47].

12.11.2.4 Murray-loop two-end method

The test equipment required for the Murray-loop two-end method is the same as that for the Murray-loop method. This method applies to the single conductor and three conductors shielded cables for locating single or double phase-to-ground faults. In addition, it can be used with belted cables to find ungrounded conductor-to-conductor faults providing that only one of the faulted conductors is grounded for the tests. However, this method will not locate three phase-to-ground faults, and it will not work on single-phase circuits unless an unfaulted conductor of known length is available for completing the bridge circuits. Moreover, the method will not locate open conductor faults. Using this method, the jumper resistance and the resistance of the leads connecting the bridge to the unfaulted and faulted conductors should be as low as possible and low concerning the phase conductor resistances [47].

12.11.2.5 Murray-Fisher loop method

With the Murray-Fisher Loop method, the length of the faulted conductor is not required, but two unfaulted conductors must be available with the length known for one of the un-faulted conductors. The equipment needed with this method is the same as that for the previous bridge methods. This method is intended to locate only single phase-to-ground faults when two unfaulted conductors, one known length, are available. It can be used with single-conductor, and three-conductor shielded cables. To locate a fault using this method, the resistance per unit length for the faulted conductor and the unfaulted conductor is not required if they are equal. If necessary, this method can be used on circuits having conductors with different resistances, provided the unfaulted conductor length and resistances per unit length of both conductors are known. However, this method will not locate open conductor faults and two conductor-to-ground faults or three conductor-to-ground faults. Also, the method cannot be used if the length of the un-faulted phase conductor is not available [47].

12.11.2.6 Varley loop method

The Varley loop method determines the distance to the fault without knowing the length of either the faulted or unfaulted conductor if the resistance per unit length of the faulted conductor is known. With this method, the bridge must be capable of operating at a high voltage to force current through the fault, and it must be capable of operating at a lower voltage. This method can be used with single-conductor, and three-conductor shielded cables to locate single and double phase-to-ground faults.

It can also locate phase-to-phase faults in belted cables, provided one of the faulted conductors is grounded. A unique aspect of this method is that it allows determining the length of the faulted conductor if the conductor resistances per unit length and the length of the unfaulted conductor are known. This method can also determine the distance to the fault without knowing circuit lengths if the faulted conductor resistance per unit length is known. As in the case with other bridge methods, the Varley loop method will not locate three phase-to-ground faults unless another conductor is run along the route of the faulted circuit for completing the bridge. Moreover, this method does not locate open conductor faults [47].

12.11.2.7 Hilborn loop method

The Hilborn loop method is very similar to the Murray loop method, except for the connection of the galvanometer in the circuit. The test equipment required for this method is essentially the same as for the other bridge methods. This method locates single phase-to-ground faults on either single-conductor or three-conductor shielded cables. The faulted and unfaulted phase conductors need not be of equal length or the same resistance per unit length. This method will locate a sheath-to-earth fault when the cable has an insulating outer jacket. Single phase-to-ground faults on single conductor circuits can be found if an over-ground twin conductor cable serves as the two un-faulted conductors. This method is of value on circuits where the conductors have large cross-sectional areas, and the conductor lengths are short since the contact resistances are included in the bridge ratio arms. This method cannot be used for three phase-to-ground and two phase-to-ground faults unless an over-ground twin-conductor cable is run for completing the circuits to the bridge. Furthermore, this method cannot be used to locate open conductor faults [47].

12.11.2.8 Open-and-closed loop method

If one unfaulted conductor is available, the Murray loop method, the Murray loop two-end method, and the Varley loop method will locate two phase-to-ground faults on a three-phase cable circuit by performing the first bridge measurement using the un-faulted conductor and one of the faulted conductors. The second set of measurements is then performed using the unfaulted conductor and the second faulted conductor. With the open-and-closed loop method, the double phase-to-ground fault is located by connecting the bridge to the two faulted conductors at the near end of the circuit. In this method, two tests are made from the near end. While the two faulted conductors at the far end are left open circuited in the first test, a jumper is connected in the second test between the two faulted conductors at the far end. With this method, an unfaulted conductor is not required. Test equipment required for this method is the bridge circuit and galvanometer. As with all other loop methods, the test leads and resistance of the jumpers should be as low as possible. This method is applied only if the fault on both conductors is at the same location. It can be used with both single-conductor and three-conductor shielded cables to locate a double phase-to-ground fault. It can also be used to locate a three-phase-to-ground fault by testing any two-phase conductors. However, this method is not intended to locate open conductor faults and cannot be used if the end of the circuit is not accessible for connecting the jumper [47].

12.11.2.9 Werren overlap method

Like the open-and-closed loop method, the Warren overlap method is intended to detect double phase to ground faults when an unfaulted conductor is unavailable. Still, it differs in that the measurements are made from both ends of the circuit rather than from just one end. Test equipment required for this method includes a bridge circuit and suitable low resistance leads and jumpers to make the connections at both ends of the circuit. This method locates two phase-to-ground faults either on three-phase circuits made with single-conductor shielded cables or on three-conductor shielded cables. For the technique to work well, phase conductors must be of the same length. Also, each phase conductor must have the same resistance per unit length, and the distance from the terminal end to the fault on each phase must be the same. Moreover, the resistances in the fault path on each phase must be constant throughout the measurement period. Furthermore, the fault resistance must be low enough to allow sufficient current flow to obtain the necessary sensitivity on the bridge null detector. This method, however, cannot be used to locate open conductor faults [47].

12.11.2.10 Impulse current method/pulse discharge detection

The impulse current method is also known as the free oscillation technique, pulse discharge detection method, the high-voltage radar method, the surge pulse method, and the pulse timing method. Test equipment required by this method includes the impulse generator or thumper, a linear coupler, and a digital recording device. Several manufacturers supply equipment for this method. The impulse current method works with both single-conductor and three-conductor cables to locate both shunt faults and open conductor faults. It locates single phase-to-ground, double phase-to-ground, and three phase-to-ground faults on three-phase circuits. When the velocity of propagation for the cable is known, the method can be applied even if the length of the cable circuit is unknown. However, the method cannot be used to determine the distance to a shunt fault by measuring the time between the pulse associated with the first transmitted current and the pulse associated with the first return. The length of cable on which this method can be used must be greater than approximately 15 meters. Results on short cable lengths have been inconclusive in many situations because of the rapidity of the pulses [47].

12.11.2.11 Pulse decay method

The pulse decay method requires a high potential tester, a voltage coupler, and a digital storage oscilloscope produced, especially for this method. The coupler acts as a voltage divider and helps damp the voltage waves returning to the hi-pot. This method can be used on both single-conductor and three-conductor cables to locate high impedance nonlinear faults breaking down at about 5 kV or higher. When the velocity of propagation for the cable is known, the method can be used even if the length of the cable circuit is unknown. It can also be used in applications where the voltage from a thumper may not be high enough to cause a breakdown at fault. The minimum length of cable on which this method can be used is generally around 15 meters. The technique is, however, not intended to locate open conductor faults. In addition, the method may not be successful if the ground return path, such as a lead jacket, has deteriorated. Finally, locating faults on tapped systems can be very difficult using this method due to wave reflections at each tap junction [47].

12.11.2.12 Standing wave differential method

The standing wave differential method is sometimes known as the quarter-wave resonance method. This method uses a variable frequency source to determine the successive frequencies which produce standing waves on the cable. Although there is no commercially available equipment for this method, oscillators and sweep generators can be used. The method can be used to detect open conductor faults. It can also be used on single-conductor and three-conductor shielded cables to detect both phase-to-ground shunt faults if the fault resistance is less than 20 ohms. Moreover, when the velocity of propagation for the cable is known, the method can be used even if the length of the cable circuit is unknown. Also, when the circuit length is known, and the measurements are made from both ends of the cable, the method can be used even if the length of the velocity of propagation is unknown. However, this method will not provide good results if discontinuities in the cable arise from different size conductors in series, taps, and changes in insulation [47].

12.11.2.13 Direct current charging method

The DC charging current method is sometimes known as the DC hi-pot method. In this method, maintenance crews isolate a cable section and apply a DC hi-pot voltage. The crews proceed to the next section if the cable holds the hi-pot voltage and continue repeating before finding a cable that cannot hold the hi-pot voltage. The cable must be isolated from the transformer because the applied voltage is DC. This method is intended for distribution primary feeder circuits with cables having separable (elbow) connectors as they allow manual sectionalizing at transformers and junction points. The DC charging current method does not prelocate the fault like the TW and bridge methods, but it can determine whether the section of cable under test faults is used in conjunction with manual sectionalizing of the circuit. Several manufacturers provide the necessary test equipment for this method, including phasing sticks, an appropriate meter, a high-voltage diode assembly for use with the phasing sticks and feed-through bushings, and parking stands for sectionalizing and making the relevant connections at each transformer. The method can be used on either single-phase or three-phase circuits. It is intended to use multigrounded neutral distribution circuits operating at voltages up to 34.5 kV using extruded dielectric concentric-neutral type cables. It is frequently applied on systems with loop-feed distribution transformers with elbow connectors. However, the method will not detect open conductor faults. In addition, it may not work on the longer-length circuits with paper-insulated cables due to the relatively high losses in the insulation.

Also, the method may not work on extruded cables where the insulation has absorbed significant moisture [42,47].

12.11.3 Less commonly used tracer methods

Tracer methods employ two pieces of apparatus namely, a transmitter and a receiver. While a transmitter produces a signal injected into a cable, a receiver senses the signal along the cable path. Detection of the applied input signal can be used to locate the cable route and identify the fault location. Some tracer methods pinpoint the fault location very accurately, usually within fractions of a meter. Other tracer methods, designed for application in duct/manhole systems, will identify the fault between two adjacent manholes. A variety of input signals can be applied to the cable conductor, such as AC, DC, high magnitude, low magnitude, high frequency, low frequency, constant frequency, and impulses. Likewise, various detecting techniques can be employed to sense the input signal, such as magnetic field pickup, electromagnetic wave reception, induced current measurements, acoustic detection, earth voltage drop, and sheath voltage drop. No technique will locate a cable fault for all possible conditions. For each type of fault, cable installation, cable construction, and environment, there may be a unique combination of the input signal and detection techniques required for a successful cable fault location. This part discusses different tracer methods less commonly used by utility fault location crews to find faults in underground cables. These methods represent alternative tracer fault location methods that can be used under particular circumstances when the equipment needed for these methods are available with the fault location crews.

12.11.3.1 Magnetic pickup method

The magnetic pickup method is also known as the tone tracing method and the search coil method. Test equipment required by this method includes the transmitter and the detector. The transmitter typically supplies a current up to 100 mA at a voltage of up to 1 kV. Audio frequency generators rated up to 200 watts output power are common. Also, impulse generators capable of supplying up to 500 A pulses are used as signal sources. In the detector, a magnetic loop, or a ferromagnetic rod antenna is generally used. In addition, a sensitive and selective low-frequency amplifier is used to amplify the detected signal. Moreover, a meter, speaker, or earphones are used to indicate the sensed signal. Some electric utilities have mounted the pickup antenna under a truck and then drive the streets to locate the fault or trace the cable route. The magnetic pickup method is widely used for cable route location. It is quite effective in finding deep-buried and submarine cables. It is used to locate faults and cable paths for nonshielded cables. However, it is generally considered a fault prelocation method since it is not as accurate as other tracer methods. This method can also be used to locate open circuit faults on nonshielded cables. It cannot be used to locate high-impedance nonlinear faults with low-voltage input signals if the source cannot establish a sufficient current flow in the fault. To locate an energized and unfaulted cable, this method does not need an additional source signal, and the cable can be traced with the detector only, sensing the fields produced by the normal load currents. Moreover, control and communications cables can be located using this method. Since this method is most useful in locating insulated wire cable faults, it is well-suited for secondary cable systems. However, this method is not applicable for fault location in the case of high resistance or high-impedance nonlinear faults or when a cable is buried in dry soil, which has high impedance to the return current [47].

12.11.3.2 Tracing current method

The tracing current method is also known as the current direction method, the directional fault indicator method, and the ballistic impulse method. In this method, an AC, pulsed DC, or pulse source can be used. The applied voltage must be high enough to cause a breakdown at fault to detect nonlinear high-impedance faults. In addition, the current transformer/meters must be sufficiently sensitive to detect the net current. The sensed signal can be displayed on an oscilloscope or a meter with associated electronics. The tracing current method is used to locate shorts and high-impedance nonlinear faults. It was successfully employed with single-conductor and three-conductor PILC cables in duct systems and on primary distribution cables of lengths ranging from 23 meters to 4.6 km. The accuracy of this method allows using it for fault prelocation for a duct/manhole system since it will indicate that the fault is between two consecutive measurement points. For direct buried cables, this technique will identify the faulted section, and further work will be required to pinpoint the fault before excavation and repair operations are started. In the case of cables with an insulating outer jacket, the sheath of the faulted cable should be grounded in each manhole and bonded to the sheaths of other cables in the manholes. Likewise, the concentric neutrals on cables in underground (URD) systems must be grounded and bonded at equipment enclosures. This method is not applicable in single-phase URD systems with transformers connected to the cable. With the transformer primary windings left directly connected from line-to-ground, the current will flow through the transformer primary windings and overload the voltage source used for testing [47].

12.11.3.3 Earth gradient method

The earth gradient method is also known as the voltage gradient method or the electric potential method. The equipment required by the earth gradient method consists of a signal source or transmitter and the detector or receiver unit. The transmitter is a continuous DC source or a DC pulse source, and the receiver is a zero-centered DC micrometer, high impedance voltmeter, or galvanometer coupled to a pair of probes or large surface area electrodes. The earth gradient method is used to locate faults on nonshielded direct-buried secondary distribution cables, shielded direct-buried cables, and street lighting cable circuits. Moreover, this method is used to find faults on bare concentric neutral direct-buried primary URD cables and coated metallic pipes to locate damaged coating. Also, it is used to locate faults on jacketed direct-buried primary cables if the fault has punctured the insulating jacket. Furthermore, this method can be used to locate shorts or high-impedance nonlinear faults and jacket faults on transmission and distribution cables. If this method is deployed on a circuit passing under a hard surface such as concrete, asphalt, or other pavement surfaces where the probes cannot penetrate that surface, large area electrodes consisting of metallic plates or wetted sponges can be used. To improve the electrical contact with the voltage probes, a conductive solution such as saltwater can be poured onto the hard surface or the sponges. The accuracy of this method is well within 0.3 meters. However, this method is not applicable to locate true open faults, and it is ineffective on most duct systems [47].

12.11.3.4 Hill-of-potential method

The hill-of-potential method is also known as the AC leakage method, the peak of potential method, and the AC earth gradient method. This method differs from the Earth Gradient method because it does not use a separate signal injection source. The test equipment for this method consists of insulated gloves or similar protective equipment, a high-impedance AC voltmeter or an auto-ranging digital VOM with high input impedance, a test cable reel, and a ground probe. This method is used on secondary distribution circuit cables that are directly buried. Also, it is applicable on circuits installed in moisture-absorbing conduit. In addition, it can be used to detect faults from the phase conductor to the earth, shorts, and some nonlinear faults by using the existing system voltage produced by the secondary of the transformer. Moreover, this method has proven effective in locating faults on street lighting circuits. The accuracy of this method is within one foot from the cable fault. However, the hill-of-potential method will not detect open conductor faults unless the source end of the conductor is also in contact with the ground and is not effective on shielded or concentric-neutral secondary cable circuits. This method is susceptible to the general earth resistivity, particularly in the area where the phase conductor is in contact with the surrounding earth. Also, this method may be challenging to use if the faulted cable is buried excessively deep or if the fault to earth on the phase conductor is located near a bare neutral conductor or metallic pipes, limiting the return current flowing in the earth. Moreover, the method cannot be used if the fault is severe enough to blow a fuse on the primary side of the transformer or the secondary cable limiters [47].

12.11.3.5 Thumper/electromagnetic wave method

The thumper/electromagnetic wave method is also known as the discharge detection method. Test equipment required for this method consists of a transmitter and a detector. A transmitter is a thumper unit comprised of a power supply, capacitor, fixed or adjustable sphere gap, and manually actuated thumper units. The detector consists of an electromagnetic wave sensor with antenna, signal amplifier, ballistic impulse detector, and a portable AM receiver tuned to an unused frequency. This method can be used on primary and secondary distribution cable systems and underground transmission cable systems. Usually, this method would not be recommended for secondary cable systems because the path of the return impulse current may not be known and may include other utilities. The method applies to cables buried two meters or more below the earth's surface. Also, it could be used on circuits with taps or those installed in nonmetallic duct lines. The faults located by this method are the high impedance nonlinear faults, opens, and low impedance faults. This technique can be used as a prelocating method or a pinpointing method. Long cable lengths should not be a problem because the attenuation of the EM wave is minimal when compared to the attenuation of an acoustic wave. This method is highly accurate due to the excellent propagation of the EM wave. Since the soil conditions will not affect the propagation of the EM wave to the same degree as they affect the acoustic wave, the depth of burial of the cable under test is not considered as a factor affecting the accuracy of this method as it is with the thumper/acoustic method. However, this method cannot be used on pipe-type cables or cables in metallic ducts since the EM wave will not propagate outside a steel pipe. Moreover, the method will not work if the magnitude of the discharge voltage is lower than the magnitude of the voltage required to arc over the gap at fault. The major drawback of the thumper/EM wave method is the complexity of the equipment required [47].

12.11.3.6 Direct current sheath potential difference method

The DC sheath potential difference method is also known as the shield potential difference method. Test equipment used by this method includes a transmitter and a detector. The transmitter is a low voltage DC power supply, and the detector is a high impedance DC voltmeter with probes. This method is used to locate shunt faults between conductor and sheath in aerial and buried metallic-sheathed cables. It can also be used to locate faults on buried street-lighting cables. The DC sheath potential difference method is valid only for conductor-to-sheath faults. This method becomes time-consuming in long lengths of cable need to be tested, but this can be alleviated by prelocating the fault with another method before employing this method. If the sheath is grounded or bonded in adjacent manholes, the parallel paths can complicate the interpretation of meter readings. Likewise, the multiple grounds or bonding on the sheath allows the flow of stray DC currents, complicating the interpretation of meter readings [47].

12.11.3.7 Sheath coil method

The sheath coil method is also known as the galvanometer method, the magnetic impulse detection method, and the hand BT method. The test equipment needed for this method includes a transmitter and a receiver. While the transmitter is an impulse current generator or thumper unit, the receiver is a sheath coil and a galvanometer or other suitable detector. This method is intended for use in duct/manhole systems having lead sheath cables. It will prelocate shorts and high impedance nonlinear faults on both single-conductor and three-conductor cables. It can also be applied on a circuit with a three-conductor cable rather than circuits with single-conductor cables. In the case of a three-conductor cable, the sheath coil is moved around the circumference of the cable, and readings are taken. When the sheath coil is located on the upstream side of the fault, the readings will change when the coil is moved around the circumference of the cable. On the other hand, when the sheath coil is located on the downstream side of the fault, the readings will not change much when the coil is moved around the circumference of the cable. This is because there is minimal phase current in the cable downstream from the fault, so the magnetic field around the circumference of the cable is due only to the sheath currents and is nearly uniform. As a result, the fault can be localized by comparing readings at adjacent manholes as the sheath coil is moved around the circumference of the cable. Following well-defined procedures, this method can also be used to identify the faulted phase and identify and match phases A, B, and C before cutting the cable length to be replaced. The sheath coil method does not pinpoint a cable fault as most tracer methods do. When applied at manhole locations, it just indicates the faulted section, with the best accuracy obtainable being the distance separating two consecutive manholes. For cables installed in ducts, such precision is considered sufficient because the faulted section between manholes, will be pulled out of the duct and repaired or replaced. For safety reasons, this method is limited to cables with a conducting sheath. The high current impulse injected into the cable conductor may be hazardous to the sheath coil operator if the cable under test does not have a conducting sheath. In addition, open conductor faults cannot generally be located with this method. Also, the method is not intended for use on URD systems using concentric neutral type cables, either jacketed or bare [47].

12.11.3.8 Pick method

The test equipment needed for the pick method includes a transmitter and a receiver. The transmitter is an interrupted DC source capable of producing a high current through fault resistance. The receiver is a pair of metallic probes, a step-up transformer, a galvanometer, or a suitable sheath voltage detector. The pick method is primarily used to locate phase-to-sheath faults on lead or metallic sheath cables in duct and manhole systems. One major urban utility applies this method in conjunction with the sheath coil method to confirm the location of the fault. The method applies only to faults where there is a reasonable return current in the conducting sheath. Thus, it applies to conductor-to-sheath faults on single conductor cables and multiconductor cables that have a conducting sheath. In addition, it can be used on three-phase feeders when the distribution transformer primary windings cannot be disconnected from the primary feeder, providing the primary windings of all transformers are connected from phase to phase. However, the Pick method will not locate open faults or phase-to-phase faults which do not involve a shield or a conducting sheath. Furthermore, it will not be successful in locating open conductor faults on any cable type, and it is not recommended for use on URD concentric neutral type cable. Moreover, it will not locate phase-to-phase faults on unshielded three-conductor cables with a lead sheath. Therefore, this tracing method is not used as a fault pinpointing technique. The best accuracy we can obtain using this method is the distance separating two consecutive manholes. Such precision is sufficient in duct systems because the complete length of cable between two manholes will be pulled out and repaired or replaced [47].

12.11.4 Practical fault location procedures

The effectiveness of fault location techniques is influenced by the various types of cable system designs. Fault location is entirely dependent on the physics of the cable system and, therefore, no method works in all applications. Before selecting the technique for locating cable faults, a good understanding of the cable and cable system characteristics are necessary. The circuit characteristics, cable type, and cables installation method will all impact fault locating attempts. In addition, they will determine which type of fault location equipment can be used and how effective the equipment will be in finding the fault. Therefore, a brief review is necessary before discussing fault locating techniques. After a faulted cable has been isolated in preparation for fault locating, it is recommended to follow a fixed plan for finding the fault. A step-by-step procedure will help in pinpointing the fault efficiently. First, it is a good idea to gather as much information as possible about the faulted cable. Some questions that will help in the fault locating process include the following:

- **Cable type**: Is the cable lead-covered, concentric neutral tape shield?
- **Insulation type**: Is the cable XLPE, EPR, paper?
- **Conductor and size**: Is the cable CU, AL, stranded, solid, 2/0, 350 MCM?
- **Length of the run**: How long is the cable?
- **Splices**: Are there splices in the cable? Are the splice locations known?
- **T-taps or wye splices**: Are there any taps? If so, are their locations known? What is the length of the branches?

This procedure consists of four main steps: test, analyze, localize, and locate (TALL). Several tests are usually performed in the "test" step, including fault resistance and loop test, time domain reflectometry (TDR) tests, and DC Hi-pot test. The second step is intended to analyze the results obtained from the aforesaid tests. The cable route should be determined or confirmed by studying good maps or physically tracing the cable route. Finally, the "localize" step is meant to prelocate the fault. The selection of a localizing method is based, at least partially, on the character of the fault. For example, bridge methods are used for single faults, TDR/low-voltage radar methods are used for faults measuring less than 200 Ω and all opens, high-voltage radar methods are used for all faults arc reflection, surge pulse reflection, and decay, and electromagnetic impulse detection is used for all shorts and some opens.

The "locate" step, also known as pinpointing the fault location, is essential before digging up the buried cable. After fault localization, a surge generator is connected to one end of the faulted cable to monitor the localized area for the telltale thump from the fault. When the thump is not loud enough to hear, a surge detector or an acoustic impulse detector may be required to localize the fault. Voltage gradient test sets can be effectively used in pinpointing faults on direct-buried secondary cables. However, the method depends on the fault existing between the conductor and earth. Therefore, a different method must be used when the cable is installed within a conduit. If the faulted cable is short, connect a surge generator, walk the cable and listen for the discharge to locate the fault. However, if the cable is very long, it might take a good deal of time to walk the cable and pinpoint the fault.

Therefore, before attempting to locate the fault, it is advisable to employ a localizing procedure to reduce overall time spent and minimize high-voltage exposure to the cable. After fault localization, a listening aid is used to zero in on the thump when the surge generator breaks down the fault. In the case of the bolted (metal-to-metal) faults on the nonburied cable, an electromagnetic impulse detector could assist in locating the faults [42,48–50].

12.11.5 Commercially available cable fault locating solutions

Nowadays, many manufacturers offer various products for cable testing, cable diagnostics, and cable fault location. In addition, cable test van systems have also been made available by many manufacturers. Furthermore, other special equipment and customized solutions have also been made available to help locate faults faster and, hence, improve system reliability, maximize uptime, and drive costs down. Some of the world's leading manufacturers in cable test, diagnostic, and fault location solutions include Megger and BAUR [48–50].

Megger fault locating products can be categorized as cable testing, cable diagnostics, cable fault location, cable test van systems, special equipment, and customized systems [48,49]. Cable test solutions focus on portability, usability, and reliability. Most of the cable test systems are large as the cables can be highly capacitive, and testing the insulation often requires a significant power output from the test equipment. Examples of this category include test systems for medium voltage cables and high power test systems for medium voltage cables, as depicted in Fig. 12.12 [51].

The prime goal of cable diagnostic solutions is to avoid service interruptions in medium, high, and extra-high voltages networks during operation. Workmanship failures on accessories, cable insulation, improper cable laying, and deterioration in joints and terminations are the leading causes of service interruptions. Cable diagnostic systems enable verifying the quality of a new cable system and assessing its condition before it is put into operation. As a result, potential issues and damage can be detected and corrected at the commissioning stage and, therefore, avoiding future network failures and the associated costs. Besides, cable fault location solutions assist the user to quickly find and localize cable faults without causing damage to fault-free parts of the cable, using well-defined fault-finding techniques and the appropriate test equipment. Selected commercially available cable fault location solutions from Megger are presented in Fig. 12.13 [51].

The concept of cable test van systems developed by Megger and other manufacturers combines cable testing, cable diagnostics, and cable fault location techniques into a van. In a test van system, all the high voltage prelocation methods are usually made available. Also, all the accessories for cable tracing, cable fault pinpointing, and cable identifying are included. Megger developed different test van systems with a variety of features. One example is a test van system designed to combine cable testing, cable diagnostics, and cable fault location techniques and fully automatic for both single- and three-phase cable fault location models. It can be customized to become

Fig. 12.12 Selected commercially available cable testing systems from Megger [51].

a complete testing and diagnostic cable solution, and with its remote-control function and GPS mapping, it enables more accurate cable fault location.

Another example is a high-performance cable test van system that can suit almost any application due to its high-power output. The third example is a manually operated, modern, modular system customized for fault location in single-phase or three-phase cables. It provides all the high voltage prelocation techniques, and it can also be set up for cable testing and diagnostics. Megger also developed a variety of special equipment and customized solutions to serve different needs and applications. Manufacturers develop the customized solutions in response to the particular needs of their customers worldwide. Selected special and customized equipment and devices are presented in Fig. 12.14 [51].

BAUR's fault locating products are designed for medium and high voltage ranges, but they can also be applied for low voltages [50]. These fault location solutions can be used to locate different types of faults such as short-circuits, earth-fault/short-circuit to earth, cable breaks, intermittent faults, and cable sheath faults. Thus, they can suit different cable designs and types. Furthermore, they are compatible with various cable installation methods and faults produced by various factors, including service life, aging, corrosion, overvoltage, thermal overload, improper

Fig. 12.13 Selected commercially available cable fault location solutions from Megger [51].

cable laying, installation errors, and damage from transportation and storage. Using BAUR's fault location solutions, fault location is conducted systematically following a four-step logical procedure. First, the fault analysis step makes it possible to determine the fault characteristics and the further procedure. Then, the fault is precisely determined by the meter during the prelocation process. The objective of the subsequent pinpointing step is to precisely pinpoint the fault to limit the ground excavation and, in turn, to minimize the repair time. The last step is cable identification which is necessary to identify the defective cable in a bundle of multiple cables at the fault location. This is very crucial if the fault is not observable from the outside. BAUR's fault locating products fall under four categories: portable devices,

Utility practices on fault location

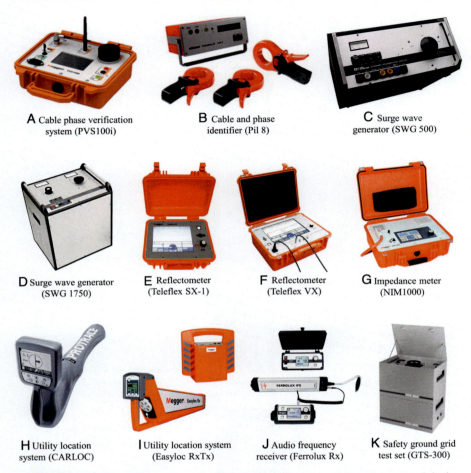

Fig. 12.14 Selected commercially available particular and customized equipment and devices from Megger [51].

high-performance modules, system solutions, and cable test vans. Examples of portable devices cable sheath testing and fault location system, phase detector, and cable identifier. Some examples of high-performance modules include surge and test generators, time domain reflectometers, and surge voltage generators. Cable fault location systems and portable cable fault location systems are some examples of the system solutions category. BAUR's cable test vans are equipped with a complete product range for cable fault location, testing, and diagnostics in one single system. They can be either fully automatic or semi-automatic systems, each with a single-phase or three phases. BAUR's fault locating products are depicted in Fig. 12.15 [52].

392 Power System Fault Diagnosis

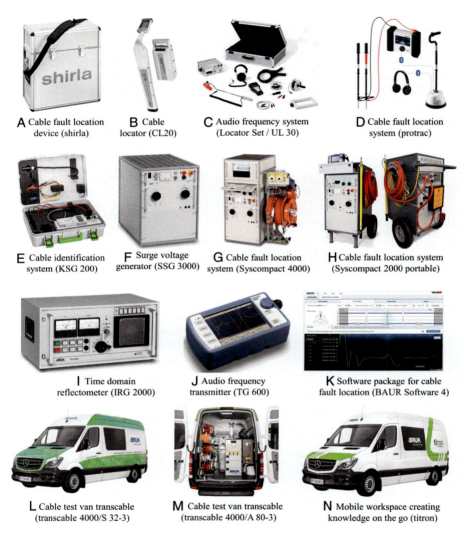

Fig. 12.15 Selected commercially available cable fault locating solutions from BAUR [52].

12.12 Summary

This chapter presented various utility practices for fault location in transmission networks, distribution systems, and underground cables. It also discussed different fault location methods along with their deployment in various commercially available fault location solutions. In addition to traditional fault location methods, some advanced fault location approaches in distribution systems have been presented. These include FLISR, PQView, VDFL, TW approach, and neutral/ground current ration method. Moreover, the chapter included some examples of different utility implementations on fault finding. Furthermore, a list of commonly used and less commonly used fault location methods for underground cables has also been detailed. Finally, the chapter provided a brief description of the selected commercially available cable fault locating solutions.

References

[1] A. Gaikwad, Transmission line protection support tools: fault location algorithms and the potential of using intelligent electronic device data for protection applications, Palo Alto, CA, 2013. https://www.epri.com/research/products/3002002381. (Accessed 29 June 2021).

[2] ABB Distribution Solutions, Fault management for grid automation (advanced fault location solution), Vaasa, Finland, 2018. https://library.e.abb.com/public/d9d1f94a3ac642f-da4d08d7b0aa65f99/Power-flow-management_broch_758910_LRENa.pdf. (Accessed 23 August 2021).

[3] M. Shafiullah, Fault diagnosis in distribution grids under load and renewable energy uncertainties, King Fahd University of Petroleum & Minerals, Dhahran, Saudi Arabia, 2018.

[4] M. Shafiullah, M.A. Abido, Z. Al-Hamouz, Wavelet-based extreme learning machine for distribution grid fault location, IET Gener. Transm. Distrib. 11 (17) (2017) 4256–4263, doi:10.1049/iet-gtd.2017.0656.

[5] S. Parmar, Fault location algorithms for electrical power transmission lines (methodology, design and testing), Delft University of Technology, 2015.

[6] M.M. Saha, R. Das, P. Verho, D. Novosel, Review of fault location techniques, Power Systems and Communications Infrastructures for the Future, Beijing, China, 2002, pp. 1–6, https://www.researchgate.net/publication/251813882. (Accessed 23 August 2021).

[7] Y.D. Mamuya, Y.-D. Lee, J.-W. Shen, M. Shafiullah, C.-C. Kuo, Application of machine learning for fault classification and location in a radial distribution grid, Appl. Sci. 10 (14) (2020) 4965, doi:10.3390/app10144965.

[8] M. Shafiullah, M. Abido, and T. Abdel-Fattah, Distribution grids fault location employing ST based optimized machine learning approach, Energies, 11 (9) (2328), doi:10.3390/en11092328.

[9] A. Al-Mohammed, M. Abido, A fully adaptive PMU-based fault location algorithm for series-compensated lines, IEEE Trans. Power Syst. 29 (5) (2014) 2129–2137. http://ieeexplore.ieee.org/xpls/abs_all.jsp?arnumber=6736121. (Accessed 19 August 2018).

[10] T.A. Short, D.D. Sabin, M.F. McGranaghan, Using PQ monitoring and substation relays for fault location on distribution systems, 2007 IEEE Rural Electric Power Conference, IEEE, 2007, pp. B4-1–B4-7.

[11] M. Mcgranaghan, T. Short, D. Sabin, Using PQ monitoring infrastructure for automatic fault location, 19th International Conference on Electricity Distribution, 2007, pp. 21–24 https://www.researchgate.net/publication/229009997_Using_PQ_Monitoring_Infrastructure_for_Automatic_Fault_Location. (Accessed 23 August 2021).

[12] M. Kezunovic, Smart fault location for smart grids, IEEE Trans. Smart Grid 2 (1) (2011) 11–22, doi:10.1109/tsg.2011.2118774.

[13] J. Izykowski, Fault location on power transmission lines. Wrocław: Oficyna Wydawnicza Politechniki Wrocławskiej, 2008.

[14] A. Bahmanyar, S. Jamali, A. Estebsari, E. Bompard, A comparison framework for distribution system outage and fault location methods, Electr. Power Syst. Res. 145 (Apr) (2017) 19–34, doi:10.1016/J.EPSR.2016.12.018.

[15] M. Shafiullah, M.A. Abido, A review on distribution grid fault location techniques, Electr. Power Components Syst. 45 (8) (2017) 807–824, doi:10.1080/15325008.2017.1310772.
[16] L. Awalin, H. Mokhlis, A. Bakar, Recent developments in fault location methods for distribution networks, Prz. Elektrotechniczny R88 (12a) (2012), pp. 206–212. http://eprints.um.edu.my/7868/. (Accessed 26 November 2015).
[17] M. Shafiullah, M.A. Abido, Distribution grid fault analysis under load and renewable energy uncertainties (Pending), US20200403406A1, 2020.
[18] S. Marx, B.K. Johnson, A. Guzmán, V. Skendzic, M.V. Mynam, Traveling wave fault location in protective relays: design, testing, and results, 16th Annual Georgia Tech Fault and Disturbance Analysis Conference, 2013, pp. 1–14. https://cms-cdn.selinc.com/assets/Literature/Publications/TechnicalPapers/6601_TravelingWave_AG_20130309_Web.pdf?v=20160427-184720. (Accessed 23 August 2021).
[19] Power Delivery Products Inc., Navigator LM overhead, 2007. http://www.powerdeliveryproducts.com/pdffiles/navlmcat.pdf. (Accessed 23 August 2021).
[20] Willfar Power, Overhead power line fault indicator, Willfar Power (2021). http://willfar-power.com/4-1-overhead-power-line-fault-indicator.html. (Accessed 30 June 2021).
[21] ADC Energy, SG203 advanced conductor mounted fault indicator, 2013. https://www.adcenergy.co.za/sites/default/files/SG203. (Accessed 23 August 2021).
[22] Four Faith, Remote overhead line fault indicator, 2019. https://en.four-faith.net/uploadfile/2019/0904/20190904034547388.pdf. (Accessed 23 August 2021).
[23] Siemens Global, Distance protection for transmission lines, 2021. https://new.siemens.com/global/en/products/energy/energy-automation-and-smart-grid/protection-relays-and-control/siprotec-4/distance-protection/distance-protection-for-transmission-lines-siprotec-7sa522.html. (Accessed 30 June 2021).
[24] HITACHI-ABB, REL650 - line distance protection, HITACHI-ABB, 2021. https://www.hitachiabb-powergrids.com/offering/product-and-system/substation-automation-protection-and-control/products/protection-and-control/line-distance-protection/rel650 (Accessed 30 June 2021).
[25] TecQuipment Academia, Distance protection relay, 2021. https://www.tecquipment.com/distance-protection-relay. (Accessed 30 June 2021).
[26] Schweitzer Engineering Laboratories, SEL-411L advanced line differential protection, automation, and control system, Schweitzer Engineering Laboratories, 2021. https://selinc.com/products/411L/. (Accessed 30 June 2021).
[27] Qualitrol Corporation, Qualitrol Traveling Wave Fault Locator Monitor, Qualitrol Corporation, New York, USA. 2021. https://www.qualitrolcorp.com/products/fault-location-monitors/tws-fl-8-and-tws-fl-1-traveling-wave-fault-locators/. (Accessed 30 June 2021).
[28] KEHUI International Ltd, XC-2100E travelling wave fault location system, 2021. https://www.kehui.com/products/xc-2100e-travelling-wave-fault-location-system/. (Accessed 30 June 2021).
[29] Schweitzer Engineering Laboratories, SEL-T400L time-domain line protection, 2021. https://selinc.com/products/T400L/. (Accessed 30 June 2021).
[30] GE and Alstom Joint Venture, Reason RPV311 digital fault recorder with fault location and PMU, 2015. https://www.think-grid.org/sites/default/files/Grid-SAS-L3-REASON_RPV311-0889-2015_10-EN.pdf. (Accessed 23 August 2021).
[31] Siemens Global, Fault recorder, 2021. https://new.siemens.com/global/en/products/energy/energy-automation-and-smart-grid/protection-relays-and-control/siprotec-5/fault-recorder.html. (Accessed 30 June 2021).

[32] GE Grid Solutions, Reason DR60 GE grid solutions, 2021. https://www.gegridsolutions.com/measurement_recording_timesync/catalog/dr60.htm. (Accessed 30 June 2021).
[33] AMETEK Power Instruments Inc., TR-3000 multi-function recorder, 2021. https://www.ametekpower.com/products/fault-recorders/tr-3000-multi-function-recorder. (Accessed 30 June 2021).
[34] SATEC Global, Second Generation Branch Feeder MonitorTM (BFM-II), 2021. https://www.satec-global.com/BFM-II. (Accessed 30 June 2021).
[35] A. Bricchi, D. Moore, J. Rostron, Fault Location Detection on Tapped Transmission Lines, CIGRE-157 Conference on Power Systems, 2010, pp. 1–10. Available: https://www.southernstatesllc.com/wp-content/uploads/2017/06/Fault-Location-Detection-on-Tapped-Transmission-Lines.pdf.
[36] J. Von Euler-chelpin, Distribution Grid Fault Location An Analysis of Methods for Fault Location in LV and MV Power Distribution Grids, Uppsala universitet, Uppsala, Sweden, 2018.
[37] G.J. Brice et al., Integrated smart distribution RD&D project final technical report, Pittsburgh, PA, and Morgantown, WV (United States) 2017. doi:10.2172/1460575.
[38] E. Personal, A. García, A. Parejo, D.F. Larios, F. Biscarri, C. León, A comparison of impedance-based fault location methods for power underground distribution systems, Energies 9 (12) (2016) 1022, doi:10.3390/en9121022.
[39] A. Maitra, Distribution fault location support tools, algorithms, and implementation approaches, Palo Alto, CA, 2013. https://www.epri.com/research/products/1024381. (Accessed 23 August 2021).
[40] Smart Grid Investment Grant Program, Fault location, isolation, and service restoration technologies reduce outage impact and duration, 2014. https://www.smartgrid.gov/files/documents/B5_draft_report-12-18-2014.pdf. (Accessed 29 June 2021).
[41] Siemens Industry Inc., Siemens distribution feeder automation (SDFA) system Wendell, North Carolina, 2014. https://assets.new.siemens.com/siemens/assets/api/uuid:91228e3d-2d52-495b-a9a7-d9cbddc70057/sdfa-brochure-flisr.pdf. (Accessed 23 August 2021).
[42] T. Short, Distribution fault location: prototypes, algorithms, and new technologies, Palo Alto, CA, 2008. https://www.epri.com/research/products/000000000001013825. (Accessed 23 August 2021).
[43] C. Melhorn, Benchmarking of fault-location technologies, Palo Alto CA, 2011. https://www.epri.com/research/products/000000000001022730. (Accessed 23 August 2021).
[44] Hydro-Québec, MILE: intelligent power line monitoring system Varennes, Québec, 2011. http://www.hydroquebec.com/innovation/en/pdf/2010G080-26A-MILES.pdf. (Accessed 23 August 2021).
[45] T. Short, Fault-location application for improving distribution system maintenance: Hydro-Quebec's experiences, Palo Alto, CA, 2011. https://www.epri.com/research/products/000000000001021999. (Accessed 23 August 2021).
[46] T.A.S. Kawady, Fault Location Estimation in Power Systems with Universal Intelligent Tuning, Technische Universität, 2005.
[47] Electric Power Research Institute (EPRI), Underground cable fault location reference and application guide: 2017 update, Palo Alto, CA, 2017. https://www.epri.com/research/products/000000003002010061. (Accessed 29 June 2021).
[48] Megger: Electrical Test Equipment, Fault finding solutions, Dover, Kent, 2003. https://www.cablejoints.co.uk/upload/Megger_Cable_Fault_Finding_Solutions.pdf. (Accessed 23 August 2021).

[49] Megger: Electrical Test Equipment, Testing, diagnostics and fault location on power cables, Dover, Kent, 2021. https://embed.widencdn.net/pdf/plus/megger/kxujlt3pck/CableCatalogue_BR_EN_V04.pdf. (Accessed 23 August 2021).
[50] BAUR, Cable fault location, Žampachova, Czech Republic, 2018. https://www.baur.cz/wp-content/uploads/2018/05/BR_821-044_Product_brochure_Cable_Fault_Location_EN.pdf. (Accessed 23 August 2021).
[51] Megger: Electrical Test Equipment, Cable fault, test and diagnostics, 2021. https://megger.com/products/cable-fault-test-and-diagnostics. (Accessed 23 August 2021).
[52] Baur GmbH, Cable fault location: the shortcut to the source of a fault, 2021. https://www.baur.eu/en/cable-fault-location. (Accessed 12 July 2021).

Appendices

A.1 Software and hardware tools

This section briefly discusses the employed software and hardware components in this book to build the intelligent fault diagnosis scheme. First, this book used RSCAD software to model the IEEE 13-node test distribution feeder and simulated the modeled distribution grid in the real-time digital simulator (RTDS) machine. In addition, the simulated data were acquired through the National Instrument (NI) Grid Automation System, known as phasor measurement units (PMUs), and the LabVIEW platform. Finally, the obtained data were analyzed to decide the simulated faults employing advanced signal processing and machine learning tools in MATLAB/ SIMULINK environment.

A.1.1 Real-time digital simulator

In 1991, RTDS Technologies Inc. introduced the most advanced, widely used, and the first commercial RTDS to analyze, design, and test electrical power systems and their components [1–4]. The RTDS solves the complex power systems via parallel computation through several digital signal processors in real-time with a time-step of order 50μs. It is used in a wide range of power system applications, including power system planning and operation, integrated protection and control, micro-grid simulation and testing, HVDC, and PMU studies to support a wide variety of built-in power system modules and user-friendly graphical interfaces [4]. Additionally, it uses a prevailing processor combined with a field-programmable gate array (FPGA) that allows the simulation for a certain number of power electronics components with a time step of 1.4-2.5 μs embedded in the 50 μs time step environment. Therefore, this technology can operate those power electronics devices accurately at higher switching frequency during the simulation process, allowing the user to incorporate physical devices into the simulation in a closed-loop environment. However, RTDS Technologies Inc. manufactures different sizes and types of racks based on customers' requirements, as shown in Fig. A.1 that comprises a combination of hardware and software components [4–6].

A.1.1.1 RSCAD software

RTDS Technologies uses RSCAD software to design, operate and troubleshoot the electric network and interface with the RTDS machine. The software is comprised of several modules, including *"File Manager" "DRAFT", "Runtime," "TLine," "Cable," "Multiplot," "CBuilder," "Help," "Convert,"* and *"Manual"* to perform the necessary steps to design and simulate the targeted power systems and to analyze their outputs.

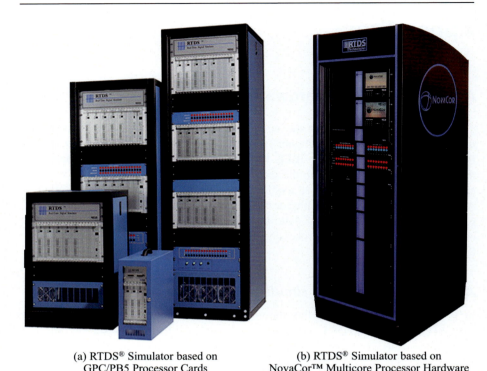

(a) RTDS® Simulator based on GPC/PB5 Processor Cards

(b) RTDS® Simulator based on NovaCor™ Multicore Processor Hardware

Fig. A.1 Real-time digital simulator (RTDS) machine [4–6].

The RTDS users organize and share their designed projects from the *"File Manager"* module, and the users launch the other RSCAD modules as shown in Fig. A.2. They use the *"Draft"* module to assemble a schematic diagram of their targeted power system graphically. The right window of the *"Draft"* module contains a comprehensive library of necessary components related to the power system, control, protection, and automation. The left window is used to model the schematic by pasting the copied components from the library, as shown in Fig. A.3. The users can change the component parameters and can group/ungroup the components according to their wishes. Besides, they can toggle from single line diagram to three-phase format and zoom in or out to view or print their modeled power systems. However, after constructing the electric networks, the users need to simulate them to create the execution code required for the RTDS machine and check preliminary errors.

The successful compilation of the modeled networks downloads the execution code onto the RTDS machine through the *"Runtime"* module. This module communicates with the simulator through GTWIF cards *via* Ethernet. This module also allows the users to control and observe the simulations in real-time shown in Fig. A.4. The users can customize the canvas by creating sliders, buttons,

Appendices

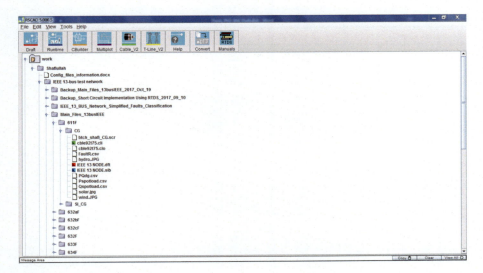

Fig. A.2 RSCAD file manager.

meters, plots, etc. The users also can plot, save and print data using this module. The users enter physical AC/DC transmission line data through the "*TLine*" module that converts the line data into a useable form to use the "*Draft*" module. The "*Cable*" module does a similar job to the physical cable data. Finally, the users can design their customized component using the "*CBuilder*" module and explore many examples through the "*Manual*" module.

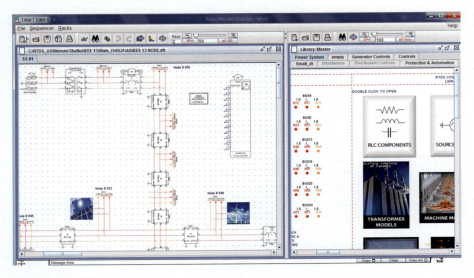

Fig. A.3 RSCAD draft file.

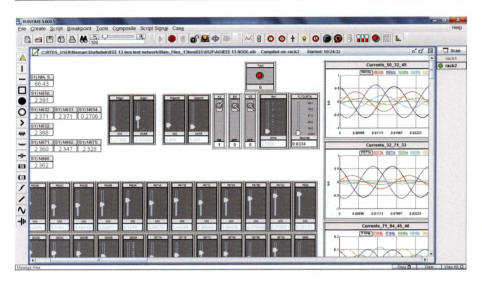

Fig. A.4 RSCAD runtime file.

A.1.1.2 RTDS hardware components

The RTDS Technology Inc. mounts the processor cards in card cages that are usually known as racks. The racks are housed in different sizes of cubicles based on customer requirements, along with necessary input/output cards, power entry components, and power supplies. The RTDS cubicles can house several racks that mean each cubicle has a different simulation capability based on several housed racks. Additionally, the RTDS hardware is modular and can gang multiple cubicles to accommodate and simulate larger sizes of power systems. Generally, the users divide the more extensive power system networks into many subsystems, and each rack simulates one subsystem. The RTDS racks are a combination of different components, as shown in Fig. A.5. The main components are listed below where some of them are accessible from the front and the rest of them from the back [4]:

- Giga Processor Card (GPC)/ PB5 Processor Card
- Giga-Transceiver Workstation Interface (GTWIF) Card
- Inter-Rack Communication (IRC) channel
- Low Voltage Digital I/O (LVDIO) Interface Panel
- High Voltage (250 V DC) Digital Output (HVDO) Interface Panel
- Giga-Transceiver Network (GTNET) Interface Card
- Giga-Transceiver Analogue Output (GTAO) Card
- Giga-Transceiver Analogue Input (GTAI) Card
- Giga-Transceiver Digital Output (GTDO) Card
- Giga-Transceiver Digital Input (GTDI) Card
- Giga-Transceiver Synchronization (GTSYNC) Card
- Giga-Transceiver Front Panel Interface (GTFPI) Card
- Global Bus Hub (GBH)

Appendices

a) PB5 Card   b) GTWIF card   c) GTNET Card   d) GTSYNC Card

e) GTDO Card   f) GTDI card   g) GTAO Card   h) GTAI Card

i) GTFPI Card   j) IRC Switch and GBH   k) LVDIO and HVDO panels

Fig. A.5 RTDS processor and related components [4].

The RTDS users the overall power system networks along with the auxiliary components (*e.g.*, transmission lines, generators, etc.) using the powerful PB5 processor card. The PB5 card contains two Freescale MC7448 RISC processors, each operating at 1.7 GHz.

It is fully compatible with the previous version of the RTDS processor card called GPC. The GTWIF card is responsible for synchronizing the racks (single cubicle or connected cubicles) through the local area network and communicates with the RSCAD software to load, start, and stop the simulation. Additionally, it is also responsible for retrieving results from the simulation and permitting the user to interact with the simulation cases. In multirack (3 or more racks) simulations, one master GTWIF card passes the time-step clock to all other racks through the GBH. However, each GTWIF has six high-speed bidirectional IRC channels that can transfer one gigabit of data per second. The RTDS cubicles can communicate with external equipment by sending and receiving digital signals through 16 digital input and 16 digital output (max 5V) ports mounted in the front of the cubicles. Besides, the users can collect a maximum of 250 V DC signals from the front of the cubicles through the 250 V DC digital output interface panel that contains 16 solid-state contacts. Both low voltage digital I/O and 250 V DC digital output interface panels are controlled from the

GTFPI card connected to PB5/GPC processor cards via fiber optic cable. The RTDS machine uses the GTNET card for Ethernet-based communication purposes installed in a slot and powered from the backplane. This card communicates with the PB5/GPC processor through a GT optical port.

The RTDS machine can be easily interfaced with external devices through Giga-Transceiver Input/output (GTIO) cards. Industry-standard optical fiber cable links these cards with the simulator, and up to eight IO cards can be daisy-chained to provide greater freedom of connection and flexibility. The GTDI card collects 64 optically isolated digital input signals from external equipment for the RTDS machine, whereas the GTDO card provides 64 optically isolated digital outputs to the external equipment from the machine. Both inputs and outputs are flexible for either a regular (50 µs) or a small (1–2 µs) time-step simulation running on the PB5/GPC card. Conversely, the GTAI card collects 12 optically isolated analog input signals from external equipment for the RTDS machine, whereas the GTAO card provides 12 optically isolated analog outputs to the external equipment from the machine. Both inputs and outputs are flexible for either a regular or a small time-step simulation running on the PB5/GPC card. The GTSYNC card synchronizes the machine time-step to an external time reference necessary for PMU benchmark testing. However, this card cannot be daisy-chained like other I/O cards rather connected to the GT port on the GTWIF card.

A.1.2 National Instrument Grid Automation Systems

The book uses NI Grid Automation System to store RTDS simulated data sent through the GTAO card. This data acquisition system combines an amplifier, global positioning system (GPS) clock, compact reconfigurable input/output (cRIO), and LabVIEW program. The amplifier receives signals from the GTAO card of the RTDS machine and conditions them before sending them to the cRIO. Finally, the LabVIEW programs store the received signals for further analysis. Additionally, this book modifies the LabVIEW program to get a synchronized timestamp for each signal using the GPS clock. However, the Grid Automation System available in 'Power System Research Lab' is limited to measure and store only twelve signals at a time.

A.1.3 MATLAB/SIMULINK

MATLAB (matrix laboratory) is a fourth-generation programming language developed by MathWorks [7]. This multiparadigm numerical computing platform allows the implementation of algorithms, plotting of data and functions and can be interfaced with other programming languages, including C/C++, Python, and Java. Scientists and engineers worldwide use this platform for many purposes, including data analytics, wireless communications, deep learning, signal processing, motor, and power control. Besides, SIMULINK is a graphical programming platform used to model, simulate and analyze multi-domain dynamic systems. MATLAB/SIMULINK can also be linked to other software and hardware, including RTDS/RSCAD, Arduino, and LabVIEW.

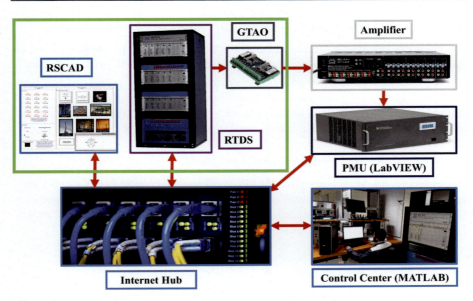

Fig. A.6 Schematic diagram of the developed laboratory prototype IFD scheme for the distribution grid. *IFD*, intelligent fault diagnosis.

A.1.4 Schematic diagram of the prototype intelligent fault diagnosis scheme

Fig. A.6 shows the complete schematic diagram for the developed laboratory prototype intelligent fault diagnosis (IFD) scheme for the distribution grids. This scheme models the targeted distribution grids in RSCAD software and simulates them in RTDS rack through the internet hub. Then, it applies different types of faults in different locations of the grids by incorporating the load demand and renewable energy generation uncertainties. For each fault, it records three-phase current signals and processed them in the MATLAB platform to train and test machine learning tools employing signal processing extracted features. Then, it collects current signals for any faulty case through physical PMU and fetches the extracted features to the trained MLT to decide on the faulty cases.

A.2 Statisctical quantities

This section presents the definitions and the mathematical expressions of the employed statistical measures in this book. It employed two different types of statistical indices, namely, to analyze the performance of the regression (fault location) problem and to extract features from the Discrete Wavelet Transform (DWT) and Stockwell transform (ST) decomposed signals [8–14].

A.2.1 Quantities related to the regression problem

This book considered the fault location problem as a regression problem. To evaluate the efficacy of the regression problem, a set of standard statistical performance indices were evaluated where $y(t)$ and $\widehat{y(t)}$ were assumed as the tth actual and predicted outputs, whereas $\bar{y}$ was the mean of actual outputs and q was the total number of outputs to be predicted. This section presents the mathematical expressions of the selected statistical performance indices used in the fault location problem.

A.2.1.1 Mean absolute percentage error

The mean absolute percentage error (MAPE) is considered one of the most prominent statistical performance indices to demonstrate the effectiveness of any regression model. The lower values of this index illustrate the strength of the developed model. The mathematical expression for MAPE is:

$$\text{MAPE} = \frac{1}{q}\sum_{t=1}^{q}\frac{\left|\widehat{y(t)}-y(t)\right|}{\left|y(t)\right|}\times 100\% \quad (A.1)$$

A.2.1.2 Root mean squared error (RMSE)

Like MAPE, the lower values of root mean squared error (RMSE) also demonstrates the strength of any regression model, whereas the higher values indicate the weakness of that model. The mathematical expression of RMSE is:

$$\text{RMSE} = \sqrt{\frac{1}{q}\sum_{t=1}^{q}\left|\widehat{y(t)}-y(t)\right|^{2}} \quad (A.2)$$

A.2.1.3 Root mean squared error-observations standard deviation ratio

RMSE-observations standard deviation ratio (RSR) is also a crucial indicator to check the effectiveness of any regression model where the lower values indicate the strength of the model and higher values illustrate the weakness of that model. The mathematical expression of the RSR index is:

$$\text{RSR} = \frac{\text{RMSE}}{\text{SD}} = \sqrt{\frac{\sum_{t=1}^{q}\left|\widehat{y(t)}-y(t)\right|^{2}}{\sum_{t=1}^{q}\left|y(t)-\bar{y}\right|^{2}}} \quad (A.3)$$

Where the SD stands for the standard deviation of the actual outputs and can be expressed mathematically as:

$$\text{SD} = \sqrt{\frac{1}{q}\sum_{t=1}^{q}\left|y(t)-\bar{y}\right|^{2}} \quad (A.4)$$

A.2.1.4 Percent bias

The ideal value of percent bias (PBIAS) is 0.0 that says the developed model can predict desired outputs accurately. However, the negative and positive values of PBIAS demonstrate the overestimation and underestimation of the predicted outputs, respectively. It can be mathematically expressed as:

$$\text{PBIAS} = \frac{\sum_{t=1}^{q}\left(y(t)-\widehat{y(t)}\right)}{\sum_{t=1}^{q} y(t)} \times 100\,\% \tag{A.5}$$

A.2.1.5 Coefficient of determination

The coefficient of determination (R^2) explains the relationship between the actual and the predicted outputs. The maximum value of R^2 can be "1" that demonstrates the strongest relationship between the predicted and the actual outputs. In contrast, the value "0" indicates the outputs can never be predicted from the given input sets. The R^2 can be expressed as:

$$R^2 = 1 - \frac{\sum_{t=1}^{q}\left|\widehat{y(t)}-\overline{y}\right|^2}{\sum_{t=1}^{q}\left|y(t)-\widehat{y(t)}\right|^2} \tag{A.6}$$

A.2.1.6 Nash–Sutcliffe model efficiency coefficient

The Nash–Sutcliffe model efficiency coefficient (NSEC) is almost identical to R^2 that evaluates the relative magnitudes of the residual variance (noise) compared to the measured data variance (information). The values of NSEC vary from $-\infty$ to 1, where the efficiency "1" indicates a perfect match between the actual and the predicted values, whereas "0" illustrates that the predicted values are as accurate as the mean of the actual values. However, the efficiency of less than zero indicates that the residual variance is larger than the data variance. The following equation provides the mathematical expression of NSEC:

$$\text{NSEC} = 1 - \frac{\sum_{t=1}^{q}\left|y(t)-\widehat{y(t)}\right|^2}{\sum_{t=1}^{q}\left|y(t)-\overline{y}\right|^2} \tag{A.7}$$

A.2.1.7 Willmott's index of agreement

The higher values of "R^2" do not necessarily mean that the predicted and the actual outputs are close to each other; instead, they demonstrate a stronger relationship between the actual and the predicted outputs. Besides, the NSEC calculates the differences between the actual and the predicted values as squared values that overestimates larger values in a time series and neglects the lower values. Consequently, the statisticians looked for another performance index that overcomes the drawbacks related to R^2 and NSEC indices. The values of Willmott's index of agreement (WIA) vary from "0" to "1" where the value "1" indicates the perfect fit and the value "0" refers to no correlation between the predicted and the actual outputs. The WIA can be expressed as:

$$\text{WIA} = 1 - \frac{\sum_{t=1}^{q}\left|\widehat{y(t)} - y(t)\right|^2}{\sum_{t=1}^{q}\left(\left|\widehat{y(t)} - \overline{y}\right| + \left|y(t) - \overline{y}\right|\right)^2} \tag{A.8}$$

A.2.2 Quantities related to features of signal processing techniques

This book decomposes electrical current signals employing DWT and ST. The DWT decomposed coefficients (detailed and approximate), including d_1, d_2, d_3, d_4, d_5, d_6, d_7, d_8, and a_8, are series of numbers as discussed in Chapter 4. This book calculated six different statistical features: entropy, energy, skewness, kurtosis, mean, and standard deviation from the mentioned coefficients. It used them as inputs of the machine learning tools. Similarly, the calculated matrices (S_{cmax}, S_{rmax}, E_{max}, and $S_{\text{c-phase-max}}$) from the ST decomposed S-matrices as discussed in Chapter 4 are also series of numbers, and the proposed approach calculates the standard deviations, maximum values, minimum values, mean values, entropy, skewness, and kurtosis from the mentioned matrices. The following part of this section presents a few selected statistical quantities as the rest are well known to researchers related to power systems. Let us assume a discrete random variable X with possible values $\{x_1, x_2, x_3, \ldots \ldots, x_n\}$.

A.2.2.1 Entropy

Generally, entropy refers to the disorder or uncertainty of the random variable X that also employs to measure the diversity of that variable. Alternatively, the average amount of information produced by a probabilistic random variable (source of data) is defined as entropy. The variable with uniform distribution contains more entropy than the variable with the non-uniform distribution. In this research, the built-in MATLAB function called *"entropy"* has been used to calculate the entropies of the coefficients obtained from DWT and the mentioned matrices obtained ST.

A.2.2.2 Energy

The percentage of energy corresponding to the DWT decomposed approximate, and detail coefficients of the current signals contain characteristic signatures related to any power system transients. Consequently, this book evaluated the percentage of energy associated with the coefficients employing a built-in MATLAB function called *"wenergy."*

A.2.2.3 Mean and standard deviation

The standard deviation measures the variation or the dispersion of a dataset from its mean. The lower values of this feature indicate that the data points are located closer

Appendices

to the mean whereas the higher values indicate that the data points are spread out over a wider range of values. However, this book used two built-in MATLAB functions, namely *"mean"* and *"std,"* to calculate the mean and standard deviation of the coefficients obtained from DWT and the mentioned matrices obtained ST in this research.

A.2.2.4 Skewness

The skewness of a dataset is the measure of lack of symmetry around the mean of that dataset. It is not necessarily related to the mean and variance of the dataset. Different datasets with the same mean and variance could have different skewness. The positive skewness refers that the more data points are located to the right of the mean than to the left. Conversely, the negative skewness refers that more data points are located to the left, whereas the zero skewness indicates that the data points are symmetric to the mean. This book calculated the skewness of the coefficients obtained from DWT and the mentioned matrices obtained ST using the built-in MATLAB function called *"skewness."*

A.2.2.5 Kurtosis

The kurtosis measures the degree of tailedness of a dataset. The negative and positive values of this feature demonstrate thinner tails and fatter tails, respectively, compared to the normal distribution. The built-in MATLAB function called *"kurtosis"* was used to calculate the skewness of the coefficients obtained from DWT and the mentioned matrices obtained ST in this research.

A.3 IEEE 13-node test distribution feeder data

This book selected the IEEE 13-node test distribution feeder that is relatively highly loaded and contains most of the inherent properties of the distribution grids, including overhead and underground cables with a variety of phases, in-line transformer, shunt-capacitor banks, unbalanced distributed, and spot loads. The IEEE-PES distribution system analysis subcommittee made the associated data available for the researchers in [15]. The following Tables A.1, A.2, A.3, A.4, A.5, A.6, A.7 and A.8 present the parameters of the IEEE 13-node test distribution feeder.

Table A.1 Overhead line configuration data [15].

Configuration	Phasing	Phase ACSR	Neutral ACSR	Spacing ID
601	B A C N	556,500 26/7	4/0 6/1	500
602	C A B N	4/0 6/1	4/0 6/1	500
603	C B N	1/0	1/0	505
604	A C N	1/0	1/0	505
605	C N	1/0	1/0	510

Table A.2 Underground line configuration data [15].

Configuration	Phasing	Cable	Neutral	Space ID
606	A B C N	250,000 AA, CN	None	515
607	A N	1/0 AA, TS	1/0 Cu	520

Table A.3 Line segment data [15].

Node A	Node B	Length (ft.)	Configuration
632	645	500	603
632	633	500	602
633	634	0	XFM-1
645	646	300	603
650	632	2000	601
684	652	800	607
632	671	2000	601
671	684	300	604
671	680	1000	601
671	692	0	Switch
684	611	300	605
692	675	500	606

Table A.4 Transformer data [15].

Item	kVA	kV-high	kV-low	R - %	X - %
Substation:	5,000	115 - D	4.16 Gr. Y	1.0	8
XFM -1	500	4.16 – Gr.W	0.48 – Gr.W	1.1	2

Table A.5 Capacitor data [15].

Node	Phase A kVAr	Phase B kVAr	Phase C kVAr
675	200	200	200
611	0	0	100
Total	200	200	300

Table A.6 Regulator data [15].

Regulator ID:	1		
Line Segment:	650 - 632		
Location:	50		
Phases:	A - B -C		
Connection:	3-Ph,LG		
Monitoring Phase:	A-B-C		
Bandwidth:	2.0 volts		
PT Ratio:	20		
Primary CT Rating:	700		
Compensator Settings:	Phase A	Phase B	Phase C
R - Setting:	3	3	3
X - Setting:	9	9	9
Voltage Level:	122	122	122

Table A.7 Distributed load data [15].

Node From	Node To	Load model	Phase A kW	Phase A kVAr	Phase B kW	Phase B kVAr	Phase C kW	Phase C kVAr
632	671	Y-PQ	17	10	66	38	117	68

Table A.8 Spot load data [15].

Node	Load model	Phase A kW	Phase A kVAr	Phase B kW	Phase B kVAr	Phase C kW	Phase C kVAr
634	Y-PQ	160	110	120	90	120	90
645	Y-PQ	0	0	170	125	0	0
646	D-Z	0	0	230	132	0	0
652	Y-Z	128	86	0	0	0	0
671	D-PQ	385	220	385	220	385	220
675	Y-PQ	485	190	68	60	290	212
692	D-I	0	0	0	0	170	151
611	Y-I	0	0	0	0	170	80
TOTAL		1158	606	973	627	1135	753

References

[1] M.D. Omar Faruque, et al., Real-time simulation technologies for power systems design, testing, and analysis, IEEE Power Energy Technol. Syst. J. 2 (2) (2015) 63–73, doi:10.1109/JPETS.2015.2427370.

[2] A. Nikander, Development and testing of new equipment for faulty phase earthing by applying RTDS, IEEE Trans. Power Deliv. 32 (3) (2017) 1295–1302, doi:10.1109/TPWRD.2015.2505779.

[3] A. Aljohani, A. Aljurbua, M. Shafiullah, M.A. Abido, Smart fault detection and classification for distribution grid hybridizing ST and MLP-NN, 2018 15th International Multi-Conference on Systems, Signals & Devices (SSD), IEEE, Hammamet, Tunisia, 2018, pp. 1–5.

[4] RTDS Technologies Inc., Home - RTDS technologies, 2021. https://www.rtds.com/. (Accessed 13 July 2021).

[5] C.M.U, Faculty of Engineering, Overview | Real Time Digital Simulator, *Faculty of Engineering, Chiang Mai University*, 2021, http://rtds.eng.cmu.ac.th/en/?page_id=141 (Accessed 24 June 2021).

[6] Center for Advanced Power Systems at Florida State University, Real Time Digital Simulator, 2021, https://www.caps.fsu.edu/about-caps/real-time-digital-simulator/. (Accessed 14 July 2021.

[7] MathWorks, MATLAB & Simulink, 2021, https://www.mathworks.com/?s_tid=gn_logo (Accessed 14 July 2021).

[8] M. Shafiullah, Fault Diagnosis in Distribution Grids under Load and Renewable Energy Uncertainties, King Fahd University of Petroleum & Minerals, Dhahran, Saudi Arabia, 2018.

[9] M.J. Rana, M.S. Shahriar, M. Shafiullah, Levenberg–Marquardt neural network to estimate UPFC-coordinated PSS parameters to enhance power system stability, Neural Comput. Appl. 31 (4) (2019), doi:10.1007/s00521-017-3156-8.

[10] M.S. Shahriar, M. Shafiullah, M.J. Rana, Stability enhancement of PSS-UPFC installed power system by support vector regression, Electr. Eng. (2017) 1–12, doi:10.1007/s00202-017-0638-8.

[11] M. Shafiullah, M.A. Abido, S-transform based FFNN approach for distribution grids fault detection and classification, IEEE Access 6 (2018) 8080–8088, doi:10.1109/ACCESS.2018.2809045.

[12] M. Shafiullah, M. Ijaz, M.A. Abido, Z. Al-Hamouz, Optimized support vector machine & wavelet transform for distribution grid fault location, 2017 11th IEEE International Conference on Compatibility, Power Electronics and Power Engineering, CPE-POWERENG). IEEE, 2017, pp. 77–82, doi:10.1109/CPE.2017.7915148.

[13] M. Shafiullah, M.A. Abido, Z. Al-Hamouz, Wavelet-based extreme learning machine for distribution grid fault location, IET Gener. Transm. Distrib. 11 (17) (2017) 4256–4263, doi:10.1049/iet-gtd.2017.0656.

[14] N. Vandeput, Forecast KPI: RMSE, MAE, MAPE & Bias, Towards Data Science, 2019, https://towardsdatascience.com/forecast-kpi-rmse-mae-mape-bias-cdc5703d242d. (Accessed 14 July 2021).

[15] IEEE PES AMPS DSAS, Test feeder working group, test feeder. In: IEEE PES AMPS DSAS Test Feeder Working Group, 2021, https://site.ieee.org/pes-testfeeders/resources/. (Accessed 13 July 2021).

Index

Page numbers followed by "*f*" and "*t*" indicate, figures and tables respectively.

A

Accuracy analysis, 172
Adaptive fault location algorithm, 167, 168*f*, 170, 200–201, 205*f*, 205, 206
Adaptive neuro-fuzzy inference system (ANFIS), 86
   first stage of, 87
   fourth layer, 88
   normalization layer, 88
   second layer of, 88
   structure, 87*f*
Advanced metering infrastructure (AMI), 1–3, 159
Alexa, 70
American Electric Power (AEP), 11
Analog-to-digital converters (ADC), 13–14
Artificial intelligence (AI) techniques, 73, 90
   artificial neural networks, 73
   with automation, 72*f*
   deep learning technique, 85
   definitions of, 69
   extreme learning machines, 78
   foundation and basic components, 71
   fuzzy logic model, 79
   genetic programming, 81
   history of, 69, 70*f*
   hybrid and ensemble, 86
      adaptive neuro-fuzzy inference system, 86
      ensemble techniques, 88
      metaheuristic algorithm, 89
   limitations of, 72
   new electricity, 71
   support vector machines, 75
Artificial neural networks (ANN), 8, 69–70, 73
   gradient descent backpropagation, 74
   Levenberg- Marquardt backpropagation, 74
   one step secant backpropagation, 74
   resilient backpropagation, 74
   scaled conjugate gradient backpropagation, 74

B

Backtracking search algorithm (BSA), 27, 44
   crossover, 45
   fitness evaluation, 45
   flowchart of, 47*f*
   initialization, 44
   mutation, 45
   pseudocode of, 46*t*
   selection I, 45
   selection-II, 46
   stopping criteria, 46
Bonneville Power Administration (BPA), 11
Branch fault location, 320–321

C

Circuit diagram, 225*f*
Classical optimization techniques, 27
   linear programming, 28
   mixed-integer linear programming, 29
   nonlinear programming, 30
Coifman wavelet, 107–108
Constriction factor particle swarm optimization (CF-PSO), 50
   flowchart of, 51*f*
Continuous wavelet transform (CWT), 103
Conventional technology, 1–3
Convolutional neural network (CNN), 85–86
Cramer's rule, 161
Current limiting damping circuit (CLDC), 225

D

Data generation and conditioning, 236
Daubechies wavelet, 107–108
Deep belief network (DBN), 86
DeepFace, 70

Index

Deep learning (DL)
　facial recognition system, 70
　technique, 85
　　convolutional neural network, 85–86
　　deep belief network, 86
　　deep stacking network, 86
　　long short-term memory, 85
　　recurrent neural network, 85
Deep stacking network (DSN), 86
Defuzzification, 88
Differential evolution (DE), 27
　algorithm, 41
　　flowchart of, 44*f*
　　pseudocode of, 43*t*
Discrete Fourier transform (DFT), 11
Discrete wavelet transform (DWT), 101–102, 104
　application of, 104
　based feature extraction, 107
　based method, 276*t*
　extracted coefficients, 111*f*
Distribution lines, 297
Distribution network observability, 133
　complete network observability, 135, 136
　phasor measurement unit channel limit, 137

E

Electric power system (EPS), 1
　distribution, 1
　fault diagnosis importance, 3
　fault diagnosis techniques, 5
　　factors affecting, 9
　　fault detection and classification techniques, 5
　　fault location techniques, 5
　　high frequency-based techniques, 9
　　impedance-based techniques, 5
　　knowledge-based techniques, 8
　　other techniques, 9
　　traveling wave-based techniques, 7
　faults, 4*f*, 4
　　open circuit, 4
　　short circuit, 4
　generation, 1
　networks, 195, 347
　structure of the modern, 2*f*
　transmission, 1
Electric transmission lines, 159

Energy-related carbon dioxide gas emission, 293
Entropy, 406
Equivalent line parameters, 228–229
Error correction (EC), 74–75
Evolutionary metaheuristic techniques, 37
　backtracking search algorithm, 44
　differential evolution, 41
　genetic algorithm, 37
External faults, 226–227
Extreme learning machines (ELMs), 73, 78, 249–250
　hidden layer output matrix, 79

F

Family of wavelets, 103–104
Fault detection, 278*t*
　and classification, 278, 314–315
　and classification technique, in EPS network, 5
　scheme, 310
Fault diagnosis, 249, 276
　results and discussions, 307
Fault inception angle, 210–211
Fault location
　algorithms, 160, 171
　　description, 230
　subroutine, 231–232, 234
Fault location technique, in EPS network, 5
　high frequency-based method, 5, 9
　impedance-based method, 5
　knowledge-based method, 5, 8
　traveling wave-based method, 5, 7
Fault modeling and feature extraction, 304
Fault point
　voltages, 170
Fault-point distance, 170
Fault point sequence
　voltages, 162*f*, 163*f*, 164*f*, 165*f*, 166*f*
Fault resistance, 172–176
Faulty section identification, 315–316
　flowchart, 202*f*
Feature extraction illustration, 107
　DWT-based feature extraction, 107
　stockwell transform, 113
Field-programmable gate array (FPGA), 293–294
First adaptive fault location algorithm (FL-1), 160–161

Index

Fourier transform (FT), 101
Four-node test distribution, 250
  feeder, 276*t*
Fuzzification, 80
Fuzzy inference system (FIS), 80
Fuzzy logic model (FLM), 72–73, 79
  structure of, 80*f*
Fuzzy membership function, 80
Fuzzy number, 81*f*

G

Gabor-Wigner transform, 101
Genetic algorithm (GA), 27, 37, 81–82
  flowchart of, 41*f*
  pseudocode of, 40*t*
Genetic programming (GP), 72–73, 81
  crossover, 83–84
  fitness evaluation, 82
  initialization, 82
  selection, 83
  termination criteria checking, 83
Giga-Transceiver Analogue Output (GTAO), 294
Global positioning system (GPS), 1–3, 121, 402
Google artificial brain, 70
Google DeepMind Challenge Match, 70
Gradient descent (GD) backpropagation, 74
Grey wolf optimization (GWO), 27, 51, 122–123
  boundary check, 53
  fitness evaluation, 52
  flowchart of, 54*f*
  initialization, 52
  pseudocode of, 54*t*
  termination criteria, 52
  updating, 53

H

Haar wavelet, 107–108
High frequency-based method, 5, 9
High voltage (HV), 1
Hilbert-Huang transform, 101

I

IEEE 13-node test distribution feeder, 298, 302
IEEE 13-node test distribution feeder data, 407
IEEE 13-node test distribution feeder modeling, 294–295

Impedance-based method, 5
Integer linear programming (ILP) problem, 29
Intelligent electronic device (IED), 11
Intelligent fault detection scheme, 276
Intelligent fault diagnosis (IFD) schemes, 249
Inter-control center communications protocol (ICCP), 121
Internal faults, 226–227
International Energy Outlook reports, 293

K

Kalman filtering, 101
Karush-Kuhn-Tucker (KKT), 27
Knowledge-based method, 5, 8
Kurtosis, 407

L

Laboratory Virtual Instrument Engineering Workbench (LabVIEW) program, 307
Lagrange multipliers, 77
Levenberg-Marquardt backpropagation, 74
Linear programming (LP), 28
Line-to-line-to-line (LLL) fault, 4
Load and renewable energy, 298
Logistic regression (LR), 74–75
Long short-term memory (LSTM), 85

M

Machine learning (ML), 69–70
Machine learning tools (MLT), 5
Mallat wavelet, 107–108
Mean absolute percentage error (MAPE), 404
Metaheuristic algorithm, 89
Metaheuristic techniques, 30, 55, 56*t*–58*t*
  evolutionary, 37
    backtracking search algorithm, 44
    differential evolution, 41
    genetic algorithm, 37
  swarm-based, 46
    constriction factor particle swarm optimization, 50
    grey wolf optimization, 51
    particle swarm optimization, 47
  trajectory-based, 31
    simulated annealing, 31
    tabu search algorithm, 34

Metal oxide varistors (MOVs), 223
Metal oxide varistor voltage-current relationship, 226
Meyer wavelet, 107–108
Micro phasor measurement unit (μPMU), 1–3
Minimum mean squared error (MSE), 307
Mixed-integer linear programming (MILP), 29, 122–123
MLP-NN parameters, 279
Morlet wavelet, 107–108
Mother wavelets, different, 108f
Multigene genetic programming (MGGP), 82
Multilayer perceptron (MLP), 73
Multilayer perceptron neural networks (MLP-NN), 73
  parameters, 310t
Multiterminal lines, 195

N

Naïve Bayes (NB), 74–75
Nash-Sutcliffe model efficiency coefficient (NSEC), 405
Network observability and measurement redundancy illustration, 126
Newton-Raphson method, 27, 159
No free lunch (NFL) theorem, 55
Nonadaptive fault location algorithm, 218
Nonlinear programming (NLP), 30
North American SynchroPhasor Initiative (NASPI), 15
Nyquist's rule, 104

O

One step secant (OSS) backpropagation, 74
Optimal phasor measurement unit placement (OPP) formulation, 123
  distribution network observability, 133
  complete network observability, 135, 136
  phasor measurement unit channel limit, 137
  general, 123
  phasor measurement units channel limit incorporation, 125
  power system contingencies incorporation, 126
Optimization processes, 27, 28f

P

Particle swarm optimization, 27
Particle swarm optimization (PSO) algorithm, 47
  fitness evaluation, 48, 49
  inertia weight, 48
  initialization, 48
  pseudocode of, 49t–50t
  termination criteria, 49
  variables boundary, 49
Phasor data concentrator (PDC), 11
Phasor measurement unit-based parameter, 227
Phasor measurement units (PMU), 159, 195
  simulation outcomes, 167
    biased and nonbiased noise, 171
    identified thevenin equivalent, 172
    noise-free measurement, 171
  thevenin equivalent identification, 160
  transmission line parameter identification, 160
Phasor measurement units (PMUs), 1–3, 10, 223–224
  application, 15
  block diagram of, 13f
  historical overview, 11
  phasor definition, 11
  phasor measurement concept, 12
  phasor measurement systems, 14
  synchrophasor and the generic phasor measurement unit, 13
PMU-based fault location scheme, 196
Power quality (PQ), 101
Prefault loading, 216–218

R

Radial basis function (RBF), 74–75
Radial buses (RB), 126
Rapid service restoration, 159
Real-time digital simulator (RTDS), 294
  hardware components, 397
  machine, 295f, 397
  software, 296f, 397
Recurrent neural network (RNN), 85
Regularization coefficients, 77
Renewable energy resources, 299
Restoration process, 223

Index

Restricted Boltzmann machine (RBM), 86
Root mean squared error (RMSE), 404
Root mean squared (RMS) value, 12

S

Saudi Electricity Company (SEC), 160
  network, 160
Scaled conjugate gradient backpropagation, 74
Selection procedure, 235
Series capacitor locations, 224
Series capacitor schemes, 197
Series compensated lines (SCLs), 223
Series faults, 4
Share coefficients, 233t
Short-time Fourier transform (STFT), 102
Signal processing techniques (SPT), 101, 102
  advanced, 101
  digital, 101
Simulated annealing (SA), 27
  algorithm, 31
    flowchart of, 34f
    pseudocode of, 33t
Simulation outcomes, 167
  biased and nonbiased noise, 171
  identified thevenin equivalent, 172
  noise-free measurement, 171
Single gene genetic programming (SGGP), 82
Single line diagram, 227f, 231f
Single-line-to-ground (SLG), 4
Sinusoidal waveform, 12f
Skewness of a dataset, 407
S-matrix, 107
Stockwell transform (ST), 106
  based approach, 276t, 315–316
  based approaches, 276
Superimposed network, 161f
Supervisory control and data acquisition (SCADA), 1–3, 121, 160
Support vector machines (SVM), 72–73, 75, 76f
Swarm-based metaheuristic techniques, 46
  constriction factor particle swarm optimization, 50
  grey wolf optimization, 51
  particle swarm optimization, 47

Symmetrical component distance relays (SCDR), 11
Synchrophasor, 13

T

Tabu search (TS), 27
  algorithm, 34
    flowchart of, 37f
    pseudocode of, 36t
Teager energy operator, 101
Thevenin equivalent (TE), 159, 160, 195–196, 223–224
  identification, 160
  model, 169f
Three-phase balanced faults, 235
Three-phase-to-ground (LLLG) fault, 4
Three-terminal transmission network, 196f
Trajectory-based metaheuristic techniques, 31
  simulated annealing, 31
  tabu search algorithm, 34
Transformers, 295
Transmission lines, 167, 195, 196–197
  parameter identification, 160
Transmission network observability, 127
  complete network observability, 130, 132
  network observability, 128
Transmission systems, 1
  extra-high voltage, 1
  high voltage, 1
  ultra-high voltage, 1
Traveling wave (TW) based method, 5, 7
  double-ended scheme, 8
  single-ended scheme, 8
Tsallis wavelet entropy, 101
Turing Natural Language Generation (T-NLG), 70
Two-terminal transmission system, 159–160

U

Uber self-driving car, 70
Underground cables (UC), 3–4, 297
  fault diagnosis in, 9
U.S. Energy Information Administration projection, 293

V

Voltage phasor measurements, 159
Voltage regulator, 297

W

Wavelet transform (WT), 101, 102
    continuous, 103
        discrete wavelet transform, 104
    stockwell transform, 106

Wide-area measurement systems (WAMS), 10
    historical overview, 11
    synchrophasor and the generic phasor measurement unit, 13
Wind turbines (WTs), 301
Winger-Ville time-frequency distribution, 101

Z

Zero injection bus (ZIB), 122–123